Molecular Mechanics
across
Chemistry

Molecular Mechanics
across
Chemistry

Anthony K. Rappé

PROFESSOR OF CHEMISTRY
COLORADO STATE UNIVERSITY
FORT COLLINS, COLORADO

AND

Carla J. Casewit

CALLEO SCIENTIFIC
FORT COLLINS, COLORADO

University Science Books
Sausalito, California

About the cover: Dynamical range of HIV protease predicted by molecular dynamics. (See Figure 3.24.)

University Science Books
55D Gate Five Road
Sausalito, CA 94965

Fax: (415) 332-5393

Production manager: *Superscript Editorial Production Services*
Copy editor: *Jeannette Stiefel*
Designer: *Robert Ishi*
Illustrator: *John Choi*
Compositor: *Coghill Composition Company*
Printer and binder: *Edwards Brothers*

This book is printed on acid-free paper.

Library of Congress Cataloging-in-Publication Data

Rappé, Anthony K., 1952–
    Molecular mechanics across chemistry / Anthony K. Rappé and Carla J. Casewit.
          p.   cm.
    Includes bibliographic references (p.   –   ) and index.
    ISBN 0-935702-77-6
    1. Molecular structure.   I. Casewit, Carla J., 1954–   .
II. Title.
QD461.R26   1996
541.2'2—dc20   96-13315
                                                    CIP

Printed in the United State of America
10 9 8 7 6 5 4 3 2 1

Dedicated to our children
Mollie, Kelly, and Charles

Contents

Chapter 8: Force Fields 385

Index 439

Preface

During our year with BioDesign, we both had the opportunity to interact with many practicing molecular modelers in industry. There we saw applications of molecular mechanics ranging from drug design to homogeneous transition metal catalysis. Inspired by this remarkable breadth, we wrote *Molecular Mechanics across Chemistry*.

This introductory molecular mechanics text uses a case-study style rather than a comprehensive review and compilation approach. The textbook thus allows readers an exposure to the relevance and utility of molecular mechanics, as well as the opportunity to study a particular chemical problem and its modeling solution in depth. Conceptual foundations and methodology are presented as needed. The desire to know the answer to a chemical question is used to motivate the understanding of the mathematical foundations. The case studies were chosen because of their pedagogical value in illustrating the concepts of molecular mechanics; it was not our intention to survey the field.

The chapters, excluding the first two chapters, are designed to be used independently to give instructors maximal flexibility. The book does not depend on a prior knowledge of elementary biochemistry, inorganic chemistry, or advanced undergraduate physical chemistry; only a good working knowledge of sophomore organic chemistry and freshman physics is mandatory. The text was written to be used independently of any software or hardware configuration. Each chapter has homework. Some of the homework problems can be solved only with a modern molecular mechanics program, while others require just pencil and paper.

Chapter 1 presents the historical and conceptual development of molecular mechanics and conformational analysis. Chapter 2 describes the determination and interpretation of preferred molecular conformations, transition state structures, and intermolecular energetics of organic systems. The modeling and design of important classes of biomolecules, and their interactions with drugs, are given in Chapters 3, 4, and 5. Chapter 6 presents modeling of polymer structure and properties. In Chapter 7 the application of molecular mechanics to inorganic molecules is described. The augmentation and design of new force fields is presented in Chapter 8. Brief appendices provide descriptions of stereochemical terminology, ideal gas thermodynamics, molecular dynamics, Monte Carlo selection procedures, conformational searching techniques, and answers to selected homework problems.

The authors are very grateful to Dr. Laurie Castonguay, Professor Marc Greenberg, Dr. Joseph Guiles, and Dr. Barry Olafson who were kind enough to read and comment on portions of the manuscript. We owe a special debt to Dr. John Briggs, Professor Robert Cave, Dr. Graham Smith, and Professor Clark Landis who read the entire manuscript and offered exceedingly helpful suggestions. We also wish to thank Professor Oren Anderson of Colorado State University, Lynn Bormann, Kathy Colwell, Dr. Georgia McGaughey, Dr. Michelle Pietsch, Dr. Sunder Ramachandran, Minna Win-Gildenmeister, and the rest of the Rappé research group for their help during the preparation of this book. Thanks are also due to Professor Jack Norton for his thoughtful support and guidance. Finally, we acknowledge the graciousness of the authors who sent us original artwork. The book is better for their generosity.

For an up-to-date snapshot of the ever-changing molecular modeling landscape, see the book's web site at http://www.chm.colostate.edu/mmac. The site also offers color images of molecules discussed in the text.

Molecular Mechanics
across
Chemistry

CHAPTER 1

Overview

1.1. INTRODUCTION

Chemistry, a largely experimental science, progresses by the development of models to explain experimental observations. Since the mid-1800s, chemists have used molecular structural models as a foundation for understanding chemical and physical properties (Riddell and Robinson, 1974). The observation that compounds with the same chemical formula, for example, β-hydroxypropionic acid (hydracrylic acid) **1.1** and α-hydroxypropionic acid (lactic acid) **1.2,** have different physical and chemical properties led

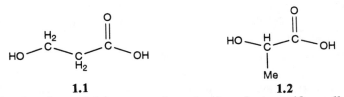

1.1   **1.2**

chemists to develop the model that atoms in molecules adopt specific, well-defined connectivities—a concept that is nearly as accepted today as gravity. Connectivities can be changed by chemical transformation but do not spontaneously interconvert, thus permitting different arrangements of the same atoms, positional isomers, to have observable differences in physical and chemical properties. The concept of positional isomers is a structural model.

The notion of molecular conformations, that is, molecular structures that are related by rotation about single bonds that do interconvert under ambient conditions, is another structural model. For example, the *trans* and *gauche* structures of hydracrylic acid, **1.1t** and **1.1g** respectively, are conformers.

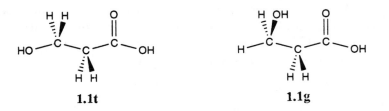

1.1t   **1.1g**

1

Distinct conformations were first suggested to exist for the molecule cyclohexane by Sache (1890, 1892), though experimental confirmation of this model did not occur until the 1940s (reviewed in Dauben and Pitzer, 1956).

In 1873, Wislicenus reported that lactic acid from sour milk, **1.2L**, (the L form) had a different optical rotatory power than lactic acid from muscle tissue, **1.2D** (the D form), yet lactic acid synthesized in the lab was not optically active.

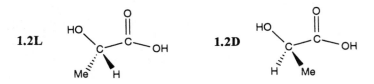

We now know that these differences are due to stereoisomerism (the synthetic compound was a racemic mixture of **1.2L** and **1.2D**). Pasteur had earlier recognized the differing optical rotatory powers of tartartic acids obtained from different sources. These results were instrumental in van't Hoff's model of a tetrahedral shape for carbon (reviewed in Riddell and Robinson, 1974). van't Hoff proposed that the only way that four substituents could be bound to a central atom and placed in two fundamentally different, nonsuperimposable, arrangements or configurations was if the central atom (carbon) were tetrahedral. Such centers are now called stereogenic. The three-dimensional (3-D) shape of carbon, taken with the concept of specific connectivities or structural formulas, helped explain why isomers possess differences in physical and chemical properties, and enabled van't Hoff to make predictions such as the number of stereoisomers present in compounds with two stereogenic carbon centers. Stereochemical terms are discussed in Appendix A.

Given the interpretive power of molecular structural models, apparent to chemists as early as the 1800s, it was only natural that chemists would develop mathematical tools to aid in understanding molecular structure and the molecular structural changes associated with chemical reactivity.

Models currently available to chemists for understanding molecular structure are numerous. They range from simple, plastic, physical molecular models to sophisticated mathematical models. Mathematical models include molecular mechanics, the extended Hückel method, Austim Model 1 (AM1), the local density functional approach, and large-scale computer intensive ab initio electronic structure procedures using extended basis sets and highly correlated electronic wave functions. Each has been usefully applied to chemical problems and each has practical limitations. For example, ab initio electronic structure procedures can be applied to any combination of atoms because the model (as the name ab initio or "from first principles" implies) is not parameterized, and hence does not require experimental data in the development of the model. But, use of ab initio electronic structure procedures requires a lot of computer time, a great deal of disk space, and a significant amount of random access memory (RAM). This means that use of the model is restricted to systems with a few to several atoms. Molecular mechanics, at the other extreme of mathematic complexity, requires several orders of magnitude less computer time to use, and so can simulate large systems, but needs to be parameterized.

Molecular mechanics is a simple, empirical "ball-and-spring" model of molecular structure. Atoms (balls) are connected by springs (bonds) that can be stretched or compressed due to intra- or intermolecular forces. The sizes of the balls and the stiffness of the springs are determined empirically, that is, they are chosen to reproduce experimental data. As with a metal spring, when a bond is pulled or "stressed," the bond will lengthen or be "strained." This change in the bond length causes the energy of the bond to rise; the increase in energy is called a bond strain energy. One important source of stress in molecules is steric interactions. Steric interactions can affect more than bond lengths; the entire structure of a molecule can change. The associated energy change is called a steric strain energy (Adams and Yuan, 1933; Brown et al., 1942).

Both suitably parameterized molecular mechanics models and ab initio electronic structure procedures can reproduce rotational barriers about single bonds and conformational energy differences (see Table 1.1), but only molecular mechanics can be used to understand the motion of segments of DNA (deoxyribonucleic acid) in solution.

Because electrons are not a part of most molecular mechanics models one cannot use these techniques to understand electronic spectroscopy and photochemistry; instead one must use electronic structure methods. Molecular mechanics models can, however, be parameterized to reproduce transition state molecular structures. The transition state mo-

TABLE 1.1. Comparison of Ab Initio and Molecular Mechanics Models of Torsional Barriers and Conformational Energy Differences[a]

Molecule Barriers	Experiment[b]	MM2	MM3	Ab Initio[c]
Me−Me	2.93	2.73[d]	2.41[d]	3.0
Me−SiH$_3$	1.7			1.4
Me−NH$_2$	2.0		1.45[e]	2.4
Me−PH$_2$	2.0		1.4[f]	2.0
Me−OH	1.1			1.4
Me−SH	1.3			1.4
trans HO−OH	1.1	1.13[g]	1.05[g]	0.9
cis HO−OH	7.0	7.04[g]	7.11[g]	9.2
trans HS−SH	6.8			6.1
cis HS−SH	7.2			8.5
Conformational energy differences				
Butane (gauche–trans)	0.97 ± 0.05[h]	0.86[d]	0.81[d]	0.8
Cyclohexane (chair–twist–boat)	5.5[i]	5.4[d]	5.8[d]	6.1[j]

[a] Energies are in kilocalories per mole (kcal/mol).
[b] From Lister (1978) unless noted.
[c] From Hehre et al., (1986) unless noted.
[d] From Allinger et al., (1989).
[e] From Schmitz and Allinger, (1990).
[f] From Fox et al., (1992).
[g] From Chen and Allenger, (1993).
[h] From Murphy et al., (1991).
[i] From Squillacote et al., (1975).
[j] From Leong et al., (1994).

lecular mechanics model does not provide information about the underlying electronic basis for low barriers in fast reactions, and high barriers in reactions that do not occur. Furthermore, the transition state modeling method relies on an electronic structure determination of the transition state geometry for use in the parameterization.

The shortcomings of molecular mechanics discussed above obscure the fact that often the act of simply visualizing a molecular structure is enough to spur new experimental research. The genetic engineering of an artifical insulin, discussed in Section 3.4, simply used a computer generated image or model of the X-ray crystal structure to select the sites of amino acid modification. The present text is primarily focused on the use of molecular mechanics models in chemistry. Most of these models are slightly more complex than a molecular image displayed on a computer screen but substantially less sophisticated than electronic structure approaches. Over the past 50 years molecular mechanics has grown into a technique used extensively by chemists to quantitate the role that steric strain plays in determining molecular structure, conformational structure, the energy differences between conformations of a molecule, and the binding interactions between molecules. In this chapter we follow the development of molecular mechanics models from the simple estimation of conformational energy differences, through the estimation of molecular geometries, to the determination of the structures of families of molecular conformations.

1.2. ENERGY EXPRESSIONS IN MOLECULAR MECHANICS

The central $C-C$ bond of lactic acid, as with other sp^3-sp^2 bonds, can adopt a conformation with either the hydrogen, the methyl, or the hydroxyl functional group of the sp^3 center eclipsing the carbonyl $C-O$ bond as indicated by the labels a, b, and c in the Newman projection shown in Figure 1.1. The three conformations are labeled **1.2H**, **1.2C**, and **1.2X** in Figure 1.2.

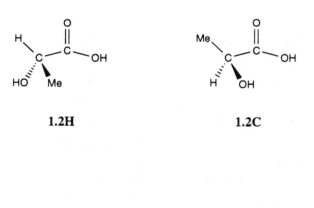

1.2H   **1.2C**

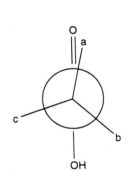

FIGURE 1.1.
Newman project of lactic acid, **1.2**.

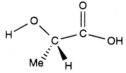

1.2X

1.2G

FIGURE 1.2.
Conformers of lactic acid, **1.2**.

In addition, if the hydroxyl group is in the eclipsing position then either an open conformation, **1.2X**, or an internally hydrogen-bonded closed conformation **1.2G** is possible (Me = CH_3). The internally hydrogen-bonded conformation, **1.2G**, is the observed gas-phase structure (van Eijck, 1983). The open eclipsed hydroxyl conformation **1.2X**, is the crystalline form (Schouten et al., 1994). Which conformation is preferred when lactic acid binds to the enzyme lactate dehydrogenase? Molecular mechanics can be used to understand the relative energies of each of the conformations for lactic acid as an isolated molecule, for lactic acid in the solid state (see Section 2.5.3), for lactic acid in solution (see Sections 3.5 and 5.2), and for lactic acid bound to the active site of lactate dehydrogenase (see Section 4.2).

To estimate the energetic difference between the various conformations of lactic acid in the gas phase, the potential energy of each conformation is calculated and compared. The potential energy for one conformer is expressed as a sum of valence or bonded interactions and nonbonded interactions, Eq. (1.1).

$$V = V_r + V_\theta + V_\phi + V_\omega + V_{vdW} + V_{el} \tag{1.1}$$

The valence interactions consist of bond stretching (V_r) and angular distortions. Included as angular distortions are bond angle bending terms (V_θ), dihedral angle torsional terms (V_ϕ), and, at times, inversion terms (V_ω). The nonbonded interactions consist of van der Waals (V_{vdW}) terms that parametrically describe the long-range induced dipole-induced dipole attraction and the short-range repulsion between any two electron densities, and electrostatic (V_{el}) terms describing the partial ionic chararacter of polar covalent bonds. At times special terms are added to describe hydrogen-bonding interactions.

The sum of all of the potential energy terms for a particular atom gives a mathematical representation for how that atom would move under the influence of the motions or displacements of all the other atoms in the system; that is, the potential energy equation is a mathematical representation for the forces experienced by that atom. Thus, the sum of the potential energy terms in a molecular mechanics model is often referred to as a force field.

The contribution that kinetic energy (temperature) makes to a molecular energy can be estimated either by carrying out a statistical thermodynamics analysis as discussed in Appendix B, a molecular dynamics study as discussed in Appendix C, or a Monte Carlo study as discussed in Appendix D.

The equations used to describe the interactions in a force field are described in more detail below. A discussion of the precise choice of parameters [natural or strainless bond distances, angles, etc., force constants, van der Waals distances and well depths, and partial charges (or bond dipoles)] has been left to Chapters 2, 3, and 8.

1.2.1. Bond Stretch

In the original Hill (1946) and Westheimer and Mayer (1946) formulations of molecular mechanics all atomic displacements from the bottom of the well were treated as harmonic. That is, for the stretching distortion of a bond between centers I and J Eq. (1.2) was used where r_{IJ} is the natural bond distance between centers I and J, and k_{IJ} is the spring or force constant for the stretch between I and J.

$$V_r = \frac{1}{2} k_{IJ}(r - r_{IJ})^2 \tag{1.2}$$

The distance r_{IJ} and force constant k_{IJ} are parameters of the force field. If a particular bond in a molecule, say the C—O double bond in **1.2**, has a bond distance, **r**, equal to r_{IJ}, then as can be seen from Eq. (1.2), the potential energy for that bond, V_r, is zero and the bond is said to be strain free. The art of force field development is primarily associated with determining the parameters such as the natural or strain-free distances, r_{IJ}, and force constants, k_{IJ} (see Chapter 8).

In the work of Dostrovsky et al. (1946) on the steric contribution to nucleophilic substitution, a Morse function (Morse, 1929) was used for the bond stretch rather than a harmonic function because bond breaking was being studied, see Eq. (1.3).

$$V_R = D_{IJ}[e^{-\alpha(r-r_{IJ})} - 1]^2 \tag{1.3}$$

In Eq. (1.3), D_{IJ} is the bond energy for the bond between centers I and J, r_{IJ} is the unstrained or natural bond distance, and α is obtained from Eq. (1.4), where k_{IJ} is the spring or force constant for the stretch between I and J.

$$\alpha = \left[\frac{k_{IJ}}{2D_{IJ}}\right]^{1/2} \tag{1.4}$$

A comparison of the Morse function with a harmonic stretch is provided in Figure 1.3. It should be apparent from Figure 1.3 that the Morse function more correctly describes dissociation. Thus this function should be used if bonds are being strained (stretched) as in bond breaking. Series approximations to the Morse function are used in some modern

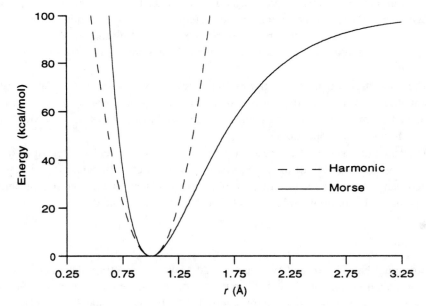

FIGURE 1.3.
Energy versus distance curves for a harmonic stretch and a Morse function. A natural distance of 1.0 Å and a force constant of 700 kcal/mol-Å² were used for each. The Morse curve assumed a bond energy of 100 kcal/mol.

force fields such as MM2 and are discussed in Chapter 2 within the context of the MM2 force field.

As shown in **1.2**, a conventional, chemist's structural formula for lactic acid, there is one C−O double bond, there are two C−O single bonds, two O−H bonds, two C−C single bonds, and four C−H bonds, totaling 11. The potential energy in a molecular mechanics description of lactic acid would reflect the structural formula and include 11 stretch terms, one for each bond.

1.2.2. Angle Bend

Angle bend interactions are still commonly taken as harmonic in θ, Eq. (1.5), though other expansions are in use and are discussed in Chapters 2, 3, 7, and 8.

$$V_\theta = \frac{1}{2}k_{IJK}(\theta - \theta_{IJK})^2 \tag{1.5}$$

In Eq. (1.5) k_{IJK}, the bending force constant, and θ_{IJK}, the strain-free bond angle, are the parameters associated with the angle between the IJ bond and the JK bond, see Figure 1.4.

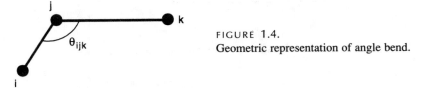

FIGURE 1.4.
Geometric representation of angle bend.

Schematic representations of the angle bend terms for lactic acid are shown in Figure 1.5. The angle bend terms include those centered at the carbonyl (C=O) carbon, at the hydroxyl (OH) oxygen atoms, and the methyl (CH₃) and methine (CH) carbon atoms.

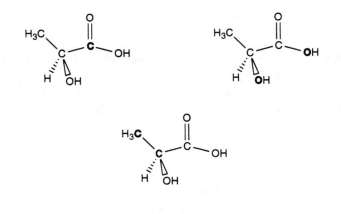

FIGURE 1.5.
Angle terms for lactic acid.

1.2.3. Torsion

As you remember from organic chemistry the substituents surrounding sp^3-sp^3 bonds adopt one of several well-defined, equivalent staggered structures. The *trans* and *gauche* conformations for hydracrylic acid are given in **1.1t** and **1.1g**. Eclipsed alternatives occur as transition states or saddle points connecting the staggered structures.

Torsional potentials are used to mimic the preference for staggered conformations about sp^3-sp^3 bonds and the preference for eclipsed conformations about sp^2-sp^3 bonds. A torsional potential is the attractive or repulsive two bond or four center interaction between the bonds between centers I and J and centers K and L that are connected by a common bond JK (see Fig. 1.6 for a geometric representation of a torsion). Torsional potential

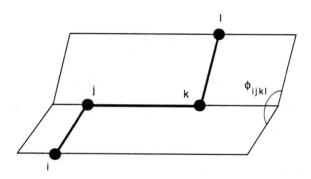

FIGURE 1.6.
Geometric representation of torsion.

models date to the work of Kemp and Pitzer (1936) on ethane. One of the first uses of torsional potentials in molecular mechanics was in the study of Pitzer and Donath (1959) on cyclopentane. The torsional potential is still almost always described with a small cosine expansion in ϕ, see Eq. (1.6), where ϕ is the torsional angle, and K_{IJKL} is the force constant.

$$V_\phi = K_{\mathrm{IJKL}} \sum_{n=0}^{m} C_n \cos n\phi \tag{1.6}$$

The coefficients C_n are determined by the rotational barrier V_ϕ, the periodicity of the potential, and the natural angle, ϕ_{IJKL}.

As shown in Figure 1.7 dihedral angle torsional terms for lactic acid would include rotations about the carbonyl carbon to methine bond, the two carbon to hydroxide bonds, and the methine to methyl bond.

1.2.4. van der Waals

In the original Hill (1946) and Westheimer and Mayer (1946) force fields, nonbonded interactions (van der Waals forces) were included using a Lennard-Jones 6-12 function, Eq. (1.7) (Lennard-Jones, 1924; Hill, 1946) or an exponential repulsion, Eq. (1.8) (West-

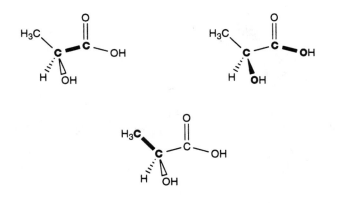

FIGURE 1.7.
Torsional terms for lactic acid.

heimer and Mayer, 1946), where ρ is the nonbonded distance, A_{IJ}, A, and B are inner-wall repulsion parameters, and B_{IJ} describes the van der Waals attraction between atoms.

$$V_{vdW} = \frac{A_{IJ}}{\rho^{12}} - \frac{B_{IJ}}{\rho^6} \tag{1.7}$$

$$V_{vdW} = Ae^{-B\rho} \tag{1.8}$$

Hill (1948) switched to an exponential-6 or Buckingham (1938) potential, Eq. (1.9), where ρ is the nonbonded distance, A and B are inner-wall repulsion parameters, and C_6 describes the van der Waals attraction.

$$V_{vdW} = Ae^{-B\rho} - \frac{C_6}{\rho^6} \tag{1.9}$$

1.2.5. Electrostatics

In the early work of Hill and Westheimer, electrostatic interactions were ignored. Kitaygorodsky (1961, 1965) proposed using either point partial charges or bond dipoles. The point charge energy is given by Eq. (1.10) and the bond dipole representation is given by Eq. (1.11).

$$V_{el} = C\frac{q_I q_J}{\varepsilon\rho} \tag{1.10}$$

$$V_{el} = C\frac{\mu_I\mu_J}{\varepsilon\rho^3}(\cos\chi - 3\cos\alpha_i\cos\alpha_j) \tag{1.11}$$

In Eq. (1.10), q_I and q_J are partial charges in electron units, ρ is the nonbonded distance in angstroms, ε is the dielectric constant, and C converts to the energy units of interest

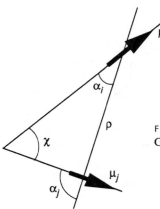

FIGURE 1.8.
Geometric representation of dipole–dipole interaction.

(332.06 for energies in kilocalories per mole). The additional variables for the bond dipole representation, Eq. (1.11), are as described in Figure 1.8.

Wiberg (1965) discussed the dilemma of including or not including nonbonded interactions between atoms connected through a central atom (1, 3 interactions). It is still the usual convention in molecular mechanics to exclude van der Waals and electrostatic interactions for atoms that are bonded to each other (1, 2 interactions) or bonded to a common atom (1, 3 interactions).

1.2.6. Other Terms

In order to reproduce molecular structures, vibrational frequencies, or intermolecular distances, other terms are sometimes used in molecular mechanics force fields. These terms include inversions (for amines and sp^2 carbon centers see Fig. 1.9) and specific hydrogen-bond potentials; these terms will be described in Chapters 2 and 3 in the context of specific force fields.

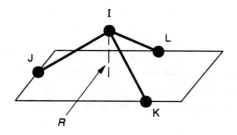

FIGURE 1.9.
Geometric representation of inversion term.

For lactic acid an inversion term centered at the sp^2 carbonyl carbon would be included, see Figure 1.10.

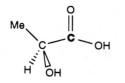

FIGURE 1.10.
Inversion term for lactic acid.

In order to properly account for the internal hydrogen bond, most molecular mechanics descriptions of **1.2G** would include a hydrogen-bond term to mathematically represent the interaction shown by a dashed line in **1.2G**.

1.2.7. Example

To see the above mathematics in action we will compare the molecular mechanics energies for four conformations of lactic acid. The total steric energies and individual term energies for four conformations of lactic acid are collected in Table 1.2. Bond distances, bond angles, and torsional angles for each of the four conformations are collected in Tables 1.3 and 1.4. The numerical results were obtained using the MM2 force field as implemented in the commercial CAChe software package.

The total energies (the first entries in Table 1.2) are used to order the conformations in terms of overall stability. The observed gas-phase conformation **1.2G** has the lowest total energy, and **1.2X**, the conformation observed in the X-ray structure, the highest energy. The stretch, bend, torsion, inversion, van der Waals, electrostatic, and hydrogen-bond terms in the energy expression can be examined to find a source of energetic differentiation. Each of the four conformations has essentially the same stretch, bend, stretch-bend cross term, and inversion energy. Conformation **1.2C** has the lowest, most favorable torsional energy. Conformation **1.2G** has the highest, most repulsive van der Waals energy and **1.2X**, the least favorable conformation, has the least repulsive van der Waals energy. Conformation **1.2G** has the least repulsive electrostatic energy; in fact, the electrostatic energy ordering parallels the total energy ordering. Each of the four conformations has a significant hydrogen-bonding contribution; for **1.2G** it is the term indicated by the dashed line in Figure 1.2c. for compound **1.2G.** For the other three conformations the methine hydroxyl group makes a weaker, but still significant, hydrogen bond to the oxygen of the carboxylic acid hydroxyl group. Given the similarity in stretch, bend, and inversion energies it is not surprising that the bond distances and angles are nearly the same for each conformation, see Table 1.3.

TABLE 1.2. Conformational Energy Comparisons for Lactic Acid[a]

Term	Conformation			
	1.2G	**1.2C**	**1.2H**	**1.2X**
Total energy	−2.374	−0.237	+0.063	+1.303
Stretch	0.117	0.095	0.082	0.077
Angle	1.208	1.222	1.200	1.195
Stretch–bend	0.054	0.067	0.057	0.060
Torsion	−2.375	−2.949	−2.298	−2.073
Inversion	0.045	0.013	0.000	0.019
van der Waals	2.095	1.596	1.705	1.064
Electrostatics	3.283	3.941	4.223	4.778
Hydrogen bonding	−6.801	−4.222	−4.907	−3.817

[a] Energies are in kilocalories per mole (kcal/mol).

TABLE 1.3. Distance and Angle Comparisons for Lactic Acid[a]

Geometric Parameter	Natural Value	Conformation			
		1.2G	**1.2C**	**1.2H**	**1.2X**
C=O	1.208	1.207	1.206	1.206	1.207
C(=O)−C	1.509	1.517	1.517	1.516	1.517
C(=O)−OH	1.338	1.331	1.333	1.333	1.334
C(Me)−OH	1.402	1.410	1.409	1.410	1.408
C−Me	1.523	1.532	1.533	1.532	1.532
C(=O)O−H	0.972	0.972	0.971	0.971	0.971
C(Me)O−H	0.942	0.942	0.945	0.945	0.943
C(Me)−H	1.113	1.117	1.116	1.115	1.116
C−H	1.113	1.114	1.115	1.114	1.114
C−H	1.113	1.115	1.114	1.114	1.114
C−H	1.113	1.115	1.114	1.114	1.114
O=C−O	122.0	121.815	120.932	121.209	120.603
C(=O)−O−H	106.1	97.960	98.549	98.301	98.457
C(Me)−O−H	106.9	108.647	109.560	109.222	108.740
C(Me)−C(=O)=O	122.5	125.117	127.853	128.315	128.019
C(Me)−C(=O)−OH	107.1	112.989	111.193	110.476	111.346
Me−C(Me)−C(=O)	109.9	112.812	113.215	111.586	112.294
Me−C(Me)−OH	107.7	108.098	108.838	108.980	108.082
Me−C(Me)−H	109.39	110.981	109.454	110.000	111.013
OH−C(Me)−C(=O)	109.5	110.666	111.448	112.400	111.192
OH−C(Me)−H	106.7	106.004	105.977	105.816	106.295
C(=O)−C(Me)−H	107.9	108.075	107.639	107.885	107.815
C(Me)−C−H	110.0	111.212	111.116	111.213	111.480
C(Me)−C−H	110.0	110.443	110.795	110.643	110.676
C(Me)−C−H	110.0	111.117	111.104	111.116	111.005
H−C−H	109.0	108.332	107.902	107.898	108.041
H−C−H	109.0	107.509	107.707	107.871	107.433
H−C−H	109.0	108.101	108.075	107.961	108.058

Bond distances are in angstroms (Å) and angles are in degree (°).

Since conformations are related by rotations about single bonds, significant differences in torsional angles are to be expected, see Table 1.4. We can find an explanation for the favorable torsional strain energy of conformation **1.2C** by examining Tables 1.4 and 1.5. The first nine torsions are for rotation about the sp^3–sp^3 C−CH$_3$ bond; in all cases the methyl hydrogen atoms adopt the expected *trans* ($\phi = 180°$) and *gauche* ($\phi = \pm 60°$) positions. The remaining torsions vary substantially. The largest difference in the torsional energies listed in Table 1.5 is for the methyl–methine–C=O torsion, where **1.2C** has a slightly favorable -0.08- kcal/mol torsional strain energy and the other three conformations have a more than 0.6-kcal/mol repulsive contribution. As discussed above, carbonyls prefer to have substituents eclipsing the C−O bond. As used by MM2, Eq. (1.6) is rearranged into an equivalent form given by Eq. (1.12).

$$V_\phi = \frac{1}{2}V_1(1 + \cos \phi) + \frac{1}{2}V_2(1 - \cos 2\phi) + \frac{1}{2}V_3(1 + \cos 3\phi) \qquad (1.12)$$

TABLE 1.4. Torsional Angle Comparisons for Lactic Acid[a]

Torsion (I−J−K−L)	Potential Parameters			Conformation			
	V_1	V_2	V_3	1.2G	1.2C	1.2H	1.2X
OH−C(Me)−Me−H	0.000	0.000	0.180	176.334	176.540	−176.833	175.568
OH−C(Me)−Me−H	0.000	0.000	0.180	−63.352	−56.636	−56.650	−64.152
OH−C(Me)−Me−H	0.000	0.000	0.180	56.609	63.307	63.250	55.835
C(=O)−C(Me)−Me−H	0.000	0.000	0.180	179.310	178.842	178.638	178.850
C(=O)−C(Me)−Me−H	0.000	0.000	0.180	−60.964	−61.215	−61.461	−61.416
C(=O)−C(Me)−Me−H	0.000	0.000	0.180	59.349	58.938	58.455	58.864
H−C(Me)−Me−H	0.000	0.000	0.237	−179.199	178.707	178.839	179.642
H−C(Me)−Me−H	0.000	0.000	0.237	−59.239	−61.139	−61.245	−60.371
H−C(Me)−Me−H	0.000	0.000	0.237	60.487	58.764	58.938	59.362
Me−C(Me)−O−H	0.800	0.000	0.090	144.621	−69.803	−85.229	−91.429
C(=O)−C(Me)−O−H	0.000	0.000	0.090	20.620	55.754	39.003	144.887
H−C(Me)−O−H	0.000	0.000	0.200	−96.326	172.570	156.519	27.809
Me−C(Me)−C(=O)=O	−0.130	0.904	0.050	−124.051	0.456	−84.442	−121.758
Me−C(Me)−C(=O)−OH	0.400	−0.300	−0.070	59.123	178.730	95.275	60.324
OH−C(Me)−C(=O)=O	0.000	0.000	0.000	−2.798	−122.636	152.780	−0.510
OH−C(Me)−C(=O)−OH	0.000	0.000	−0.110	−179.624	55.638	−27.503	−178.427
H−C(Me)−C(=O)=O	−0.167	0.000	−0.100	112.865	121.564	36.499	115.637
H−C(Me)−C(=O)−OH	0.000	0.000	−0.016	−63.961	−60.163	−143.784	−62.280
C(Me)−C(=O)−O−H	0.000	0.500	0.000	176.731	−177.884	179.957	178.376
O=C(=O)−O−H	−3.285	5.600	0.000	−0.213	0.526	−0.303	0.283

[a] Angles are in degree (°).

TABLE 1.5. Torsional Strain Energy Comparisons for Lactic Acid[a]

Torsion (I−J−K−L)	Potential Parameters			Conformation			
	V_1	V_2	V_3	1.2G	1.2C	1.2H	1.2X
Me−C(Me)−O−H	0.800	0.000	0.090	0.131	0.544	0.467	0.438
C(=O)−C(Me)−O−H	0.000	0.000	0.090	0.066	0.001	0.025	0.057
H−C(Me)−O−H	0.000	0.000	0.200	0.133	0.007	0.067	0.111
Me−C(Me)−C(=O)=O	−0.130	0.904	0.050	0.641	−0.080	0.842	0.673
Me−C(Me)−C(=O)−OH	0.400	−0.300	−0.070	0.082	0.000	−0.160	0.073
OH−C(Me)−C(=O)=O	0.000	0.000	0.000	0.000	0.000	0.000	0.000
OH−C(Me)−C(=O)−OH	0.000	0.000	−0.110	0.000	−0.001	−0.062	0.000
H−C(Me)−C(=O)=O	−0.167	0.000	−0.100	−0.148	−0.140	−0.184	−0.146
H−C(Me)−C(=O)−OH	0.000	0.000	−0.016	0.000	0.000	−0.011	0.000
C(Me)−C(=O)−O−H	0.000	0.500	0.000	0.002	0.001	0.000	0.000
O=C(=O)−O−H	−3.285	5.600	0.000	−3.285	−3.284	−3.285	−3.285

[a] Energies are in kilocalories per mole (kcal/mol).

13

The MM2 force field accounts for the preference for sp^3–sp^2 bond eclipsing by the cos 2ϕ term; when ϕ is near zero, cos 2ϕ is near one, so $1 - \cos 2\phi$ is roughly zero. Plugging the data for the four conformations of lactic acid given in Tables 1.4 and 1.5 for the methyl–methine–C=O torsion term into Eq. (1.12) yields Eqs. (1.13–1.16).

$$V_{1.2G} = \frac{1}{2}\{-0.130[1 + \cos(-124.051)] + 0.904[1 - \cos(2 \times -124.051)]$$
$$+ 0.05[1 + \cos(3 \times -124.051)]\} \quad (1.13)$$

$$V_{1.2C} = \frac{1}{2}\{-0.130[1 + \cos(0.456)] + 0.904[1 - \cos(2 \times 0.456)]$$
$$+ 0.05[1 + \cos(3 \times 0.456)]\} \quad (1.14)$$

$$V_{1.2H} = \frac{1}{2}\{-0.130[1 + \cos(-84.442)] + 0.904[1 - \cos(2 \times -84.442)]$$
$$+ 0.05[1 + \cos(3 \times -84.442)]\} \quad (1.15)$$

$$V_{1.2X} = \frac{1}{2}\{-0.130[1 + \cos(-121.758)] + 0.904[1 - \cos(2 \times -121.758)]$$
$$+ 0.05[1 + \cos(3 \times -121.758)]\}(1.16)$$

Evaluation of the cosine terms yield Eqs. (1.17–1.20).

$$V_{1.2G} = \frac{1}{2}[-0.130 \times 0.4401 + 0.904 \times 1.3730 + 0.05 \times 1.9776) =$$
$$\frac{1}{2}(-0.0572 + 1.241 + 0.099) \quad (1.17)$$

$$V_{1.2C} = \frac{1}{2}(-0.130 \times 1.9999 + 0.904 \times -0.0001 + 0.05 \times 1.9997) =$$
$$\frac{1}{2}(-0.260 + 0.000009 + 0.10) \quad (1.18)$$

$$V_{1.2H} = \frac{1}{2}(-0.130 \times 1.0969 + 0.904 \times 1.9812 + 0.05 \times 0.7131) =$$
$$\frac{1}{2}(-0.143 + 1.7911 + 0.0356) \quad (1.19)$$

$$V_{1.2X} = \frac{1}{2}(-0.130 \times 0.4737 + 0.904 \times 1.4459 + 0.05 \times 1.9958) =$$
$$\frac{1}{2}(-0.0616 + 1.307 + 0.099) \quad (1.20)$$

The small magnitude of the second term in Eq. (1.18) is responsible for the small torsional energy for **1.2C**.

In addition to determining the relative ordering of the conformations of lactic acid the individual terms of a molecular mechanics energy expression can be used to provide a fairly detailed explanation for the relative ordering. At this point a relevant question arises: Where did the geometric coordinates of the conformations come from? The answer to this question is the subject of Section 1.3.

1.3. MINIMIZATION TECHNIQUES

As shown in Table 1.3 the carbonyl $C-O$ bond distance was slightly different for each of the four conformations of lactic acid and each $C-O$ bond was slightly perturbed from the strain-free or natural value of 1.208 Å. The angle strain energy was also slightly different for each of the four conformations. The only angle that was perturbed more than a degree from the strain-free value was the carboxylic acid $C-O-H$ angle. In contrast, most of the torsion angles deviated substantially from their strain-free values of $0°$, $\pm60°$, or $180°$. These distortions are due to the small magnitudes of single-bond torsion potentials and the presence of comparable magnitude nonbonded interactions. The 3-D shape of a molecule, even a molecule as simple as lactic acid, is dictated by a summation of a large number of competing weak interactions including torsional potentials, van der Waals forces, electrostatic interactions, and hydrogen-bonding interactions. Assessing the relative importance of each of these competing, energetically small interactions visually from a physical model is rather difficult, and energy minimization of a computer model is usually resorted to in order to determine the overall structure of a molecule. The equilibrium, minimum, or ''best'' structures of lactic acid referred to in Section 1.2 were structures that had been minimized. These minimized structures were those for which all forces were balanced. There are a number of procedures for obtaining ''best'' or minimized molecular structures.

We begin a discussion of energy minimization by considering a one-dimensional (1-D) harmonic oscillator where the energy, V, is expressed in terms of a single variable, x, and two constants, k and x_0, see Eq. (1.21).

$$V = \frac{1}{2}k(x - x_0)^2 \tag{1.21}$$

The objective, given a trial structure, say x_1, is to determine how to move from x_1 to the geometry, that is, the value of x, that minimizes the energy V. Differentiation of the function given by Eq. (1.21) with respect to the variable x twice yields Eqs. (1.22) and (1.23) for the first and second derivatives.

$$\frac{dV}{dx} = k(x - x_0) \tag{1.22}$$

$$\frac{d^2V}{dx^2} = k \tag{1.23}$$

Combination of Eqs. (1.22) and (1.23), rearrangement of the resulting equation, and substitution of x_1 for x results in an estimate for the distance from our current point x_1 to x_0 (the optimum or lowest energy structure) given by Eq. (1.24).

$$x_0 - x_1 = -\frac{1}{\frac{d^2V}{dx^2}}\frac{dV}{dx} \tag{1.24}$$

For this 1-D, harmonic example, the function will be minimized (have $\dfrac{dV}{dx} = 0$ and $\dfrac{d^2V}{dx^2} > 0$) in a single step. For an anharmonic potential it may take several steps to arrive at the minimum.

1.3.1. Newton–Raphson

The multidimensional analog of the above procedure is referred to as Newton–Raphson minimization. In a Newton–Raphson minimization the $3n$ coordinates to be minimized form a vector $\mathbf{x}$, the derivative of the energy with respect to the $3n$ coordinates also forms a vector $\mathbf{g}$, and the second derivatives of the energy with respect to the coordinates forms a matrix $\mathbf{A}$. The resulting multidimensional analog to Eq. (1.24) is Eq. (1.25).

$$\mathbf{x}_0 - \mathbf{x}_1 = -\mathbf{A}^{-1}\mathbf{g} \tag{1.25}$$

The practical difficulties of a Newton–Raphson minimization include determination of the second derivatives [for n atoms there are $(3n)^2$ second derivatives], storage of the $(3n)^2$ second derivative matrix elements, and obtaining the inverse of the $3n \times 3n$ second derivative matrix needed for Eq. (1.25). Three minimization methods more commonly used include the steepest descent method, the conjugate gradient method, and the Fletcher–Powell method. These methods are discussed below.

1.3.2. Steepest Descents

The steepest descent method, which uses the least information of the procedures described, is the least efficient but is the simplest to implement. In steepest descents $\dfrac{d^2V}{dx^2}$ (the force constant) is assumed to be a universal constant γ, and Eq. (1.26) is used to estimate the distance from x_1 to the minimum x_0.

$$x_0 - x_1 = -\gamma \frac{dV}{dx} \tag{1.26}$$

In Eq. (1.26) γ, the universal scaling constant, is used to make the geometric step of proper size.

1.3.3. Fletcher–Powell

In the Fletcher–Powell method the force constant $\left(\dfrac{d^2V}{dx^2}\right)$ is initially set to a constant and subsequently estimated from finite differences of the first derivative from one step to the next using Eq. (1.27).

$$\frac{d^2V}{dx^2} \cong \frac{\left(\dfrac{dV}{dx_{old}}\right) - \left(\dfrac{dV}{dx_{new}}\right)}{x_{old} - x_{new}} \tag{1.27}$$

Because the Fletcher–Powell method uses second derivative information, like the Newton–Raphson procedure, the difficulties are nearly the same. The Fletcher–Powell method does not have to incur the expense of evaluating the second derivative matrix but does suffer imprecision in the estimation of second derivative matrix.

1.3.4. Conjugate Gradient Method

The most common minimization technique used is the conjugate gradient method. Whereas the steepest descent method chooses the descent direction based on the gradient of the current step and makes a single step in this direction, the conjugate gradient method starts along the steepest descent direction, continues along this direction until a minimum in this direction is attained, and then proceeds along a direction perpendicular or conjugate to this current direction. The rationale is that once the current direction has been minimized, any improvement will occur in orthogonal directions. This process is continued until the energy is minimized below a preset threshold.

1.3.5. Comparison of the Methods

The above discussion becomes concrete when steepest descent, conjugate gradient, and Newton–Raphson minimizations of the same, crude trial structure of lactic acid are compared (see Table 1.6). The total energies and norm of the gradient, $|\text{grad}|$, of the energies (see Eq. 1.28) are reported for each minimization step.

$$|\text{grad}| = \sqrt{\sum_i \left(\frac{dV}{dx_i}\right)^2} \tag{1.28}$$

A small $|\text{grad}|$ is an indication that the overall derivative of the energy with respect to all of the coordinates is small. This quantity, in conjunction with the change in energy with each step, are often used to determine if the minimization procedure has converged or found an answer. The relative times for each procedure are also reported in Table 1.6. The attributes of a minimization procedure that determine its utility are the number of steps or guesses needed to find the ''best'' or lowest energy structure, and the time associated with each step. For the crude trial structure of lactic acid, the Newton–Raphson procedure required 15 steps of minimization, and the conjugate gradient required 72 steps to reach a $|\text{grad}|$ of 0.5 kcal/mol-Å or a root mean square (rms) gradient of 0.14 kcal/mol-Å (rms $= \dfrac{|\text{grad}|}{\sqrt{n}}$, where n is the number of variables, here n is $3 \times$ the number of atoms in the molecule). Because each Newton–Raphson step took more computer time than each conjugate gradient step, each of the two procedures required the same amount of computer time to find the minimum. Note that the steepest descents procedure still had not converged after 500 steps. From a comparison of energy sequences for the conjugate gradient and Newton–Raphson procedures, one can see that the conjugate gradient procedure reached a relative energy of 14.00 after 14 iterations ($\sim \frac{1}{5}$ of the total number of iterations) and the Newton–Raphson reached the same relative energy after 5 steps ($\sim \frac{1}{3}$ of the total). These observations support the common practice of using either steepest descents or conjugate gradient minimization for the initial ''rough'' minimization of a crude trial struc-

TABLE 1.6. Minimization Comparisons[a]

Iteration	Steepest Descents CPU time: 41.08 (in s)		Conjugate Gradient CPU time: 15.77 (in s)		Netwon Raphson CPU time: 14.84 (in s)	
	Energy	|grad|	Energy	|grad|	Energy	|grad|
1	1541.51	2363.2	1541.51	2363.2	1541.51	2363.2
2	709.34	1633.6	586.36	1023.6	409.50	985.9
3	456.35	1226.9	257.37	501.9	76.54	214.2
4	294.83	893.4	137.73	280.7	23.08	62.8
5	218.34	683.8	102.66	261.8	14.00	125.9
6	168.73	528.2	68.53	185.5	7.41	51.4
7	137.86	419.4	45.55	151.7	10.77	186.4
8	116.23	336.2	32.09	134.9	2.27	37.6
9	100.20	271.6	25.06	111.8	2.61	84.8
10	87.99	221.7	18.17	83.8	0.96	40.9
11	78.29	182.9	15.95	40.6	0.44	3.5
12	70.42	153.6	15.37	38.2	1.89	89.5
13	63.88	131.5	14.61	36.4	0.07	1.4
14	58.34	115.1	14.00	20.5	0.15	25.4
15	53.57	102.8	13.78	18.7	0.00	0.2
16	49.41	93.6	13.59	13.5		
17	45.76	86.3	13.47	11.8		
18	42.51	80.4	13.40	15.0		
19	39.62	75.5	13.29	12.6		
20	37.04	71.1	13.15	14.0		
21	34.72	67.3	13.07	21.9		
22	32.64	63.7	12.93	23.2		
23	30.76	60.4	12.69	45.2		
24	29.07	57.4	12.04	45.8		
25	27.54	54.5	11.25	57.9		
26	26.16	51.9	10.25	73.9		
27	24.91	49.3	7.76	63.6		
28	23.78	46.9	5.91	53.4		
29	22.76	44.7	3.90	45.8		
30	21.83	42.5	2.76	38.2		
31	20.99	40.5	1.73	29.7		
32	20.22	38.6	1.22	21.1		
33	19.53	36.8	0.87	23.0		
34	18.90	35.1	0.54	15.5		
35	18.33	33.5	0.36	12.4		
36	17.80	32.0	0.28	11.1		
37	17.33	30.6	0.21	9.1		
38	16.89	29.2	0.15	7.0		
39	16.49	27.9	0.11	4.0		
40	16.13	26.8	0.10	4.4		
42	15.79	25.6	0.09	2.6		
44	15.49	24.6	0.08	3.0		
45	15.21	23.6	0.07	2.6		
46	14.95	22.6	0.06	2.8		

TABLE 1.6. (*continued*)

	Steepest Descents			Conjugate Gradient			Netwon Raphson	
CPU time: (in s)	41.08			15.77			14.84	
Iteration	Energy	|grad|		Energy	|grad|		Energy	|grad|
47	14.71	21.8		0.06	2.7			
48	14.49	20.9		0.05	3.2			
49	14.28	20.2		0.04	4.1			
50	14.09	19.5		0.03	2.9			
51	13.91	18.8		0.03	2.7			
52	13.75	18.2		0.03	3.2			
53	13.59	17.6		0.02	3.0			
54	13.44	17.1		0.02	3.0			
55	13.31	16.6		0.01	1.8			
56	13.18	16.1		0.01	1.5			
57	13.06	15.7		0.01	1.7			
58	12.94	15.3		0.01	1.5			
59	12.83	14.9		0.01	1.7			
60	12.73	14.6		0.01	1.3			
61	12.63	14.3		0.01	1.5			
62	12.53	14.0		0.01	1.1			
63	12.44	13.8		0.00	1.0			
64	12.35	13.6		0.00	1.2			
65	12.26	13.4		0.00	1.1			
66	12.17	13.2		0.00	1.0			
67	12.10	13.1		0.00	0.8			
68	12.01	12.9		0.00	0.8			
69	11.93	12.8		0.00	0.8			
70	11.85	12.7		0.00	0.5			
71	11.78	12.6		0.00	0.6			
72	11.70	12.5		0.00	0.5			
74	11.41	12.3						
76	11.26	12.2						
80	10.98	12.2						
90	10.27	12.2						
100	9.56	12.4						
120	8.11	12.2						
130	7.42	11.8						
140	6.78	11.3						
150	6.19	10.8						
175	4.61	16.5						
200	3.81	11.9						
250	2.87	25.0						
300	2.30	17.8						
350	1.90	11.2						
400	1.63	20.8						
500	1.14							

[a] Energies are in kilocalories per mole (kcal/mol) and |grad| are in kilocalories per mole per angstrom (kcal/mol-Å).

ture, as these procedures are computationally more efficient at improving a rough guess, and then follow this by a few steps of Newton–Raphson minimization to "finish off" the process. Of the techniques discussed here, the Newton–Raphson method uses the most precise description of the molecular potential energy surface, and hence should take the fewest number of steps to yield a minimized structure. However, each step also takes the longest to perform of the methods presented.

1.4. CONFORMATIONAL SEARCHING

The presence of a large number of potential energy terms in a force field, including a number of weak interactions consisting of van der Waals forces, electrostatic interactions, and small barrier torsional potentials suggests that there may be more than one combination of atomic positions that gives a minimum on the potential energy surface. We have already discussed four conformations for lactic acid. Minimization procedures can only guide a molecule from the starting conformation to the nearest minimum on the potential energy surface. This local minimum is not necessarily, or in fact even usually, the best or lowest energy conformation; it is simply the minimum closest to the trial structure. The four conformations of lactic acid discussed in Section 1.1 were obtained by starting from four separate trial structures, each with torsional angles appropriate for the conformation being sought. There are other conformational possibilities for lactic acid involving rotations about the $C(=O)-OH$ bond. The lowest energy rotamer of this bond type has $\phi = 0°$ (already considered), but there is a secondary, higher energy minimum with $\phi = 180°$. A complete conformational analysis would include these $\phi = 180°$ conformations as well as the $\phi = 0°$ possibilities since intramolecular nonbonded interactions might make this a preferred conformation. A complete analysis by trial and error conformation construction will get tedious quickly. As the number of rotatable bonds increases trial and error will also become faulty because low-energy combinations of medium energy torsional minima are likely to be missed. Fortunately, in addition to assessing discrete molecular structures, molecular mechanics can be used to systematically determine the number of minima and the energetic differences between these minima.

   Determination of molecular shape or structure is thus a problem of finding the global or lowest energy minimum and the relatively small set of low-energy conformations present in a large family of viable structures. Insuring that one has obtained the best solution is, in fact, an insoluble problem. Though the global minimum problem has yet to be solved in mathematics, many conformational searching techniques have been developed and used successfully in molecular mechanics. These techniques are overviewed in Appendix E. Exhaustive grid searching, Monte Carlo searching, and molecular dynamics searching are outlined here.

1.4.1. Grid Searches

Grid search methods systematically vary each of several geometric variables in a molecule, typically the torsional angles, by some increment, while keeping the remaining coordinates (bond lengths and angles) fixed. Alternatively, the geometric variables of interest can be incrementally changed while all of the remaining coordinates are independently optimized; this generates an adiabatic grid surface. Either way one simply looks at the set

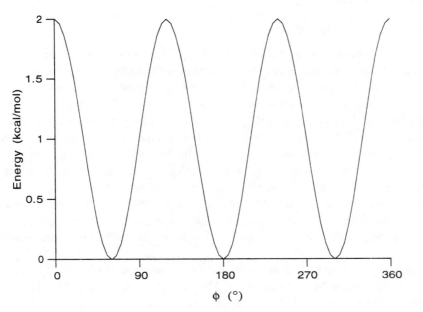

FIGURE 1.11.
Energy as a function of torsional angle for a sp^3 center bonded to a sp^3 center.

of resulting structures and energies and picks out the low-energy conformations. Unfortunately, the number of possible conformations grows exponentially with the number of torsions.

Typical sp^3–sp^3 torsional potentials have barriers of about 2 kcal/mol, and three minima, see Figure 1.11.

Thus, for each sp^3–sp^3 bond in a molecule there are likely three minima (ignoring cases were atom–atom collisions exclude a conformation). For n such bonds in a molecule there are 3^n possible minima with a corresponding exponential growth in complexity (3, 9, 27, 81, 243, 729, 2187, 6561, ...).

Without a procedure to select in advance only those conformers that are likely to be relatively low-energy local minima, the generation and energy evaluation of conformations becomes intractable for large molecules.

1.4.2. Monte Carlo

Within the framework of a Monte Carlo (MC) simulation (Wood and Parker, 1957) the dynamic behavior of a molecule is simulated by making random changes to the system, such as random changes in dihedral angles. The energy of a trial conformation is calculated and accepted if the energy has decreased or meets the requirement of a particular algorithm. Conventional MC methods are rather inefficient in exploring the configurational space of large molecules because the vast majority of the simulation time is spent sampling high-energy conformations. More modern approaches that focus on specific coordinates have been developed (Saunders, 1987, 1989; Ferguson and Raber, 1989; Chang et al., 1989) and a comparison of Monte Carlo searching procedures for a small organic molecule, cycloheptadecane has appeared (Saunders et al., 1990). An overview of MC sampling is provided in Appendix D.

1.4.3. Molecular Dynamics

The third major conformational searching technique is generically referred to as molecular dynamics (MD). Here heat is used as a means of surmounting potential energy barriers, and hence changing from one conformation or local minimum to another. As discussed here MD is used as a conformational searching tool, but MD is also used to study the dynamical properties of a system (Chapters 3 and 5) or to generate statistical mechanical average properties of a system. Above 0 K molecules possess kinetic energy. In molecular mechanics, heat is provided in the form of kinetic energy. Newton's equations of motion are used to describe the time evolution of a molecular system. That is, from any given molecular geometry intra- and intermolecular velocities, rather than the logic of a minimization algorithm, are used to decide what change in coordinates will be made. To properly determine the change in coordinates due to molecular motions, an incredibly short time scale is needed, on the order of 1 fs (10^{-15} s). An overview of MD is provided in Appendix C.

Three types of MD calculations are used for conformational searching: (1) conventional microcanonical MD, (2) quenched MD, and (3) annealed MD (see Fig. 1.12). In conventional microcanonical MD, the dynamical behavior of a molecular system is monitored as a function of time at a "constant" temperature. In microcanonical MD the total energy of the system is conserved, so the kinetic energy (or temperature) will fluctuate. This type of dynamics is used to look at the motion of a molecule. This time course of molecular motion is referred to as a "trajectory"; it can be stored on disk by a computer and examined to understand the dynamical properties of a molecular system. This approach samples relatively high potential energy structures due to the presence of significant kinetic energy in the system (0.894 kcal/mol per atom at 300 K). In quenched MD, structures are periodically extracted from the microcanonical MD time progression and minimized. The configurations found in this set of minimized structures are analyzed for uniqueness and the low-energy subset of unique structures said to represent the "structure" of the molecular system. In annealed MD, the temperature of the system is incrementally increased and then decreased between low and high temperatures for a number of designated cycles with the lowest energy structures being saved for further minimization and analysis. Although annealed MD does not actually minimize the energy, if the low temperature is small (i.e., 0 K) and the temperature step size is small, then the system will cool slowly enough to find a low-energy minimum on the potential energy surface without getting trapped in a high-energy local minimum. This process is the computational analog of metallurgical annealing, where a mixture of metals is subjected to several cycles of heating and slow cooling to ensure the sample has reached a strain-free, thermodynamic condition.

1.5. ENERGETICS

The term molecular conformations is used to describe molecular structures that interconvert under ambient conditions. This implies that several conformations may be present, in differing concentrations, under ambient conditions. A correct description of "the" molecular structure, "the" molecular energy, or "the" spectrum for a molecule with several conformations must comprise a proper weighting of all of the conformations. For example,

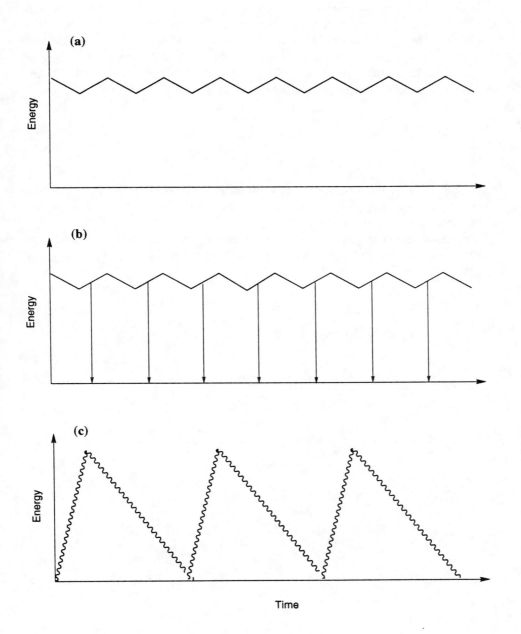

FIGURE 1.12.
Energy as a function of time curves for (a) conventional microcanonical MD, (b) quenched MD, and (c) annealed MD.

the gas-phase structure of lactic acid discussed in Section 1.2, obtained at 298 K, was developed from a microwave spectrum that was a superposition of the microwave spectra of each of the conformations of lactic acid present in significant concentration at that temperature. The MM2 energies of Table 1.2 can be used to obtain a theoretical estimate of to what extent higher energy conformations contributed to the observed microwave spectrum of lactic acid. Statistical mechanics provides the Boltzmann equation, Eq. (1.29), which is used to obtain the probability or population of each conformation i, P_i.

$$P_i = \frac{f_i e^{\frac{-E_i}{RT}}}{\sum_j f_j e^{\frac{-E_j}{RT}}} \tag{1.29}$$

In Eq. (1.29) f_i is the number of states or conformations of energy E_i (or the degeneracy of conformation i), R is 1.98 cal/mol-K (the ideal gas constant), T is the temperature, and the j summation is over all the conformations.

The theoretical probability of each conformation of lactic acid at 298 K can be obtained by plugging the energetic data found in Table 1.2 into Eq. (1.29), giving Eqs. (1.30–1.33).

$$P_{1.2G} = \frac{1}{1 + \exp\left(-\frac{2.137}{0.590}\right) + \exp\left(-\frac{2.437}{0.590}\right) + \exp\left(-\frac{3.677}{0.590}\right)} = 0.957 \tag{1.30}$$

$$P_{1.2C} = \frac{\exp\left(-\frac{2.137}{0.590}\right)}{1 + \exp\left(-\frac{2.137}{0.590}\right) + \exp\left(-\frac{2.437}{0.590}\right) + \exp\left(-\frac{3.677}{0.590}\right)} = 0.0256 \tag{1.31}$$

$$P_{1.2H} = \frac{\exp\left(-\frac{2.437}{0.590}\right)}{1 + \exp\left(-\frac{2.137}{0.590}\right) + \exp\left(-\frac{2.437}{0.590}\right) + \exp\left(-\frac{3.677}{0.590}\right)} = 0.0154 \tag{1.32}$$

$$P_{1.2X} = \frac{\exp\left(-\frac{3.677}{0.590}\right)}{1 + \exp\left(-\frac{2.137}{0.590}\right) + \exp\left(-\frac{2.437}{0.590}\right) + \exp\left(-\frac{3.677}{0.590}\right)} = 0.0019 \tag{1.33}$$

In Eqs. (1.30–1.33) the E_i values are taken relative to the lowest energy conformation. From this analysis, we can see that the population of the lowest energy conformation is 95.7%; thus assigning the microwave spectrum to a single conformation was quite reasonable.

In addition to considering the population of higher energy conformations, a detailed study of the temperature dependence of the thermodynamics of a molecular system must include the population of vibrational, rotational, and translational states, as well the effect of zero-point motion. These topics are discussed in Appendix B and are used in Sections 2.5.2, 6.5, and 7.4.

Given this discussion of thermodynamics it might be tempting to compare the steric or strain energies of molecules to see which molecule is thermodynamically preferred. Consider using molecular mechanics steric energies to calculate an enthalpy for the combustion of methane (CH_4), the reaction given in Eq. (1.34).

$$CH_4 + 2O_2 \rightarrow CO_2 + 2H_2O \qquad (1.34)$$

Because 1,2 and 1,3 nonbonded interactions are not included in molecular mechanics, the steric energy for CH_4 and each of the other molecules in Eq. (1.23) is zero, predicting an exothermicity of zero for the combustion of methane (not much of a fuel!). The experimental ideal gas exothermicity (at 0 K) is 192 kcal/mol. Molecular mechanics fails to correctly predict combustion thermodynamics because molecular mechanics steric energy expressions do not account for bond energies and atomic heats of formation. Additional terms can be added to the steric energy expression in order to be able to estimate heats of formation, and hence molecular enthalpy differences. The procedures used in MM2(3) to obtain heats of formation are discussed in Section 2.2.

Can the steric energies of molecular mechanics be used to predict the thermodynamics of a simple tautomerization such as the keto–enol equilibrium given in Eq. (1.35)?

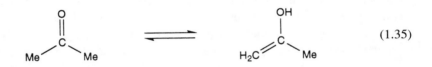

$$(1.35)$$

The experimental heat of formation difference for the reaction in Eq. (1.35) is approximately 8 kcal/mol, with acetone more stable. The MM2 steric energy for acetone, as implemented in CAChe, is -0.48 kcal/mol and the steric energy for vinyl alcohol 1.83 kcal/mol, yielding a steric energy difference of 2.3 kcal/mol, with acetone more stable. Again using the MM2 steric energy differences to predict thermodynamics is substantially in error because of the bond energy terms; the bond energies of the CH, C=O, and C−C bonds broken do not precisely balance the OH, C=C, and C−O bonds formed. The number of bonds is retained in Eq. (1.35) but the types of bonds change.

How about using molecular mechanics to predict energy differences between positional isomers, where the number and precise types of bonds are retained? Consider the energy difference between hydracrylic acid, **1.1** and lactic acid, **1.2**. From Table 1.2 the MM2 strain or total energy of the lowest energy conformation of lactic acid, **1.2**, is -2.37 kcal/mol. The MM2 energy for hydracrylic acid, **1.1**, is -13.23 kcal/mol, predicting a difference in stability of 10.86 kcal/mol, with hydracrylic acid the more stable isomer. Experimentally, the heat of formation difference is approximately -4.4, with lactic acid more stable.

The problem with using molecular mechanics to predict positional isomer energy differences is more subtle than tautomeric energy differences. The probable source of error in MM2 is the use of bond dipoles to represent electrostatic interactions. When molecules are far apart, electrostatic interactions can be described in terms of a series in charge moments: monopoles or partial charges, dipole moments, quadrapole moments, and higher moments. As the molecules come closer together electrostatics need to be described in terms of nuclear charges and electron density distributions that are centered at atom positions. Errors in intermolecular binding energies have previously been attributed to a breakdown in the dipole–dipole approximation used by MM2 (Lipkowitz et al., 1989).

Most other force fields use partial charges for electrostatics, which also causes problems for estimating the energy differences between positional isomers. In force fields that use partial charges, 1,4 nonbonded interactions are included, but 1, 3 nonbonded interactions are left out. The attractive interaction between the hydroxyl oxygen and the carboxylic acid carbon is absent in lactic acid because in lactic acid it is a 1,3 interaction, (see **1.3**), while in hydracrylic acid it is a 1,4 interaction, (see **1.4**).

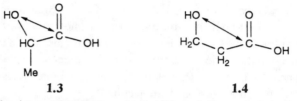

1.3	**1.4**

Molecular mechanics can be employed to predict relative energies of conformers and configurational isomers, but not positional isomers or the relative energies of different molecules. Successful examples of using MM2 to understand the relative energies of conformers and configurational isomers are described in Sections 2.3. Molecular mechanics can also be used to study the binding between molecules if intermolecular interactions have been appropriately parameterized. A successful example of using MM2 to understand the binding energies between organic molecules is described in Section 2.5.2.

1.6. APPLICATIONS

In the following chapters, we will describe the application of molecular mechanics to a wide variety of chemical disciplines. Some, like organic chemistry, have a long history of molecular mechanics use, while others, such as inorganic chemistry, are just starting to exploit molecular mechanics. The case studies included were chosen to illustrate how molecular mechanics can be used to contribute to the solution of real-world problems. The case studies also demonstrate, by inference, what is not possible. For example, there is a misconception that a computer can convert an amino acid sequence into a nicely folded 3-D protein structure. This is not possible now, nor is it likely to be possible in the near future. However, through homology modeling, one can, given enough information about evolutionarily related proteins, accomplish this goal (see Section 3.8).

A cautionary note regarding the use of scientific software is in order. If you intend to publish the results of a molecular mechanics study you need to make sure that the software package you are using will permit you to comply with the following statement written by the editorial board of the Journal of the American Chemical Society: "When computational results are an essential part of a manuscript, sufficient detail must be given, either within the paper or in supplementary material, to enable readers to reproduce the calculations. This includes, for example, force field parameters and equations defining the model or references to where such information is available in the open literature (Bard, 1994)." You should verify that **all** the parameters and equations being used by your study have been published; for example, a 1976 reference to a 1994 method is probably not complete.

Homework

1.1. Which of the pairs of molecules shown in Figure 1.13 are positional isomers, stereoisomers, and conformers? Which are none of the above?

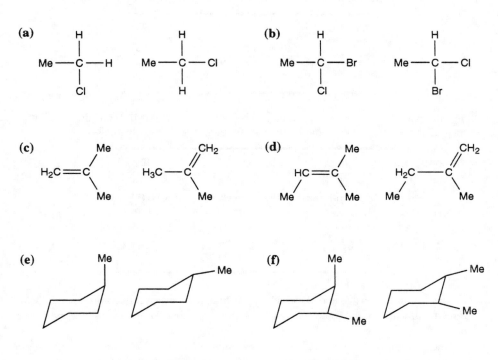

FIGURE 1.13.
Positional isomers, stereoisomers, and conformers.

1.2. What does "steric energy" mean?

1.3. Is the carbonyl carbon-to-methine carbon bond distance of 1.517 Å obtained by MM2 the natural carbonyl carbon-to-methine carbon bond distance? If not, what is the MM2 natural carbonyl carbon-to-methine carbon bond distance?

1.4. Do you think that natural bond distances vary from force field to force field?

1.5. With the use of Eq. (1.2) and the following data: k_{IJ} = 700 kcal/mol-Å^2, r_{IJ} = 1.1 Å, and r = 1.2 Å, what is the numerical value of V_r? If an r of 1.0 Å is used instead of 1.2 Å, what is V_r? If k_{IJ} is 1000 kcal/mol-Å^2 instead of 700 kcal/mol-Å^2, what is the strain energy (V_r)?

1.6. With the use of Eq. 1.3 and the following data: k_{IJ} = 700 kcal/mol-Å^2, D_{IJ} = 100 kcal/mol, r_{IJ} = 1.1 Å, and r = 1.2 Å, what is the numerical value of V_r? If an r of 1.0 Å is used instead of 1.2 Å, what is V_r?

1.7. In Section 1.3 Newton–Raphson minimization was discussed in terms of a harmonic potential [Eq. (1.2)]. Use a Morse potential [Eq. (1.3)] and Eq. (1.24) for a 1-D Newton–Raphson optimization to minimize V_r $\left(\dfrac{dV}{dx} < 1.0 \ kcal/mol \right)$ starting from r = 1.3 Å with r_{IJ} = 1.1 Å, k_{IJ} = 720 kcal/mol, and D_{IJ} = 110 kcal/mol. *(Remember:* $\left(\dfrac{de^x}{dx} \right) = e^x \ dx$.]

1.8. With the use of the table of coordinates given below and the following force field parameters: k_{IJ} = 700 kcal/mol-Å^2, r_{IJ} = 0.93 Å, k_{IJK} = 100 kcal/mol-rad^2, θ_{IJK} = 104.5°, what is the stretch–strain energy for water at this geometry? What is the bend–strain energy for water at this geometry? What is the total strain energy for H_2O at this geometry?

Atom	X	Y	Z
O	0.0	0.0	0.0
H$_1$	0.0	−0.65	0.60
H$_2$	0.0	0.65	0.60

1.9. For the two lowest energy conformations of *n*-butane, the MM2 force field (as implemented in CAChe) obtained the individual strain energy terms listed below.

Term	A	B
Stretch	0.15	0.16
Stretch-bend	0.05	0.07
Bend	0.29	0.63
Torsion	0.01	0.44
vdw	1.68	1.75
Total	?.??	?.??

Of the conformations **A** and **B**, which is the lowest in energy? According to the MM2 force field, what are the sources of the energetic preference for the lowest energy conformation?

The remaining problems are written assuming that you have access to a molecular mechanics software package with a graphical user interface.

1.10. The amino acid alanine is quite similar to lactic acid; the hydroxyl group is simply replaced by an amino functional group. Sketch in and use the minimization facility of your MM software package to obtain the steric strain energies of the four conformations of alanine analogous to **1.2G**, **1.2C**, **1.2H**, and **1.2X**. Which conformation is the most stable and why?

1.11. Use the conformational searching facility of your MM software package to see if there are other low–energy conformations of alanine.

1.12. Use the data you obtained in Problems 1.10 and 1.11 and the Boltzmann equation, Eq. (1.29), to estimate the populations of the four lowest conformations of alanine.

1.13. Sketch in a trial structure for lactic acid and use the various minimization options available in your MM software package to see which approach is most time efficient at converging the same trial structure.

1.14. Sketch in [using a two-dimensional (2-D) sketcher] a trial structure for the six carbon ring skeleton of cyclohexane, making sure that you keep the ring flat. Minimize the structure with an available minimizer. If the minimizer utilized first or higher order derivative information during the minimization you should see a near perfect, flat hexagon on the screen. This occurs because, for each carbon atom, the derivatives of the steric energy associated with moving out of the plane of the screen forward or backward are the same, and in fact are both zero. The molecule is in a position not unlike standing a pencil on its point. A small nudge would cause the pencil to fall over and a small nonplanar molecular distortion should cause the ring to pucker. If your software package permits individual atomic distortions, move one atom out of the molecular plane and reminimize the structure. The molecule should now have adopted either a chair or twist–boat conformation. If you move carbon atoms at opposite ends of the hexagon out of the plane in the same direction you should be able to preferentially form a twist–boat structure. If you instead move carbon atoms at opposite ends of the hexagon out of the plane in opposite directions you should be able to preferentially form the chair structure. If you have been able to trap the molecule in a twist–boat structure, test the conformation searching aspects of your molecular mechanics software package to see how fast it can find the lower energy chair structure.

References

Adams, R.; Yuan, H. C. (1933), The stereochemistry of diphenyls and analogous compounds, *Chem. Rev.* **12**, 261.

Allinger, N. L. (1977), Conformational Analysis. 130. MM2. A hydrocarbon force field utilizing V_1 and V_2 torsional terms, *J. Am. Chem. Soc.* **99**, 8127.

Allinger, N. L.; Miller, M. A.; VanCatledge, F. A.; Hirsch, J. A. (1967), Conformational analysis. LVII. The calculation of the conformational structures of hydrocarbons by the Westheimer–Hendrickson–Wiberg method, *J. Am. Chem. Soc.* **89**, 4345.

Allinger, N. L.; Tribble, M. T. (1972), Conformational analysis—LXXX The hydrinanone ring system, *Tetrahedron* **28**, 1191.

Allinger, N. L.; Yuh, Y. H.; Lii, J.-H. (1989), Molecular mechanics. The MM3 force field for hydrocarbons. 1, *J. Am. Chem. Soc.* **111**, 8551.

Bard, A (1994), *J. Am. Chem. Soc.* **116**, 9A.

Brown, H. C.; Schlesinger, H. I.; Cardon, S. Z. (1942), Studies in stereochemistry. I. Steric strain as a factor in the relative stability of some coordination compounds of boron, *J. Am. Chem. Soc.* **64**, 325.

Buckingham, R. A. (1938), The classical equation of state of gaseous helium neon and argon, *Proc. R. Soc. London, Ser. A* **168**, 264.

Chang, G.; Guida, W. C.; Still, W. C. (1989), An internal coordinate monte carlo method for searching conformational space, *J. Am. Chem. Soc.* **111**, 4379.

Chen, K.; Allinger, N. L. (1993), A molecular mechanics study of alkyl peroxides, *J. Comp. Chem.* **14**, 755.

Dauben, W. G.; Pitzer, K. S. (1956), Conformational analysis. In *Steric Effects in Organic Chemistry*, Newman, M. S., Ed., Wiley, New York, p. 1.

Dostrovsky, I; Hughes, E. D.; Ingold, C. K. (1946), Mechanism of substitution at a saturated carbon atom. Part XXXII. The role of steric hindrance. (Section G) magnitude of steric effects, range of occurrence of steric and polar effects, and place of the Wagner rearrangement in nucleophilic substitution and elimination, *J. Chem. Soc.* 173.

Ferguson, D. M.; Raber, D. J. (1989), A new approach to probing conformational space with molecular mechanics: random incremental pluse search, *J. Am. Chem. Soc.* **109**, 4371.

Fox, P. C.; Bowen, J. P.; Allinger, N. L. (1992), MM3 molecular mechanics study of alkylphosphines, *J. Am. Chem. Soc.* **114**, 8536.

Hehre, W. J.; Radom, L.; Schleyer, P. v. R.; Pople, J. A. (1986), *Ab Initio Molecular Orbital Theory,* Wiley, New York.

Hill, T. L (1946), On Steric Effects, *J. Chem. Phys.* **14**, 465.

Hill, T. L. (1948), Steric effects. I. Van der Waals potential energy curves, *J. Chem. Phys.* **16**, 399.

Kemp, J. D.; Pitzer, K. S. (1936), Hindered rotation of the methyl groups in ethane, *J. Chem. Phys.* **4**, 749.

Kitaygorodsky, A. I. (1961), The interaction curve of non-bonded carbon and hydrogen atoms and its application, *Tetrahedron* **14**, 230.

Kitaygorodsky, A. I. (1965), The principle of close packing and the condition of thermodynamic stability of organic crystals, *Acta Crystallogr,* **18**, 585.

Lennard-Jones, J. E. (1924), On the determination of molecular fields.-II. From the equation of state of a gas, *Proc. R. Soc. London, Ser. A* **106**, 463.

Leong, M .K.; Mastryukov, V. S.; Boggs, J. E. (1994) Structure and conformations of six-membered systems A_6H_{12} (A = C, Si): *Ab initio* study of cyclohexane and cyclohexasilane, *J. Phys. Chem.* **98**, 6961.

Lister, D. G.; Macdonald, J. N.; Owen, N. L. (1978), *Internal Rotation and Inversion,* Academic Press, London.

Lipkowitz, K. B.; Baker, B.; Zegarra, R. (1989), Theoretical studies in molecular recognition: enantioselectivity in chiral chromatography, *J. Comp. Chem.* **10**, 718.

Morse, P. M. (1929) Diatomic molecules according to the wave mechanics. II. Vibrational levels, *Phys. Rev.* **34**, 57.

Murphy, W. F.; Fernandez-Sanchez, J. M.; Raghavachari, K. (1991), Harmonic force field and raman scattering intensity parameters of *n*-butane, *J. Phys. Chem.* **95**, 1124.

Pitzer, K. S.; Donath, W. E. (1959), Conformations and strain energy of cyclopentane and its derivatives, *J. Am. Chem. Soc.* **81**, 3213.

Riddell, F. G.; Robinson, M. J. T. (1974), J. H. van't Hoff and J. A. Le Bell–Their historical context, *Tetrahedron* **30** 2001.

Sache, H. (1890), Ueber die geometrischen isomerien der hexamethylenderivate, *Chem. Ber.* **23**, 1363.

Sache, H. (1892), Uber die konfigurationen der polymethyleneringe, *Z. Phys. Chem.* **10**, 203.

Saunders, M. (1987), Stochastic exploration of molecular mechanics energy surfaces. Hunting for the global minimum, *J. Am. Chem. Soc.* **109**, 3150.

Saunders, M. (1989), Stochastic search for the conformations of bicyclic hydrocarbons, *J. Comp. Chem.* **10**, 203.

Saunders, M.; Houk, K. N.; Wu, Y.-D.; Still, W. C.; Lipton, M.; Chang, G.; Guida, W. C. (1990), Conformations of cycloheptadecane. A comparison of methods for conformational searching, *J. Am. Chem. Soc.* **112**, 1419.

Schmitz, L. R.; Allinger, N. L. (1990), Molecular mechanics calculations (MM3) on aliphatic amines, *J. Am. Chem. Soc.* **112**, 8307.

Schouten, A.; Kanters, J. A.; van Krieken (1994), Low temperature crystal structure and molecular conformation of L-(+)-lactic acid, *J. Mol. Struct.* **232**, 165.

Squillacote, M. S; Sheridan, R. S.; Chapman, O. L,; Anet, F. A. L (1975), Spectroscopic Detection of the Twist–Boat Conformation of Cyclohexane. A Direct Measurement of the Free Energy Difference between the Chair and the Twist-Boat, *J. Am. Chem. Soc.* **97**, 3244.

van Eijck, B. P. (1983), The microwave spectrum of lactic acid, *J. Mol. Spectrosc.* **101**, 133.

Westheimer, F. H.; Mayer, J. E. (1946), The theory of the racemization of optically active derivatives of diphenyl, *J. Chem. Phys.* **14**, 733.

Wiberg, K. B. (1965), A scheme for strain energy minimization. Application to the cycloalkanes, *J. Am. Chem. Soc.* **87**, 1070.

Wood, W. W.; Parker, F. R. (1957), Monte Carlo equation of state of molecules interacting with the Lennard-Jones potential. I. A supercritical isotherm at about twice the critical temperature, *J. Chem. Phys.* **27**, 720.

Further Reading

Berg, U.; Sandström, J. (1989), Static and dynamic stereochemistry of alkyl and analogous groups, *Adv. Phys. Org. Chem.* **25**, 1.

DeKock, R. L.; Madura, J. D.; Rioux, F.; Casanova, J. (1994), Computational chemistry in the undergraduate curriculum. In *Reviews in Computational Chemistry,* Vol. 4, Lipkowitz, K. B.; Boyd, D. B., Eds., VCH, New York, p. 149.

Kollman, P. A.; Merz, K. M. (1990), Computer modeling of the interactions of complex molecules, *Acc. Chem. Res.* **23**, 246.

Schlick, T. (1993), Optimization methods in computational chemistry. In *Reviews in Computational Chemistry,* Vol. 3, Lipkowitz, K. B.; Boyd, D. B., Eds., VCH, New York, p. 1.

Ugi, I.; Bauer, J.; Bley, K.; Dengler, A.; Dietz, A.; Fontain, E.; Gruber, B. Herges, R.; Knauer, M.; Reitsam, K.; Stein, N. (1993), Computer-aided solution of chemical problems—The historical development and present state of the art of a new discipline of chemistry, *Angew. Chem. Int. Ed. Engl.* **32**, 201.

van Gunsteren, W. F.; Berendsen, J. C. (1990), Computer simulation of molecular dynamics: methodology, applications, and perspectives in chemistry, *Angew. Chem. Int. Ed. Engl.* **29**, 992.

Warshel, A. (1991), *Computer Modeling of Chemical Reactions in Enzymes and Solutions*, Wiley, New York.

Weber, J.; Morgantini, P. -Y. (1990), Molecular computer graphics and visualization techniques in chemistry. In *Scientific Visualization and Graphics Simulation* Thalmann, D., Ed., Wiley, New York, p. 203.

Westheimer, F. H. (1956), Calculation of the magnitude of steric effects. In Newman, M. S., Ed., *Steric Effects in Organic Chemistry*, Wiley, New York, Chapter 12, p. 523.

CHAPTER 2

Organics

2.1. INTRODUCTION

Molecular mechanics found its early and natural home within the general discipline of organic chemistry beginning with Hill (1946) and Westheimer and Mayer (1946, reviewed in Westheimer, 1956). Molecular mechanics has typically been used to probe the isolated molecule (or gas phase) three-dimensional (3-D) structure of small-to-medium sized organic molecules, and to understand the energetic differences between molecular conformations. Three conformational studies are discussed in Section 2.3.

The second and third case studies in this chapter rely on a combination of molecular mechanics and structural information provided by nuclear magnetic resonance (NMR) spectroscopy. Aspects of the NMR experiment relevant to these investigations are discussed in Boxes 2.1–2.3.

In large measure, molecular mechanics was created to quantitate the contribution that sterics makes to the transition state structures and energetics of organic reactions rather than ground-state structures. The early molecular mechanics work focused on reactions where the barrier to reaction was associated with an internal rotation instead of bond making and breaking (reviewed in Westheimer, 1956). Houk extended molecular mechanics to treat reactions where chemical bonds are made and broken (reviewed in Houk et al., 1990). A transition state modeling study on the facial selectivity of the addition of nucleophiles to carbonyls is presented in Section 2.4.

Molecular mechanics has also been used in organic chemistry to examine the nature and magnitude of intermolecular interactions important in diastereoselectivity. Applications of molecular mechanics to chiral chromatography and diastereomeric salt formation are discussed in Section 2.5. The diastereomeric salt formation case study is a solid state simulation. Special technologies have been developed for simulation in the solid state, and they are reviewed in Box 2.4.

The preeminent force field in organic chemistry is MM2, developed by Allinger and numerous co-workers. For organic molecules, MM2 (or its successor MM3) is the force field of choice because it has been extensively benchmarked, it yields good structural results, and experimental heats of formation are well reproduced. Because most of the case studies in this chapter utilize MM2, this force field and its successor MM3 are discussed in Section 2.2.

33

2.2. ALLINGER'S MM2 AND MM3

MM2 (Allinger, 1977; Burkert, 1982), and MM3 (Allinger et al., 1989), are by far the most popular force fields for small molecule, organic, molecular mechanics. These force fields have been developed to reproduce experimental structure, conformational energy differences, and heats of formation. Reproduction of vibrational frequencies is only fair by MM2; and heats of sublimation are substantially in error (Allinger, 1989; Lii et al., 1989a; Lii and Allinger, 1989b). In the successor, MM3, reproduction of vibrational frequencies (Lii, 1989a), and for the solid state, heats of sublimation and crystal packing geometries (Lii, 1989b) were also included in the parameterization. Agreement with experiment has been substantially improved. In this discussion we focus on the more recent MM3 force field, although a description of MM2 is provided because it is still in common use.

The MM3 energy expression includes the bond stretching (V_r), bond angle bending (V_θ), dihedral angle torsion (V_ϕ), inversion (V_ω), van der Waals (V_{vdw}), and electrostatic (V_{el}) terms discussed in Section 1.2. In contrast to the harmonic force fields detailed in Sections 1.2 and 3.2, harmonic stretch and harmonic bend potentials were not used in MM2 and MM3. The reason is harmonic potentials could not adequately describe the structural distortions found in highly strained organic molecules. Since a detailed reproduction of strained molecular structures was sought in MM2 and MM3, stretch–bend, stretch–torsion, bend–torsion, and bend-bend cross terms were used to provide a good description of small variations in structure as a function of steric interactions and intramolecular strain. Each of these terms is described below.

2.2.1. Bond Stretch

In Allinger's force fields, bond stretching interactions are described as Taylor series approximations to a Morse function [see Eq. (1.3)]. For MM3, a quartic approximation was used, Eq. (2.1). For MM2, a cubic approximation was used, Eq. (2.2).

$$V_r = \frac{1}{2}k_{IJ}(r - r_{IJ})^2\{1 - B(r - r_{IJ})[1 - C(r - r_{IJ})]\} \tag{2.1}$$

$$V_r = \frac{1}{2}k_{IJ}(r - r_{IJ})^2[1 - B(r - r_{IJ})] \tag{2.2}$$

In Eqs. (2.1) and (2.2), k_{IJ} is the force constant in units of kcal per mole per squared angstrom (kcal /mol-Å^2) and r_{IJ} is the standard or natural bond length in angstroms. In Eq. (2.1), B is set to 2.55 Å^{-1} and to 2 Å^{-1} in Eq. (2.2). For MM3, C in Eq. (2.1) is set to $7/12$ Å^{-1}. B and C are cubic and quartic correction constants.

A harmonic potential does not describe the general shape of a bond stretch when compared to a Morse function (see Fig. 2.1). The harmonic potential does not dissociate correctly: As the bond distance increases, the potential energy rises to infinity rather than leveling off to the bond energy. Furthermore, the asymmetry of the Morse (or real) potential near the bottom of the well is not included. This asymmetry is referred to as anharmonicity. Addition of a cubic term, as in MM2, greatly improves the shape of the potential near the bottom of the well, but at long r, the cubic term swamps the harmonic potential

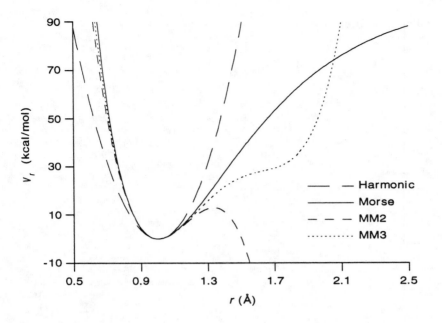

FIGURE 2.1.
Energy versus distance curves for harmonic, Morse, cubic (MM2), and quartic (MM3) stretch functions. A natural distance of 1.0 Å and a force constant of 700 kcal/mol-Å² were used for each. The Morse curve assumed a bond energy of 100 kcal/mol. For the cubic function, B was set to 2 Å^{-1} and for the quartic function, B was set to 2 Å^{-1} and C was set to $\frac{7}{12}$ Å^{-1}.

causing the potential to turn over and approach $-\infty$. This means that if a trial bond distance is longer than the maximum in the potential curve, the two bonded atoms will fly apart during minimization. For the typical potential shown in Figure 2.1, the barrier to explosion is only 13 kcal/mol. This difficulty was addressed in MM3 by the addition of a quartic term. The new bond stretch equation, Eq. (2.1), which retained the improved description of the bottom of the well, tracked the Morse potential at longer r than the cubic potential, and did not explode at long r.

2.2.2. Angle Bend

Angle bend interactions are expanded in θ, through sixth order for MM3, see Eq. (2.3).

$$V_\theta = \frac{1}{2}K_{IJK}(\theta_0 - \theta_{IJK})^2\{1 - B(\theta_{IJK} - \theta_0) + C(\theta_{IJK} - \theta_0)^2 -$$
$$D(\theta_{IJK} - \theta_0)^3 + E(\theta_{IJK} - \theta_0)^4\} \quad (2.3)$$

The parameter B is set to 1.4×10^{-2}, C is set to 5.0×10^{-5}, D is set to 7×10^{-7}, and E is set to 9×10^{-10}. Angle bend interactions are also expanded through sixth order in θ for MM2, though only the harmonic and sextic terms are used, see Eq. (2.4).

$$V_\theta = \frac{1}{2}K_{IJK}(\theta_{IJK} - \theta_0)^2\{1 + C(\theta_{IJK} - \theta_0)^4\} \quad (2.4)$$

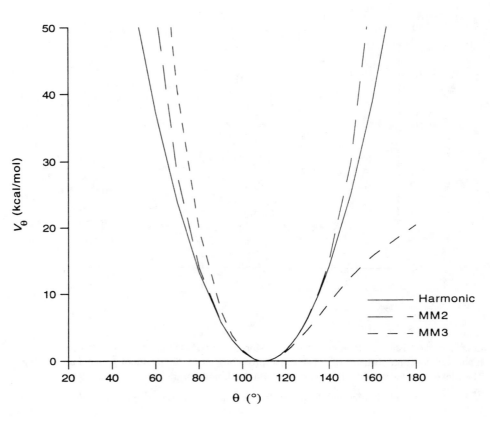

FIGURE 2.2.
Energy versus angle curves for harmonic, harmonic plus sextic (MM2), and full sextic (MM3) bend functions. A natural angle of 109.471° and a force constant of 100 kcal/mol-rad² were used for each. For the harmonic plus sextic potential, C of Eq. (2.4) was set to 7×10^{-8}. For the full sextic potential, Eq. (2.3), B was set to 1.4×10^{-2}, C to 5.0×10^{-5}, D to 7×10^{-7}, and E to 9×10^{-9}.

For MM2, C is set to 7×10^{-8}. The MM2 and MM3 potentials are compared to a harmonic angle bend potential in Figure 2.2. The abrupt change in slope in the harmonic potential at 180° has been remedied by inclusion of higher order terms in the sextic potentials.

2.2.3. Torsion

The torsional potentials in MM2 and MM3 for two bonds IJ and KL connected via a common bond JK are described with a three term cosine expansion in ϕ, where the coefficients in Eq. (2.5), V_n, are determined to reproduce the complex shapes of torsional curves.

$$V_\phi = \frac{V_1}{2}(1 + \cos \phi) + \frac{V_2}{2}(1 - \cos 2\phi) + \frac{V_3}{2}(1 + \cos 3\phi) \qquad (2.5)$$

2.2.4. Inversion

An out-of-plane bending term is used for planar systems in MM2 and MM3. For a central atom I, surrounded by atoms J, K, and L, the inversion is defined in terms of the distance R between atom I and the plane containing J, K, and L [see Fig. 2.3 and Eq. (2.6).

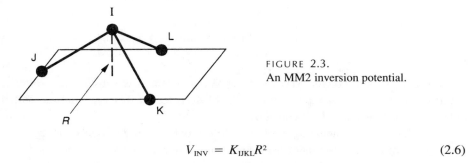

FIGURE 2.3.
An MM2 inversion potential.

$$V_{INV} = K_{IJKL}R^2 \tag{2.6}$$

For the harmonic potential in Eq. (2.6) the natural bond distance is zero.

2.2.5. Cross Terms

In addition to the conventional functional forms discussed in Chapter 1, MM3 and MM2 both use interaction cross terms including, for MM3, stretch–bend, stretch–torsion, and bend–bend interactions.

For an angle involving the r_{IJ} and r_{JK} bonds, the stretch–bend cross term is given by Eq. (2.7).

$$V_{r\theta} = K_{r\theta}[(r - r_{IJ}) + (r - rr_{JK})](\theta - \theta_{IJK}) \tag{2.7}$$

The stretch–torsion cross term for a torsion about a central JK bond involving two sp^3 centers is described by Eq. (2.8)

$$V_{r\phi} = K_{r\phi}(r - r_{JK})(1 + \cos 3\phi) \tag{2.8}$$

The bend–bend cross term, included for angles centered at the same atom, is given in Eq. (2.9).

$$V_{\theta\theta'} = K_{\theta\theta'}(\theta - \theta_{IJK})(\theta - \theta_{LJM}) \tag{2.9}$$

2.2.6. van der Waals

Nonbonded interactions (van der Waals forces) are included in MM3 and MM2 force fields using a two-term Hill (1948) representation of an exponential-6 or Buckingham potential, see Eq. (2.10).

$$V_{vdW} = Ae^{-B\rho} - \frac{C_6}{\rho^6} \tag{2.10}$$

For MM3, the van der Waals equation is given in Eq. (2.11) and for MM2 it is given in Eq. (2.12).

$$V_{vdW} = D_{IJ}\left[1.84 \times 10^5 e^{-12(\rho/\rho_{IJ})} - 2.25\left(\frac{\rho_{IJ}}{\rho}\right)^6 \right] \tag{2.11}$$

$$V_{vdW} = D_{IJ}\left[2.90 \times 10^5 e^{-12.5(\rho/\rho_{IJ})} - 2.25\left(\frac{\rho_J}{\rho}\right)^6 \right] \tag{2.12}$$

As shown in Figure 2.4, the MM3 van der Waals potential is substantially softer (the repulsive inner wall is less repulsive) than the MM2 potential. The potential was softened to better reproduce short H−H contacts.

Most force fields, including MM2 and MM3, obtain van der Waals parameters (D_{IJ} and ρ_{IJ}) for hetereonuclear pairs from the homonuclear parameters through the use of combination rules. Both MM2 and MM3 use a geometric combination rule for energy, D_{IJ}, see Eq. (2.13),

$$D_{IJ} = \sqrt{D_I \times D_J} \tag{2.13}$$

and an arithmetic combination rule for distance, ρ_{IJ}, see Eq. (2.14).

$$\rho_{IJ} = \frac{1}{2}(\rho_I + \rho_J) \tag{2.14}$$

To account for the observed displacement of electron density in H−X bonds towards X, hydrogen van der Waals interactions are evaluated with the hydrogen coordinates displaced towards X (Allinger et al., 1971).

2.2.7. Electrostatics

Unlike the harmonic force fields, MM2 and MM3 describe electrostatic interactions in terms of bond dipoles rather than point charges, see Eq. (2.15), where C is used to convert to kilocalories per mole, ε is the dielectric constant, and the remaining variables are as described previously in Figure 1.8.

$$V_{el} = C\frac{\mu_i \mu_j}{\varepsilon \rho^3}(\cos \chi - 3 \cos \alpha_i \cos \alpha_j) \tag{2.15}$$

The bond dipoles are set by an atom pair.

The force fields MM2 and MM3 follow the usual convention in molecular mechanics of excluding nonbond interactions for atoms that are bonded to each other (1, 2 interactions) or bonded to a common atom (1, 3 interactions).

2.2.8. Hydrogen Bonding

Hydrogen-bonding (HB) interactions were originally dealt with through modified van der Waals parameters in MM3 (Schmitz and Allinger, 1990). More recently, the hydrogen-bond interaction was made angularly dependent (Allinger et al., 1992), see Eq. (2.16),

(a)

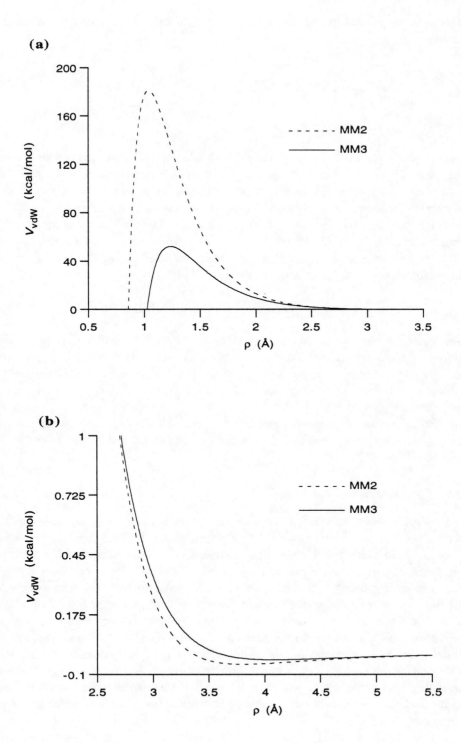

(b)

FIGURE 2.4.
The C−C van der Waals potentials for MM2 and MM3.

where ω is the angle between the hydrogen-bond donor and acceptor centered at the hydrogen.

$$V_{HB} = D_{IJ}\left[1.84 \times 10^5 e^{-12(\rho/\rho_{IJ})} - 2.25\left(\frac{\rho_{IJ}}{\rho}\right)^6 \right](1 - \cos \omega) \qquad (2.16)$$

2.2.9. Heats of Formation

As developed so far, the energies obtained from MM2 and MM3, like the harmonic force fields, are only strain energies. Strain energies cannot be compared between molecules. This limitation has been overcome in MM2 and MM3 by adding "heat of formation" parameters to the force field (Allinger, 1989). These new terms do not affect minimized molecular structures but do change relative energetics. The basic scheme for heats of formation at 25°C, in MM2 and MM3, starts with a set of bond terms for bond A−B, which roughly correspond to the difference between the bond energy D_{AB} and the atomic heats of formation $\Delta H_{A\text{ atom}}$ and $\Delta H_{B\text{ atom}}$, see Eq. (2.17).

$$\Delta H_{AB} \approx D_{AB\text{ bond}} - \Delta H_{A\text{ atom}} - \Delta H_{B\text{ atom}} \qquad (2.17)$$

A set of group increments for entities such as methyl, secondary, or tertiary carbon atoms and four- or five-membered rings are added to the bond terms. To account for the presence and population of higher energy states, corrections are added, one to approximate the contribution due to higher energy molecular conformations and one to account for the population of higher energy vibrational states due to low-frequency torsions. The heat of formation calculation is completed by adding a term (2.4 kcal/mol) to approximate the pressure–volume product term and the change in translational and rotational energy in going from 0 to 298 K.

2.2.10. Transferability of Parameters

In all molecular mechanics force fields a degree of transferability of parameters within a force field is assumed. Rather than have different parameters for every possible combination of atoms in every possible chemical environment, it is assumed that atoms in the same general environment are basically the same. For example, for carbon there are sp^3, sp^2, and sp environments. Each hybridization is assigned a unique label, with associated force field parameters. The unique labels, defining a particular environment for a particular element, are referred to as atom types. For example, the same sp^3 carbon atom type would be used to describe the carbon atoms in both ethane and cyclohexane. That means that the natural C−C r_{IJ}, and k_{IJ} are the same for ethane and cyclohexane. Because different force fields use a variety of functional forms, one force field might use a cubic stretch and another a harmonic stretch, and because force fields are parameterized from different data the parameters of a force field cannot be transferred from one force field to another without care.

2.2.11. Examples

Let us consider the energy expressions for a few hydrocarbons. For methane, at its equilibrium, minimized, or "best" geometry, the sum of all included interactions in MM2 or

MM3 is 0 kcal/mol. This occurs because all individual terms in the energy expression can be simultaneously satisfied and as can be seen from Eq. (2.1), when the C−H distance, r, is equal to the natural distance, r_{U}, the stretch strain energy, V_r, is zero.

For ethane at its MM2 minimized geometry, the MM2 force field (as implemented in CACHe, PC-Model, and CERIUS[2]), yields an energy of 0.82 kcal/mol for the sum of all of the force field energy terms rather than the 0 kcal/mol of methane. This nonzero value is due to the presence of repulsive 1,4 H−H van der Waals interactions in ethane. The resulting "best" bond distances and bond angles of the MM2 minimized geometry for ethane are thus not the strain-free or natural values for the individual C−C and C−H stretch and C−C−H and H−C−H bend terms. For example, the MM2 minimized C−C distance of ethane is 1.531 Å, whereas the strain-free or natural value is 1.523 Å. With the use of a MM2 C−C stretching force constant of 4.4 mdyn/Å or 633.072 kcal/mol-Å², and the MM2 stretch potential energy expression, Eq. (2.2), we obtain Eq. (2.18), which gives a C−C bond strain energy of 0.02 kcal/mol.

$$V_{r(C-C)} = \frac{1}{2}633.072 \text{ (kcal/mol-Å}^2)(1.531 \text{ Å} - 1.523 \text{ Å})^2$$
$$[1 - 2\text{Å}^{-1}(1.531 \text{ Å} - 1.523 \text{ Å})] \quad (2.18)$$

The energy is conventionally referred to as a strain energy because V_r for the bond without external influences would be zero. The MM2 equilibrium C−C−H angles of ethane are 110.99°, whereas the strain-free values are 110.0°. This leads to a total C−C−H angle strain energy of 0.05 kcal/mol. The summed stretch energy is 0.03 kcal/mol and the total bend energy is 0.10 kcal/mol. The stretch–bend cross term energy is 0.01 kcal/mol. Because the equilibrium geometry is a perfectly staggered conformation the torsion strain energy is 0.0 kcal/mol. The van der Waals energy is 0.68 kcal/mol.

For butane there are two conformations (*trans* and *gauche*) that are experimentally within 0.7 kcal/mol of each other (Murphy et al., 1991), see Figure 2.5. Molecular mechanics can provide an explanation for this energy difference. By using the MM2 force field, the *trans* conformation has a total strain energy of 2.18 kcal/mol and the *gauche* conformation a strain energy of 3.05 kcal/mol, yielding an energy difference between the two conformations of 0.87 kcal/mol. The individual term energies suggest that, at least for the MM2 force field, the *gauche* conformation is destabilized relative to the *trans* conformation by bend and torsional strain (see Table 2.1). Also collected in Table 2.1 are the individual term energies obtained with the harmonic Dreiding force field, at the geom-

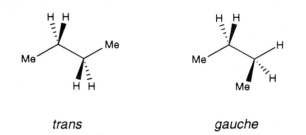

trans   gauche

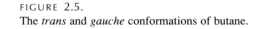

FIGURE 2.5.
The *trans* and *gauche* conformations of butane.

TABLE 2.1. Butane *trans* and *gauche* Individual Strain Energy Terms (kcal/mol)

Term	MM2			Dreiding		
	trans	*gauche*	ΔE	*trans*	*gauche*	ΔE
Stretch	0.15	0.16	0.01	0.33	0.38	0.05
Stretch–bend	0.05	0.07	0.02			
Bend	0.29	0.63	0.34	0.51	1.15	0.64
Torsion	0.01	0.44	0.43	0.01	0.11	0.01
vdW	1.68	1.75	0.07	3.59	3.59	0.00
Total	2.18	3.05	0.87	4.44	5.23	0.79

etry optimized with the Dreiding force field (Mayo et al., 1990). While both force fields provide nearly the same energy difference between the two conformations (0.87 kcal/mol for MM2 and 0.79 kcal/mol for Dreiding) the energetic sources of the conformational energy difference are somewhat different. The MM2 attributes the energy difference to bend and torsional strain, whereas the Dreiding force field lumps the entire effect into bend strain. That the two force fields attribute conformational ΔE values to different terms in the energy expression is due to the force fields using alternative functional forms and parameterization procedures (Burkert and Allinger, 1982). Neither force field has the ''correct'' answer. While it is comforting to be able to provide a simple physical explanation for an energetic difference, it may not always be warrented and due caution is in order.

For a conformationally constrained ring system such as cyclohexane there are two low lying conformations, a ground-state chair conformation and a twist–boat conformation, 5.5 kcal/mol higher in energy (Squillacote et al., 1975), see Figure 2.6. For the MM2 force field, the chair conformation has a strain energy of 6.56 kcal/mol and the twist–boat conformation has a strain energy of 11.94 kcal/mol, yielding an energy difference between the two conformations of 5.38 kcal/mol. The individual MM2 term energies for these two conformations are listed in Table 2.2. Again, at the MM2 equilibrium geometry of the lower energy chair conformation, all of the various terms are not at strain–free values. The

Chair

Twist–boat

FIGURE 2.6.
Chair and twist–boat conformations of cyclohexane.

TABLE 2.2. Cyclohexane Chair and Twist–Boat
Individual Strain Energy Terms for the MM2 Force Field

Term	Chair	Twist–boat	ΔE
Stretch	0.32	0.40	0.08
Stretch–Bend	0.08	0.12	0.04
Bend	0.36	0.74	0.38
Torsion	2.16	5.59	3.43
vdW	3.64	5.09	1.45
Total	6.56	11.94	5.38

total stretch–strain energy is 0.32 kcal/mol because the equilibrium C–C bond distances of 1.535 Å are 0.012 Å longer than the MM2 strain-free C–C distance of 1.523 Å and the C–H bond distances of 1.117 Å are 0.004 Å longer than the strain-free distance of 1.113 Å (0.043 × 6 kcal/mol for the C–C bonds and 0.005 × 12 kcal/mol for the C–H bonds). The bend energy is 0.36 kcal/mol. The total stretch–bend cross-term energy is 0.08 kcal/mol. Due to 1,3 axial H–H interactions, the molecule has distorted from a purely all staggered conformation so now the molecule has torsional strain. One pair of 1,3 axial H–H interactions is shown in Figure 2.7. The total torsional strain energy is

FIGURE 2.7.
The 1, 3 axial interactions in the chair conformation of cyclohexane.

2.16 kcal/mol and the van der Waals energy is 3.64 kcal/mol. Comparison of the individual term energies in Table 2.2 suggests that, at least for the MM2 force field, the twist–boat conformation is destabilized relative to the chair conformation by torsional strain and van der Waals repulsions.

A rather dramatic example of molecular distortion and strain is tri-*tert*-butylmethane, (Bartell and Bürgi, 1972). The tertiary to quaternary C–C bonds of tri-*tert*-butylmethane (indicated by boldface lines in Fig. 2.8), are lengthened from a normal or ideal C–C distance of 1.53 to 1.61 Å in the experimental structure. The quaternary-to-methyl C–C

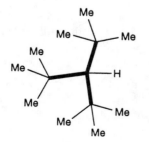

FIGURE 2.8.
Tri-*tert*-butylmethane. The tertiary to quaternary C–C bonds are indicated by boldface lines.

bonds, in contrast, only lengthen to 1.55 Å. The MM2 minimized geometry of this mole-cule has been calculated. Tertiary-to-quaternary C—C bond distances of 1.59 Å and qua-ternary-to-methyl C—C bond distances of 1.55 Å are observed. These distortions of the tertiary-to-quaternary bonds lead to a C—C stretch–strain energy of 6.3 kcal/mol. In order to relieve the repulsive van der Waals interactions among the *tert*-butyl groups, the ter-tiary-to-quaternary C—C bonds lengthen. But, there is a limit to the lengthening because there is a restoring force that opposes bond stretching, as given by the stretch energy. The fundamental tenant of molecular mechanics is that the balance of all such forces plays a major role in determining the shapes and energetics of molecules.

2.3. CONFORMATIONAL ANALYSIS

2.3.1 Background

Some of the most commonly asked structural questions in organic chemistry and other disciplines are What does the molecule look like? What is it's three-dimensional shape? Is the shape different in solution than in the solid state? Why is a particular conformation preferred over another? Molecular mechanics has been and is being used to address these questions. Three examples are provided in this section. A review of a classic study on the conformational preferences of hydrindanes is discussed in Section 2.3.2. A characteriza-tion of the solution conformations of cinchona alkaloids is described in Section 2.3.3; and a more detailed investigation of the conformation and relative configuration of seven stereogenic centers in a 2-methoxy-3-sulfonyl-1,3-oxazolidine is presented in Section 2.3.4.

2.3.2 Isomer Energies of Hydrindanes and Hydrindanones

Fused rings, often the building blocks of natural products such as steroids, provide a classic example of the use of molecular mechanics in organic chemistry (Allinger and Tribble, 1972). The fused five- and six-membered ring system bicyclo[4.3.0]nonane, hy-drindane **2.1**,

2.1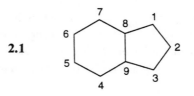

serves as a model of the steroid C/D ring system, indicated with boldface lines in **2.2**.

2.2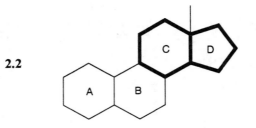

There are two important configurational isomers of this ring system, a trans fused ring, **2.1t** and a cis fused ring, **2.1c**.

2.1t   **2.1c**

The energetic ordering of the two isomers is dependent on the substitution pattern on the rings. Understanding the configurational preference of fused ring systems is important in explaining the stereochemistry of asymmetric ring-forming transformations (see, e.g., Meyers et al., 1992). For the parent system, **2.1**, the experimental ΔH_f of trans bridgehead linkage is 1.07 kcal/mol less than the cis linkage (the trans isomer is more stable). For derivatives with a methyl at sites 8 and 9 of **2.1** the cis forms are lower in energy. These junctions between the rings are generally called bridgehead positions. Placement of a methyl in the 8 position is analogous to the C/D ring system in steroids, see **2.2**.

In order to understand the source of energetic differentiation in fused five- and six-membered rings, Allinger and Tribble (1972) of the University of Georgia carried out molecular mechanics minimizations on the cis and trans configurations for a series of substituted hydrindanes and hydrindanones using a predecessor (Allinger et al., 1971; Allinger et al., 1972) to the popular MM2 and MM3 force fields discussed in Section 2.2.

METHODS

Bond deformations were described with a harmonic potential. A harmonic potential plus a cubic correction term described angular deformations. A stretch–bend cross term was included to account for the lengthening of bonds for strained ring systems. Torsions were described with a threefold Fourier expansion. Nonbonded interactions were described with a Hill potential (Hill, 1948), a two parameter representation of an exponential-6 or Buckingham potential (Buckingham, 1938). The van der Waals interactions for hydrogen atoms were evaluated with the hydrogen atom displaced to 92% of the bond distance.

RESULTS

The primary goal of Allinger and Tribble (1972) was to understand the configurational preferences of fused ring systems. This required reasonable estimates of the molecular structures as well as the molecular strain energies. As a first calibration of the force field, Allinger and Tribble compared the calculated structure of androsterone, **2.3**, with the X-ray crystal structure.

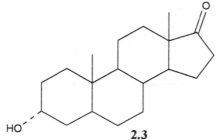

2.3

The overall root-mean-square (rms) bond distance discrepancy between calculated and experimental bond distances was 0.010 Å, which is quite good.

The use of rms errors is rather common in structural analyses. The rms error is a measure of the deviation of a calculated quantity from an observed quantity X (bond distance, bond angle, coordinate, vibrational frequency, etc.). The rms error or difference is given by Eq. (2.19), where X_i^{obs} is the ith observed quantity, X_i^{calc} is the calculated value and N is the number of terms included in the rms analysis. The corresponding rms for bond angles on **2.3** was 1.3°.

$$error_{rms} = \sqrt{\left(\frac{\sum_i^N (X_i^{obs} - X_i^{calc})^2}{N} \right)} \qquad (2.19)$$

In addition to structural calibration, Allinger and Tribble (1972) compared experimental and calculated configurational energy differences for a series of substituted hydrindane derivatives. The results are collected in Table 2.3. The force field correctly reproduced the sign of the differentiation (e.g., compounds known to prefer cis are calculated to be more stable as cis configurations). The magnitude of the error was less than 1 kcal/mol with the exceptions of **2.4**, R = H and **2.4**, R = Me, where the errors were 1.1 kcal/mol.

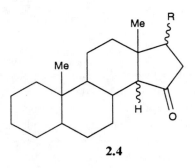

2.4

After verifying that the force field could reproduce the structure and relative energetics of substituted hydrindanes, Allinger and Tribble analyzed the individual stretch, bend, torsion, and van der Waals energy contributions. The intent was to understand why partic-

TABLE 2.3. Experimental and Calculated Free Energy Differences (kcal/mol) for Substituted Hydrindanes[a]

Compound	Experimental $\Delta G°$	Calculated $\Delta G°$
2.1	+0.4	+0.7
2.6 8-Methylhydrindan-2-one	−2.8	−2.1
2.4 [b]	−1.2	−2.3
2.4 [c]	−0.3	−1.4
2.4 [d]	+0.7	+0.7

[a] A minus sign (−) indicates cis is lower in energy.
[b] Where R = H.
[c] Where R = Me (Me = CH$_3$).
[d] Where R = i−Pr (i−Pr = isopropyl).

TABLE 2.4. Relative Strain
Energies (kcal/mol) for **2.1t** and
2.1c[a]

Term	
Bend	−0.6
Torsion	+0.6
vdW	+1.1
Total	+1.1

[a] Where a minus sign (−) indicates cis is
lower in energy.

ular substitutions caused a change in the preference of cis over trans. For the parent hydrindane, **2.1**, the trans isomer **2.1t** had 0.6 kcal/mol more bending strain than the cis, **2.1c**
(see Table 2.4), because for the trans isomer two equatorial substituents on the cyclohexane ring were being twisted together to form a five-membered ring. The dihedral angle
between an axial substituent and an equatorial substituent is about 55°, whereas the dihedral angle between two equatorial substituents is about 63°. This leads to an axial–
equatorial combination of substituents being closer together and hence more consistent
with the smaller size of a cyclopentane ring. As shown in Figure 2.9(a), the two equatorial
substituents will have to swing around to be a part of the five-membered ring. The cis
isomer had more torsional strain, 0.6 kcal/mol, and a higher, repulsive, van der Waals
energy by 1.1 kcal/mol. This differential was explained in terms of the five-membered
ring being flatter, and hence having more eclipsed C−H interactions, see Figure 2.9(b).
The net energetic difference of 1.1 kcal/mol in favor of the trans isomer is in agreement
with the experimental difference in ΔH_f of 1.07 kcal/mol. (These enthalpic differences
should not be confused with the free energy differences presented in Table 2.3; see Appendix B for a discussion of free energy estimation.)

(a)   **(b)**

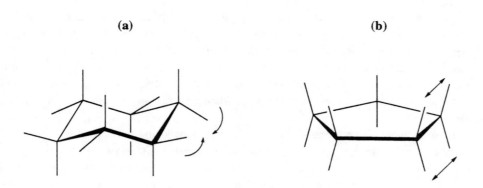

FIGURE 2.9.
(a) Ring strain induced in cyclohexane by five-membered ring formation. The necessary distortions
are indicated by arrows. (b) Five-membered ring eclipsing C–H interactions, indicated by arrow.

TABLE 2.5. Strain Energies (kcal/mol) of Substituted Hydrindanes[a]

| Term | Hydrocarbon, **2.5** | | | 1-Keto, **2.6** | | | 2-Keto, **2.7** | | | 3-Keto, **2.8** | | |
| | | *cis* | | | *cis* | | | *cis* | | | *cis* | |
	trans	eq	ax	*trans*	eq	ax	*trans*	eq	ax	*trans*	eq	ax
Bend	1.8	0	0	2.4	0	−0.3	2.2	0.2	0	2.6	0.2	0
vdW	0.4	0	0.6	0.5	0	0.4	0.4	−0.2	0	0.4	−0.2	0
Torsion	0	0.7	0.5	−0.4	0	0.1	−0.5	0.2	0	−0.4	1.1	0.2
Total	1.5	0	0.4	2.5	0	0.2	2.1	0.2	0	2.6	1.1	0

Axial = ax and equatorial = eq.

Allinger and Tribble (1972) also investigated methyl substitution at the bridgehead. The energy terms for the parent hydrocarbon, 8-methylhydrindane, **2.5**, as well as the three possible ketones 8-methylhydrindan-1-one, **2.6**, 8-methylhydrindan-2-one, **2.7**, and 8-methylhydrindan-3-one, **2.8**, are collected in Table 2.5.

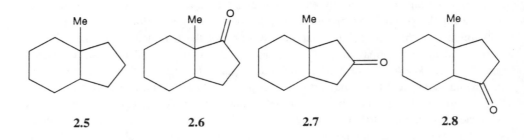

2.5   **2.6**   **2.7**   **2.8**

For each system, angle strain dominated the differentiation and caused the *trans* isomer to be the least stable. For the cis bridgehead configuration the 8-methyl substituent could either be axial or equatorial. Axial placement of the methyl group corresponds to the configuration adopted in a steroid, so a comparison between the two cis configurations was of interest. For the ketones the two configurations (axial and equatorial methyl) were within 0.2 kcal/mol except for the 3-keto system, **2.8**. Here torsional strain in the equatorial configuration caused the steroid configuration to be the lowest. In the nonsteroid form the carbonyl approximately eclipsed a hydrogen atom, whereas in the steroid form the carbonyl eclipsed a methylene group. The preferred conformations in 2-butanone and propionaldehyde place a methyl group rather than hydrogen in a position eclipsing the carbonyl.

In summary, Allinger and Tribble (1972) reproduced the experimentally observed structure of androsterone, **2.3**, and experimental energetic differences between configurational isomers for a series of fused rings. In addition, analysis of the individual terms in the molecular mechanics energy expression showed that the smaller size of the five-membered ring, in comparison to the more strain-free six-membered ring, was responsible in large measure for the observed configurational preferences.

2.3.3 Solution Conformations of Cinchona Alkaloids

Cinchona alkaloids, **2.9**, are important pharmacological agents for the treatment of malaria, heart arrhythmias, and nocturnal leg cramps (Bigger and Hoffman, 1990; Webster, 1990).

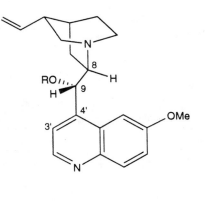

2.9

Cinchona alkaloids such as quinidine (2.9, R = H), and its diastereomer quinine are used as chiral resolving agents (Jacques et al., 1981) and in the chromatographic separation of enantiomers (Salvadori et al., 1987). Cinchona alkaloids have also been employed quite successfully as chiral adjuvants. Chiral adjuvants are auxiliary reagents added to a reaction to increase stereoselectivity. Catalytic amounts of quinine and quinidine have been used as the source of chirality in an asymmetric Michael addition (Wynberg and Helder, 1975) and in the [2 + 2] cycloaddition between chloral and ketene (Wynberg and Staring, 1984). Catalytic quantities of quinine and quinidine have been used as chiral phase-transfer catalysts (Pluim and Wynberg, 1980), and as excellent chiral ligands in the asymmetric dihydroxylation of alkenes (Sharpless et al., 1992).

In order to probe the intimate details of the mechanism of action of the cinchona alkaloids as chiral adjuvants, Dijkstra, Kellogg, and Wynberg of the University of Groningen, and Svenden, Marko, and Sharpless of the Massachusetts Institute of Technology (Dijkstra et al., 1989) undertook a thorough investigation of the preferred conformations of cinchona alkaloids in solution. As should be apparent from the molecular structure **2.9**, conformational flexibility about the interring linkage single bonds ($C_{4'}-C_9$ and C_9-C_8) would be expected, with the possibility of "closed" and "open" conformations, see Figure 2.10. In the open conformation the $C_4'-C_9-C_8-N$ torsional angle is near 180° (a *trans* arrangement), whereas in the closed conformation the $C_4'-C_9-C_8-N$ torsional angle is near −60° (a *gauche* arrangement). Furthermore, the alkaloid ring can either lie above or below the quinoline ring, depending on whether the $C_{3'}-C_{4'}-C_9-C_8$ angle is near −90° or near 90°, see Figure 2.10.

Proton NMR (^{1}H NMR) spectroscopy was used as a tool to understand the solution conformations for a series of substituted dihydroquinidines. The (*p*-chlorobenzoyl)dihydroquinidine (R = *p*-ClBz, where B_z = benzoyl) derivative was known to adopt a closed

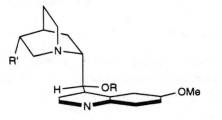
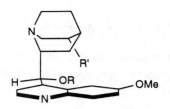

Closed-above   Open-above

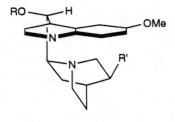

Closed-below   Open-below

FIGURE 2.10.
Conformations of quinidines.

conformation in the solid state, but its conformation in solution as well as the solution conformations for other substituted dihydroquinidines were not known. This result led to the discovery of differences in nuclear Overhauser effect (NOE) intensities, and hence molecular structures, as a function of substitution at C9 and as a function of solvent. In order to help understand the source of these conformational differences they turned to molecular mechanics to determine how many low-energy conformations there were and whether or not the conformational energy differences were dependent on the substituent at C9. This application of molecular mechanics is typical of its present day usage in organic chemistry. That is, molecular mechanics plays a small role, being used to provide support for an experimental finding or providing a means of rationalizing an observation or supposition.

METHODS

Torsional profiles about the $C_{4'}-C_9$ and C_9-C_8 bonds were generated for the parent dihydroquinidine, R = H (open conformation observed), acetyldihydroquinidine, R = Ac, where Ac = acetyl (closed conformation observed), and methoxydihydroquinidine, R = Me (open and closed conformations observed) using the CHEMX modeling program. The geometries of the resulting conformations were minimized using the MM2 force field (see Section 2.2) as implemented in Allinger's MMP2 program.

Interring NOEs were used to establish nonbonded proton–proton distances, and the Altona extension of the Karplus equation was used with measured $^3J_{HH}$ coupling constants to establish dihedral angles. The NOE spectroscopy and the Karplus equation are outlined below in Boxes 2.1 and 2.2, respectively.

BOX 2.1   NUCLEAR OVERHAUSER EFFECT

The coupling between NMR methods and molecular modeling in this example of a combined molecular mechanics–NMR study is qualitative. The observation of an NOE effect suggests that a particular pair of nuclear spins are in close proximity but specific distance information is neither generated nor used. The example in Section 2.3.4 and case studies in Chapters 3 and 7 use NOE derived distance information more quantitatively as distance constraints in molecular dynamics (MD) simulations. Before discussing the results of this cinchona alkaloid work we briefly review how structural information is obtained from NOE and $^3J_{HH}$ coupling constant data. A good presentation of the NOE experiment can be found in the book by Neuhaus and Williamson (1989).

In NMR spectroscopy, a signal is observed when radiation is absorbed or emitted due to a nuclear spin flap. This requires the nucleus to have a nonzero nuclear spin and for there to be a difference in the populations of the up (α) and down (β) nuclear spin states (for spin $1/2$ nuclei). Since both absorption and emission can be induced by the incident radiation, if nonradiative mechanisms are not present for the decay from the higher energy spin state to the lower energy state, the signal would rapidly go away or become saturated.

In the NOE experiment the nonradiative process that permits signal detection (and actually leads to signal enhancement in some cases) is due to the interaction between two nuclear spin dipoles. The magnitude of this dipole–dipole interaction is dependent on the distance between the two dipoles. If a particular spin is irradiated and the intensity for another spin is enhanced or partially suppressed the two spins are likely geometrically close, even though they may be far apart in the connectivity of the molecule. The precise distance dependence of this dipolar interaction is discussed in Section 2.5.

BOX 2.2   Three Bond Proton–Proton Couplings: The Karplus Relations

Based on a valence bond treatment of the contact electron spin–nuclear spin interaction of ethane, Karplus proposed that $^3J_{HH}$ should be proportional to the square of the cosine of the dihedral angle relating to two protons ($^3J_{HH}$ α $\cos^2\phi$) for the $H-C-C-H$ coupling paths in ethane and ethylene (Karplus, 1959). This led to the Karplus relation, Eq. (2.20) (Karplus, 1963).

$$^3J_{HH} = A \cos^2\phi + B \cos \phi + C \qquad (2.20)$$

If the three bond scalar coupling, $^3J_{HH}$, is measured and the proportionality constants A, B, and C are known for related molecules, then the dihedral angle ϕ can be estimated.

Karplus pointed out that scalar coupling and, hence, the parameters in Eq. (2.20) are dependent on the interactions between electron densities. The coupling should therefore be dependent on the $C-H$ and $C-C$ bond distances, the $C-C-H$ bond angles, and electronegativities of substituents

(continued)

BOX 2.2 *(continued)*

bound to the carbon atoms. Because terms in the electronic wave function that contribute to the scalar coupling constant are quite small ($< 0.4\%$) they are quite difficult to calculate using electronic structure methods. The parameters in Eq. (2.20) are typically obtained from fitting experimental data (see, e.g., Haasnoot et al., 1980). There are uncertainties in the exerimentally determined A, B, and C constants because the experimental data base is heavily weighted towards $180°$ and $60°$ for sp^3–sp^3 centers. Furthermore, unless the molecular framework is rigid, the precise experimental dihedral angle that corresponds to J is not known. A flexible molecule adopts numerous configurations in solution. For flexible molecules an energy weighted average should be used, but the energy weighting factors in solution are not easily determined. In addition, there are experimental uncertainties in the measured J values.

When Eq. (2.20) was used with $A = 7.76$, $B = -1.0$, and $C = 1.40$ for a set of 315 coupling constants, an rms error in J of 1.2 Hz was obtained (Haasnoot et al., 1980). The error was substantially reduced when the effect of electronegative substituents was included. The improved form (Haasnoot et al., 1980) is given in Eq. (2.21)

$$^3J_{HH} = P_1 \cos^2\phi + P_2 \cos\phi + \sum_i \Delta\chi_i \{P_4 + P_5 \cos^2(\xi_i \cdot \phi + P_6 \cdot |\Delta\chi_i|)\} \qquad (2.21)$$

where $\Delta\chi_i$ is the difference in Huggins electronegativity (Huggins, 1953) between substituent i and hydrogen, ξ_i is ± 1 depending on the orientation of the substituents, and $P_1 = 13.86$, $P_2 = -0.81$, $P_4 = 0.56$, $P_5 = -2.32$, and $P_6 = 17.9°$. Inclusion of the electronegativity effect reduced the rms error in J to 0.511 Hz for a set of 315 coupling constants.

For the chair conformation of cyclohexane, coupling constants of 15.9, 3.48, and 2.89 Hz are obtained from Eq. (2.21) (using H−C−C−H dihedral angles of $\phi_{ax,ax} = 175°$, $\phi_{ax,eq} = 57.4°$, and $\phi_{eq,eq} = -60°$ and a carbon substituent $\Delta\chi$ of 0.4. Experimentally, the scalar couplings are $^3J_{ax,ax} = 13.12$ Hz, $^3J_{ax,eq} = 3.65$ Hz, $^3J_{eq,eq} = 2.96$ Hz (Garbisch and Griffith, 1968). Enhancements to this approach are still being developed (Diez et al., 1989; Donders et al., 1989; Altona et al., 1989).

RESULTS

Before using NOE information as a structural tool one must know the identity of the individual peaks in the ^{1}H spin system. The authors used a combination of the chemical shift positions, $^3J_{HH}$ coupling constants, and the presence of NOEs between known and unassigned protons to map out the complete ^{1}H spin system. With this knowledge in hand the observation of a NOE between H_5, H_8, and H_{18} for (p-chlorobenzoyl)dihydroquinidine indicated that these protons were in close proximity and suggested that the (p-chlorobenzoyl)dihydroquinidine adopted a closed conformation in solution as well as in the solid state, see Figure 2.11(a). Additional support for this conformational assignment came from the $^3J_{H_8H_9}$ coupling constant of 7.5 Hz, which was consistent with a torsional angle of about 155°, using Eq. (2.20). Because the ^{1}H NMR spectra of the R = CONMe$_2$ and R = Ac derivatives were very similar to the p-chlorobenzoyl derivative, these compounds were also assigned closed conformations (they had $^3J_{H_8H_9}$ coupling constants between 7.5 and 8.3 Hz). Interestingly, the parent dihydroquinidine, R = H, and methoxydihydroquinidine, R = Me, had substantially different $^3J_{H_8H_9}$ coupling constants, 3.5 and 3.9 Hz,

(a)   **(b)**

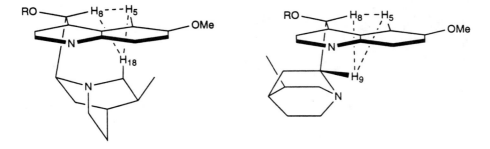

FIGURE 2.11.
The NOE active nonbond distances in the (a) closed-below and (b) open-below conformations of
quinidines.

respectively, which were consistent with either a 120° or 60° dihedral angle. This dihedral
angle ambiguity was resolved by irradiation of H8 and observation of strong NOEs both
for H5 and H9 for the methoxydihydroquinidine, see Figure 2.11(b). These observations
were consistent with a *trans* conformation for the $C_{4'}-C_8-C_9-N$ dihedral angle and an
overall open conformation for the molecule. Given the similarity of ^{1}H NMR spectra for
the methoxy and parent dihydroquinidines the parent was also assigned an open conformation.

In the authors, Dijkstra et al., also used ^{1}H NMR spectroscopy to investigate the dependence of the molecular conformation on solvent. For (*p*-chlorobenzoyl)dihydroquinidine
there were only minor differences changing from chloroform-d_1 to acetone-d_6, acetonitrile-
d_3, dichloromethane-d_2, or toluene-d_8 (for toluene the conformation was thought to open
up slightly from a $H_8-C_9-C_8-H_9$ dihedral angle of ~155° to ~140°). For methoxydihydroquinidine, however, changing from chloroform-d_1 to dichloromethane-d_2 caused substantial changes in the ^{1}H spectrum: $^3J_{H_8H_9}$ changed from 3.9 to 6.6 Hz and irradiation of
H8 yielded an NOE between H5 and H18. This analog changed from an open conformation to a closed conformation upon changing the solvent from chloroform to dichloro-
methane!

In order to understand the dependence of the overall molecular conformation on substitution at C9 and solvent, the authors used molecular mechanics. Torsional profiles about
the $C_{4'}-C_9$ and C_9-C_8 bonds were generated for the parent dihydroquinidine (open conformation experimentally observed), acetyldihydroquinidine (closed conformation observed), and methoxydihydroquinidine (open and closed conformations observed
depending on solvent). For all three molecules four low-energy conformations were found.
See Figure 2.12 for the structures for the parent dihydroquinidine. The torsional angles
about the $C_{4'}-C_9$ and C_9-C_8 bonds are collected in Table 2.6. In calculated conformations
1 and 2, the $C_{4'}-C_8-C_9-N$ dihedral angles were $-51.6°$ and $-54.6°$, respectively, making them both closed conformations. The difference between conformations 1 and 2 was
whether the alkaloid ring was above the quinoline ring (conformation 2) or below it (conformation 1) due to $C_{3'}-C_{4'}-C_9-C_8$ dihedral angles of 104.4° (conformation 1) and

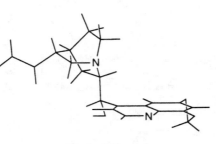

Quinidine-1

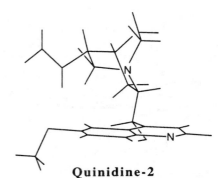

Quinidine-2

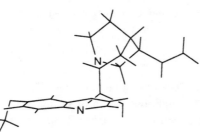

Quinidine-3

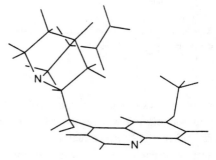

Quinidine-4

FIGURE 2.12.
Four low-energy conformations of the parent dihydroquinidine. [Reproduced with permission from Dijkstra, G. D.; Kellogg, R. M.; Wynberg, H.; Svendsen, J. S.; Marko, I.; Sharpless, K. B. (1989), *J. Am. Chem. Soc.* **111,** 8069. Copyright © 1989 American Chemical Society.]

TABLE 2.6. Calculated Relative Energies and Torsional Angles for the $C_{4'}-C_9$ and C_9-C_8 Bonds for a Set of Dihydroquinidines, **2.9**

	Conformation	$C_{3'}-C_{4'}-C_9-C_8$	$C_{4'}-C_9-C_8-N_1$	Energy (kcal/mol)[a]
Parent dihydroquinidine, R = H	1	104.4	− 51.6	0.0
(open conformation observed)	2	− 71.3	− 54.6	0.11
	3	− 93.7	− 155.3	1.18
	4	87.4	− 150.7	5.30
Methoxydihydroquinidine, R = Me	1	104.6	− 51.5	0.0
(open and closed	2	− 69.8	− 54.8	0.24
conformation observed)	3	− 92.8	− 155.6	1.98
	4	85.1	− 176.9	4.33
Acetyldihydroquinidine, R = Ac	1	106.2	− 42.1	0.0
(closed conformation observed)	2	− 64.6	− 41.1	0.33
	3	− 91.9	− 155.5	0.57
	4	83.7	− 172.8	2.43

[a] Relative to conformation 1.

$-71.3°$ (conformation 2). For the remaining two calculated conformations (3 and 4) the $C_{4'}-C_8-C_9-N$ dihedral angles were $-155.3°$ and $-150.7°$, respectively, making them both open conformations. As for conformations 1 and 2, the difference between conformations 3 and 4 was whether the alkaloid ring was above the quinoline ring (conformation 3) or below it (conformation 4) due to $C_{3'}-C_{4'}-C_9-C_8$ dihedral angles of $-93.7°$ and $87.4°$.

Conformations 1–3 were found to be energetically close (within ~ 1 kcal/mol) and conformation 4 was found to be higher in energy (~ 3 kcal/mol) for each of the three compounds studied. Conformations 2 and 3 correspond to the experimentally observed conformations. There is no experimental evidence for conformation 1. Furthermore, the authors, Dijkstra et al., report that the molecular mechanics calculations did not show a gradual change in energy difference between conformations 2 and 3 as a function of benzylic substituent. This is not in agreement with their experimental observations. Dijkstra et al. suggested that this disagreement might be due to a solvation effect as the molecular mechanics work was done in the gas phase, while the NMR work was done in solution. The preferred conformation of methoxydihydroquinidine (R = Me) was experimentally found to depend on solvent.

Since the solvents and the quinidines were all polar the differential quinidine–solvent interaction missing in the molecular mechanics study was possibly a dipole–dipole interaction. Dijkstra et al. calculated the dipole moments of each of the conformations for each of the molecules at their molecular mechanics optimized geometries. Interestingly, the conformation with the lowest dipole moment for each of the molecules was the one found in solution. Another suggested possibility for their disagreement with experiment was specific quinidine–quinidine interactions in solution. Alkaloid–alkaloid interactions had previously been observed for cinchona alkaloids.

Molecular mechanics was used to establish that there were four low-energy conformations for a set of three quinindines. Unfortunately, the calculated energy differences were not consistent with the NMR-derived conformations. It is likely that a detailed understanding of the conformational preferences of cinchona alkaloids in solution will require molecular mechanics simulations in solution.

2.3.4. Assignment of Stereogenic Centers by a Combination of NOESY and Restrained Molecular Dynamics

Establishing the relative and absolute configuration of stereogenic centers in synthetic intermediates, as well as the study of the conformational preferences of such molecules, is of major importance for both the synthesis of pharmacologically active molecules and the development of new synthetic methodologies.

X-ray crystallography is the most precise method of assigning the relative configurations of a set of stereogenic centers, but it requires the preparation of a crystalline sample. In addition, X-ray crystallography does not provide information regarding the conformation of the molecule in solution.

Nuclear magnetic resonance spectroscopy has seen extensive use in the structural characterization of oligopeptides, proteins, and oligonucleotides in solution. Reggelin et al. (1992) of the Technische Universität München demonstrated that NOE spectroscopy, when combined with restrained molecular dynamics, can be applied to the characterization of stereogenic centers in conventional organic molecules.

The molecule chosen for this initial study was 2-methoxy-3-sulfonyl-1,3-oxazolidine, **2.10**.

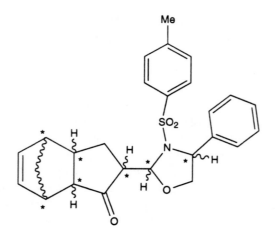

2.10

This compound is an intermediate in the diastereoselective transformation of a vinyl sila-nol into a β-keto aldehyde as shown in Figure 2.13 (Hoppe et al., 1991). This particular molecule was of interest because it had a limited degree of conformational flexibility but contained seven stereogenic centers, as indicated by the stars in **2.10**.

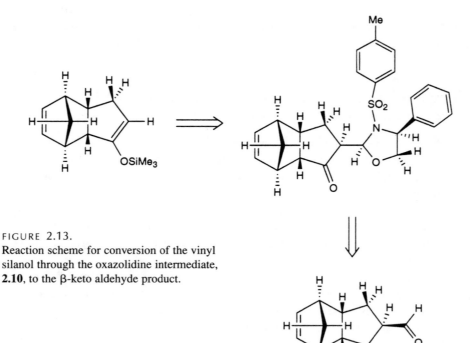

FIGURE 2.13.
Reaction scheme for conversion of the vinyl silanol through the oxazolidine intermediate, **2.10**, to the β-keto aldehyde product.

METHODS

The initial molecular structure of **2.10** was sketched in using Insight, from Biosym; the unpublished CFF91 force field was augmented to include terms for the sulfonamide, and the structure was minimized. From this initial structure a restrained, quenched MD simulation was carried out. That is, a 100 picosecond (ps) (1 ps $= 10^{-12}$ s) MD simulation, with a 1 femtosecond (fs) (1 fs $= 10^{-15}$ s) time step, was carried out at 500 K, structures being extracted and minimized every 50 fs, with terms added to the force field to cause the molecule to distort to a conformation (and stereoconfiguration) consistent with the NMR data. The terms added to the force field from NMR data come from the distance dependence of NOE interactions, which are discussed in Box 2.3.

The NOE derived distances for **2.10** are collected in Table 2.7. They were used as the r_0 values in a biharmonic constraint function, Eq. (2.22) (Clore, 1985),

$$V_{\text{NOE}} = \frac{1}{2}k_{\text{inner}}(r - r_0)^2 \qquad \text{if } r < r_0$$

$$V_{\text{NOE}} = \frac{1}{2}k_{\text{outer}}(r - r_0)^2 \qquad \text{if } r > r_0 \qquad (2.22)$$

with an inner-force constant of 5 kcal/mol·Å^2 and an outer-force constant of 50 kcal/mol·Å^2. These additional "bonds" provide an energy penalty to structures that do not satisfy the NMR distance data. If Reggelin et al. had simply carried out a minimization

---

BOX 2.3.   **Distance Dependence of NOE**

As discussed in Box 2.1 the effectiveness of the nonradiative relaxation pathway due to dipolar coupling is related to how far apart the dipoles are. The distance dependence of the dipole–dipole coupling interaction goes as $\frac{1}{r^3}$, so the rate constant for signal enhancement is proportional to $\frac{1}{r^6}$, barring complications due to internal molecular motion. If one monitors the NOE effect between a pair of nuclei as a function of time, the slope of the intensity–time curve is proportional to the distance between the nuclei to the inverse sixth power. Linearity in this curve implies that the enhancement observed is due to a first-order process, quite likely NOE. The extrapolation of the full set of lines for each pair of interactions in the system back to zero time gives "initial build-up rates." The build-up rates provide relative distances between the nuclei in a molecule. These relative distances can be converted to real distances by using an internal calibration; a geometrically defined pair of spins (such as methylene protons) is assigned a reasonable, known, distance, r_{AB}, and all other distances, r_{CD}, are defined in terms of this distance. In Eq. (2.23) τ_{AB} is the initial build-up rate or enhancement for the AB spin pair and τ_{CD} is the initial build-up rate or enhancement for the CD spin pair.

$$\frac{\tau_{\text{AB}}}{\tau_{\text{CV}}} = \left(\frac{r_{\text{CD}}}{r_{\text{AB}}}\right)^6 \qquad (2.23)$$

(continued)

BOX 2.3.   Distance Dependence of NOE *(continued)*

The NOE distance information can be collected in either a conventional intensity versus chemical shift [one-dimensional (1-D)] experiment where saturation of a particular spin causes a change in the intensity of other spins (as shown in Fig. 2.14) or in the more modern two-dimensional Nuclear Overhauser Effect SpectroscopY (2-D NOESY) experiment, where the NOE information is contained in the intensity of cross peaks. An idealized NOESY spectrum is shown in Figure 2.15. Since chemical shift information is supplied along both the *x* and *y* axes the NOESY spectrum is symmetric, and since the 2-D *x* and *y* drawing space is used up providing chemical shift information, the height of the peaks or amplitudes of the signals are displayed as contours of increasing intensity (shown as concentric circles in this idealized spectrum). As can be seen in Figure 2.15(b) the conventional 1-D spectrum, in Figure 2.15(a), appears along the diagonal of the 2-D spectrum. There is additional information contained in Figure 2.15(b); these additional features (cross peaks), labeled A and B, provide the NOE data.

The cross peaks in Figure 2.16(b) labeled *A* (connected to the first peak and the fourth peak by horizontal and vertical dashed lines) indicate that protons 1 and 4 are close enough to have a strong NOE. The cross peaks labeled *B* (connected to the second peak and the fourth peak by horizontal and vertical dashed lines) indicate that protons 2 and 4 are close enough to have a weak NOE. The distance between 1 and 4 is likely shorter than the distance between 2 and 4 due to the larger amplitude (more concentric circles) for the cross peak connecting 1 and 4 than the cross peak connecting 2 and 4. The lack of additional cross peaks suggest that the proton pairs 1 and 3, 2 and 3, 3 and 4, and 1 and 2 are too far apart to have a significant NOE.

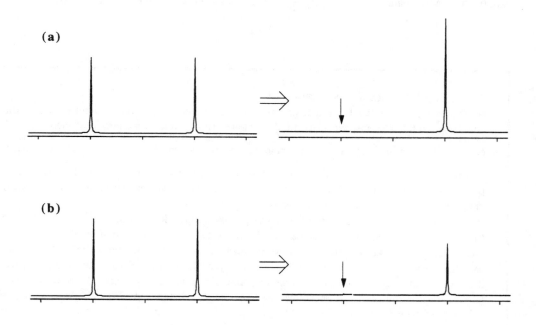

FIGURE 2.14.
(a) The NOE enhancement due to a two quantum transition. (b) The NOE inversion due to a zero quantum transition.

(a) 1-D Spectrum

(b) 2-D Spectrum

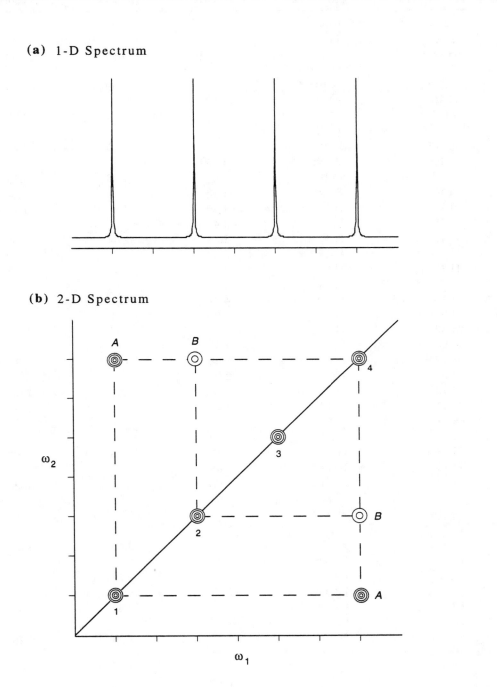

FIGURE 2.15.
(a) Idealized 1-D spectrum. (b) Idealized four proton NOESY spectrum.

TABLE 2.7. Interproton Distances (Å) from NOESY and MD Simulation for **2.10**

Diagonal Peak	Cross Peak	Distance[a]	Distance[b]	Distance[c]
T2,T6	H-6		3.38	3.70
H-6	T2,T6	3.30		
H-10[pro-S]	P2,P6	3.10	3.08	4.33
P2,P6	H-10[pro-S]			
P2, P6	H-10[pro-R]		2.89	3.32
H-10[pro-R]	P2, P6	2.79		
T3, T5	T7		2.24	2.99
T7	T3, T5	2.17		
P3, P5	H-10[pro-R]		3.90	4.70
H-10[pro-R]	P3, P5	3.23		
P3, P5	H-10[pro-S]		4.18	5.55
H-10[pro-S]	P3, P5	3.48		
H-12	H-15[pro-R]	3.39	3.60	3.49
H-15[pro-R]	H-12	3.46		
H-12	H-10[pro-S]	2.73	2.84	3.04
H-10[pro-S]	H-12	2.87		
H-13	H-15[pro-R]	3.37	3.58	3.49
H-15[pro-R]	H-13	3.50		
H-13	H-10[pro-S]	4.57	4.69	4.44
H-10[pro-S]	H-13	4.90		
H-12	H-9	4.28	4.47	4.29
H-9	H-12	4.40		
H-12	H-6	2.88	3.12	3.58
H-6	H-12	2.95		
H-12	H-11	2.58	2.70	2.68
H-11	H-12	2.63		
H-13	H-6	2.87	3.06	3.63
H-6	H-13	2.99		
H-13	H-14	2.57	2.73	2.69
H-14	H-13	2.65		
H-12	H-14	4.46	4.61	4.29
H-14	H-12	4.53		
T2, T6	H-3		2.49	4.92
H-3	T2, T6	2.46		
T2, T6	H-1		2.48	3.82
H-1	T2, T6	2.45		
P2, P6	H-4[pro-R]		2.79	3.45
H-4[pro-R]	P2, P6	2.73		
P2, P6	H-3		2.82	3.04
H-3	P2, P6	2.75		
P3, P5	H4[pro-R]		3.79	5.28
H-4[pro-R]	P3, P5	3.67		
P3, P5	H-3		3.47	5.09
H-3	P3, P5	3.67		
H-1	H-10[pro-R]	3.47	3.48	3.72
H-10[pro-R]	H-1	3.44		
H-1	H-6	2.31	2.34	2.53
H-6	H-1	2.31		
H-1	H-4[pro-S]	2.58	2.60	3.08
H-4[pro-S]	H-1	2.58		
H-1	H-4[pro-R]	4.44	3.67	3.86
H-4[pro-R]	H-1	4.45		

Diagonal Peak	Cross Peak	Distance[a]	Distance[b]	Distance[c]
H-1	H-3	3.42	3.19	3.55
H-3	H-1	3.43		
H-3	H-4*pro-S*	2.30	2.31	2.23
H-4*pro-S*	H-3	2.29		
H-3	H-4*pro-R*	2.90	2.88	2.86
H-4*pro-R*	H-3	2.90		
H-4*pro-R*	H-10*pro-R*	3.25	3.21	3.38
H-10*pro-R*	H-4*pro-R*	3.21		
H-4*pro-R*	H-8	3.64	3.59	4.38
H-8	H-4*pro-R*	3.63		
H-4*pro-R*	H-4*pro-S*	1.78	1.78	1.79
H-4*pro-S*	H-4*pro-R*	1.78		
H-14	H-15*pro-S*	2.66	2.70	2.68
H-15*pro-S*	H-14	2.66		
H-14	H-15*pro-R*	2.58	2.62	2.64
H-15*pro-R*	H-14	2.60		
H-11	H-10*pro-R*	4.57	3.79	3.71
H-10*pro-R*	H-11	4.55		
H-11	H-10*pro-S*	2.80	2.81	3.00
H-10*pro-S*	H-11	2.89		
H-11	H-15*pro-R*	2.59	2.62	2.65
H-15*pro-R*	H-11	2.60		
H-11	H-15*pro-S*	2.70	2.71	2.68
H-15*pro-S*	H-11	2.68		
H-8	H-15*pro-S*	2.52	2.54	2.70
H-15*pro-S*	H-8	2.50		
H-6	H-10*pro-S*	2.31	2.32	2.30
H-10*pro-R*	H-6	2.29		
H-6	H-10*pro-R*	2.90	2.66	2.97
H-10*pro-R*	H-6	2.87		
H-9	H-15*pro-S*	2.43	2.42	2.51
H-15*pro-S*	H-9	2.39		
H-9	H-10*pro-S*	2.96	2.83	2.87
H-10*pro-S*	H-9	3.04		
H-9	H-10*pro-R*	2.27	2.27	2.25
H-10*pro-R*	H-9	2.24		
H-14	H-8	2.43	2.50	2.62
H-8	H-14	2.45		
H-11	H-9	2.45	2.53	2.52
H-9	H-11	2.47		
H-10*pro-R*	H-10*pro-S*	1.79	1.82	1.74
H-10*pro-S*	H-10*pro-R*	1.86		
H-15*pro-R*	H-15*pro-S*	1.87	1.85	1.78
H-15*pro-S*	H-15*pro-R*	1.85		
T2, T6	H-4*pro-S*		3.86	5.88
H-4*pro-S*	T2, T6	3.83		
H-1	H-10*pro-S*	4.63	4.42	4.14
H-10*pro-S*	H-1	4.76		
H-3	H-10*pro-R*	4.59	4.55	4.11
H-10*pro-R*	H-3	4.53		

[a] Intensity ratio method. [b] Cross-peak analysis. [c] The parameter r_{MD}.

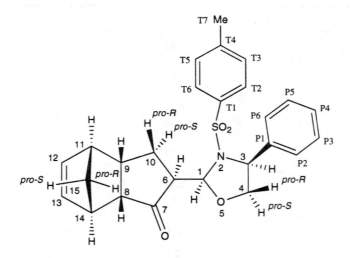

FIGURE 2.16.
The NOE plus energy restrained-MD derived stereochemistry for **2.10**.

with these constraints, the molecule would simply have fallen into the nearest, high-energy minimum. Carrying out MD permits the molecule to climb over low-energy potential barriers. For the molecule at hand, with 7 stereogenic centers, the MD simulation could have been carried out for each of the 16 possible diastereomers, ultimately choosing the structure that best satisfied the NOE constraints and/or had the lowest constrained energy; instead the authors (Reggelin et al.) chose to use restraining force constants sufficiently large so as to swamp out the angle and torsional potentials that would enforce a particular chirality at a given center. The NOE restraining ''bond'' terms were stiff enough to cause centers of the wrong chirality (inconsistent with the NOE distance data) to invert even though there are four bonds to carbon.

The first step in a structural NMR study consists of a complete assignment of the ^{1}H spin system. The nonaromatic protons of **2.10** were assigned using the conventional 1-D ^{1}H NMR spectrum in conjunction with a *TO*tal *C*orrelation *S*pectroscop*Y* (TOCSY) spectrum. Total correlation spectroscopy provides cross peaks between all members of a coupled spin system, and hence defines which protons are coupled to which other protons in the spin system. Although the proton pairs at carbon atoms 4, 10, and 15 have different chemical environments, and hence are diastereotopic, their assignments could not be made at this point, see Figure 2.16 for atom numbers. The dependence of $^3J_{CH}$ upon the amplitude of the proton detected heteronuclear long range correlation (HMBC) signal, along with the dependence of $^3J_{CH}$ on dihedral angle [Eq. (2.17)], was used to assign the methylene protons attached to carbon atoms 10 and 15. The relationship between the two rings, connected by the bond between C1 and C6, was also established by the HMBC spectrum. The remaining two protons (those bound to C4) remained unassigned until the 2-D NOE experiment (NOESY) was carried out.

Interproton distances were obtained from the NOESY spectrum in two ways. The authors (Reggelin et al.) first used the conventional approach of carrying out the experiment with several mixing times. Recall the proton–proton distance is proportional to the slope of the intensity versus mixing time curves. The relative distances so obtained were calibrated assuming the H—H distance of the geminal protons at C4 to be 1.78 Å, Eq. (2.23). The second technique for obtaining proton–proton distances rests on the relationship between the cross and diagonal peaks in the NOESY spectrum, and is independent of the initial build-up rate approximation. The distance between a pair of protons r, is related to the diagonal and cross-peak intensities (a_{AA} and a_{AB}), see Eq. (2.24), where τ_C (3.56×10^{-11} s) is obtained from the assumed distance between the methylene or geminal proton pair at C4.

$$r = \left\{ \frac{-2q\tau_{mix}}{\ln\left[\dfrac{a_{AA} + a_{AB}}{a_{AA} - a_{AB}}\right]} \left(\frac{6\tau_C}{1 + 4\omega^2\tau_C^2} - \tau_C \right) \right\}^{1/6} \tag{2.24}$$

The proton–proton distances from this analysis ($\tau_{mix} = 600$ ms) and the more conventional analysis are collected in Table 2.7.

RESULTS

In order to determine the solution structure of **2.10**, including an assignment of the stereochemistry of seven stereogenic centers, the authors (Reggelin et al.) used NOESY NMR spectroscopy to estimate intramolecular nonbonded distances. These NOE derived distances, tabulated in Table 2.7, were used as restraints in an MD simulation. The final stereochemistry of the molecule is illustrated in Figure 2.16. The proton–proton distances from the lowest energy restrained-MD structure are collected in Table 2.7 for comparison to the NMR derived distances. The overall agreement is reasonable (within 1 Å), with the exceptions of the distances between H-10$^{pro\text{-}S}$ and P2, P6 (should be 3.1 Å, calculated to be 4.33 Å); T2, T6 and H-3 (should be 2.49 Å, calculated to be 4.92 Å); P3, P5 and H-10$^{pro\text{-}S}$ (should be 4.18 Å, calculated to be 5.55 Å); T2, T6 and H-1 (should be 2.48, calculated to be 3.82); P3, P5 and H-4$^{pro\text{-}R}$ (should be 3.79, calculated to be 5.28), P3, P5 and H-3 (should be 3.47, calculated to be 5.09); and T2, T6 and H-4$^{pro\text{-}S}$ (should be 3.86, calculated to be 5.88).

The discrepancy between these calculated and observed distances is a consequence of the conformational flexibility of the phenyl and toluyl aromatic rings. Since there is near free rotation about the single bonds connecting the aromatic rings to the rest of the molecule, P2 and P6, P3 and P5, T2 and T6, and T3 and T5 cannot be distinguished and contribute to the same cross-peak intensity. The NOE experiment weights "conformationally equivalent" distances differently than would a simple arithmetic average (Güntert et al., 1989). This can be seen from a rearrangement of Eq. (2.23). For **2.10**, the H—H distance of the geminal protons at C4 and corresponding NOE enhancement are taken as constant. Therefore the "NOE" distance, r_{NOE}, is given by Eq. (2.25) where τ is the cross-peak intensity for the spin pair of interest.

$$\tau = \frac{\text{const}}{r_{NOE}^6} \tag{2.25}$$

For a cross-peak intensity due to two protons (normalized to a single proton) the observed enhancement is the average of the two separate enhancements, Eq. (2.26) for peaks in the slow-exchange regime.

$$\tau_{ave} = \frac{1}{2}(\tau_a + \tau_b) \qquad (2.26)$$

Expressing the enhancements in terms of the distances, that is, combining Eq. (2.25) and (2.26) yields Eq. (2.27).

$$\frac{const}{r_{NOE}^6} = \frac{1}{2}\left\{\frac{const}{r_a^6} + \frac{const}{r_b^b}\right\} \qquad (2.27)$$

The constant terms in the numerator drop out of Eq. (2.27) and rearrangement yields Eq. (2.28) which relates the observed "NOE" distance to the individual conformationally "averaged" distances.

$$r_{NOE} = 2^{1/6}\frac{r_a r_b}{(r_a^6 + r_b^6)^{1/6}} \qquad (2.28)$$

If r_a is equal to r_b, then Eq. (2.28) simplifies to the result that r_{NOE} is also equal to r_a. If $r_a << r_b$, then the r_a^6 term in the denominator of Eq. (2.28) can be ignored and Eq. (2.29) is obtained.

$$r_{NOE} = 2^{1/6}r_a \qquad (2.29)$$

There are geometric conditions where the "NOE" derived distance would be independent of r_b.

To illustrate the dependence of "distance" on the averaging procedure, consider the H-10^{pro-S}, P2,P6 case and assume that the average, computed, distance of 4.33 Å is the observed average distance. The results of applying Eqs. (2.28) and (2.29) for a range of r_a values are collected in Table 2.8. Given the observed r_{NOE} of 3.1 Å, individual r_a and r_b distances of 2.75 and 5.91 Å are consistent with the NOE data and the computed average distance. The NOE derived "average" distance is substantially skewed towards the shorter of the two individual distances. This conformational effect is quite common and

TABLE 2.8. The NOE Derived Distances for an Average Distance of 4.33 Å

r_a	r_b	r_{NOE}	$2^{1/6}r_a$
2.0	6.66	2.245	2.245
2.5	6.16	2.804	2.806
2.75	5.91	3.082	3.087
3.0	5.66	3.355	3.367
3.5	5.16	3.868	3.928
4.0	4.66	4.245	4.480
4.33	4.33	4.33	4.860

could be included in a restraining function by using Eq. (2.28) to determine the restraining distance r in Eq. (2.22). For three "equivalent" distances, such as between the protons of a methyl group and an external proton, Eq. (2.29) can be generalized to Eq. (2.30).

$$r_{NOE} = 3^{1/6} \frac{r_a r_b r_c}{(r_a^6 r_b^6 + r_a^6 r_c^6 + r_b^6 r_c^6)^{1/6}} \tag{2.30}$$

In summary, MD in conjunction with NOE data can be used to determine the structure of a complex organic molecule with several stereogenic centers. However, care must be taken for systems where multiple conformations are likely to be populated in solution.

2.4. TRANSITION STATE MODELING

2.4.1 Background

Structural data for the ground states of molecules can be obtained straightforwardly from X-ray diffraction for crystalline substances, electron diffraction and microwave spectroscopy for gas-phase substances, and NMR spectroscopy for molecules in solution. No experimental method exists for determining the molecular structures of transition states or saddle points. From the early days of quantum mechanics, chemists hoped that theoretical chemistry could provide structural insight into the transition state or saddle point region of the potential surface. Molecular mechanics was, in fact, developed by Westheimer and Mayer (1946) to rationalize the rates of racemization of substituted biphenyls in terms of steric effects in the transition state. With the advent of analytic gradient methodologies and the general accessibility of commercial software, such as the Gaussian series of programs, ab initio electronic structure techniques have become a relatively routine method for obtaining structural information about transition states (Hehre et al., 1986). This methodology, unfortunately, is limited to small molecular systems (up to 20 atoms or so). DeTar (1974) and later Houk et al. (1990) devised molecular mechanics force fields modified to treat the saddle point regions of the potential energy surface for individual reactions. Houk's transition state modeling approach begins with an ab initio determination of the saddle point structure for a small model system. Given this structure, the atoms that define the saddle point structure are either held fixed, or molecular mechanics bond distance and angle bend terms are estimated to reproduce the ab initio geometry for those centers directly involved in the reaction. The molecular framework is then elaborated and steric contributions to the saddle point structure and energetics are assessed by molecular mechanics.

2.4.2 Carbonyl Reduction

The reduction of carbonyls by nucleophilic reagents such as lithium aluminum hydride (LiAlH$_4$ or LAH), Eq. (2.31), is one of the cornerstone reactions of organic chemistry (House, 1972).

$$\tag{2.31}$$

R and R' in Eq. (2.31) are alkyl substituents. If the carbonyl is unsymmetrically substituted or contained in a ring, the two carbonyl π faces react differently, either due to an electronic effect or steric discrimination. Numerous models have been proposed to rationalize the facial selectivity of this general class of reaction (Cram and Abd Elhafez, 1952; Dauben et al., 1956; Cherest et al., 1968). In order to quantitatively probe the steric contribution to this reaction, Wu and Houk (1987) of the University of California, Los Angeles, reported a transition state modeling study of acyclic and cyclic ketones.

METHODS

Wu and Houk (1987) began the simulation by determining the transition state structures for the addition of LiH across the $C-O$ π bond of acetone using a 3–21G basis set and an ab initio Hartree–Fock wave function. This ab initio transition state structure served as the natural transition state geometry for the molecular mechanics studies. Both the $C-O$ and $C-H$ natural bond distances of 1.269 and 2.035 Å, respectively, were obtained from this structure. In addition, a $O-C-H$ natural bond angle of 96.2° was found. They also determined the transition state structures and relative energetics for the addition of NaH across the $C-O$ π bond of acetaldehyde and three conformations of proprionaldehyde using a 3–21G basis and a Hartree–Fock wave function. This set of relative energetics was used to modify the torsional parameters (angle and force constant) for the $C_\alpha-C_{CO}$ bond of the MM2 force field (see Section 2.2) as implemented in a local version of the MM2 program.

RESULTS

Wu and Houk (1987) evaluated the energy differences at the transition state structure for hydride approach, and hence carbonyl facial preference for several acyclic as well as cyclic ketones. The transition state force field was found to reproduce experimental isomer ratios (facial preference) rather well. A subset of the published data is collected in Table 2.9.

After the experimental observations were reproduced Wu and Houk (1987) examined their results and explained the facial selectivity in terms of torsional strain, consistent with Felkin's model (Cherest et al., 1968).

To understand this torsional strain, consider acetone as a prototype carbonyl; in the ground state one of the hydrogen atoms (H^a) on each of the methyl groups is parallel to the $C-O$ bond vector, see **2.11a** in Figure 2.17.

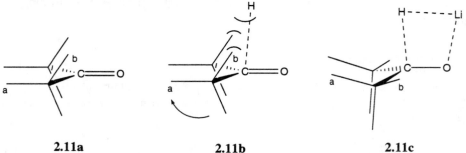

2.11a   **2.11b**   **2.11c**

FIGURE 2.17.
Acetone and hydride addition to acetone transition state models.

TABLE 2.9. Stereoselectivity of Hydride Addition

			Stereochemistry	
Compound[a]	Substituents	Nucleophile[b]	Experimental	Calculated

| | R=H | LAH | 88–91:12–9 | 88:12 |
| | R=Me | LAH | 95:5 | 90:10 |

	R_1=Me, R_2=R_3=R_4=R_5=H	LAH	60–82:40–18	82:18
	R_1=R_2=Me, R_3=R_4=R_5=H	LAH	62:38	73:27
	R_3=Me, R_1=R_2=R_4=R_5=H	LAH	84–87:16–13	88:12
	R_3=R_4=R_5=Me, R_1=R_2=H	LAH	20:48:80–52	30:70

	X=CH_2	LAH	91:9	89:11
	X=O	LAH	15:85	9:91
	X=CH_2	MeMgI	45:55	68:32
	X=O	MeMgI	98:2	94:6
	X=S	MeMgI	7:93	3:97

[a] Nucleophile=Nu, tert−butyl=t−Bu, and phenyl=Ph.
[b] Lithium aluminum hydride=LAH.

As the hydride approaches, the methyl groups rotate to relieve a repulsive interaction between the approaching hydride and the hydrogen atoms of each of the methyl groups labeled b, see **2.11b** of Figure 2.17. The transition state for attack by lithium hydride is shown in **2.11c** of Figure 2.17 (Wu and Houk, 1987).

The facial selectivity for the cyclic ketones in Table 2.9 was explained by replacing pairs of hydrogen atoms in a second view of the acetone transition state model, **2.11d** in Figure 2.18(a), by a trimethylene fragment. Replacement of the hydrogen atoms labeled a, which were nearly parallel in the transition state, yielded a transition state for axial hydride attack that was achieved without introducing significant ring strain, see the transformation from **2.11d** to **2.12** shown in Figure 2.18(a).

68

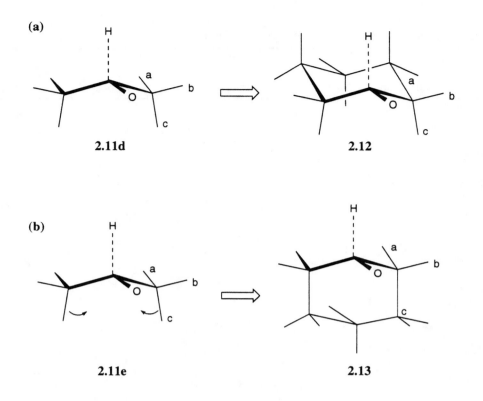

FIGURE 2.18.
Cyclic modifications to the hydride addition to acetone transition state model. (a) Axial hydride attack and (b) equatorial hydride attack.

Replacement of the hydrogen atoms labeled c corresponded to equatorial attack by the nucleophile, see the transformation from **2.11e** to **2.13** shown in Figure 2.18(b). As indicated by the arrows in **2.11e**, the two hydrogen atoms labeled c were not parallel in the transition state for lithium hydride addition to acetone. Addition of a trimethylene fragment required bending the $C-H^c$ bond vectors inward, introducing strain into the transition state. According to Wu and Houk (1987) this strain manifested itself as torsional strain. This torsional strain as well as conventional steric interactions were included in the force field and reasonable agreement with experiment was found.

The Houk transition state modeling approach has been criticized by Menger and Sherrod (1990) who questioned the uniqueness of the transition state model structures and parameters. They found that a variety of transition state parameterizations could reproduce trends in reactivity and that even a ground-state model could work. Menger and Sherrod (1990) concluded that it was not wise to correlate molecular mechanics derived transition state structural information with reactivity as the structural information was somewhat arbitrary. Eurenius and Houk (1994) recently responded to this criticism pointing to an improper oxygen van der Waals radius in the Menger and Sherrod analysis. This interchange is reminiscent of the uniqueness of force field parameters discussion in Section 2.2. Care must always be taken to obtain parameters in as rational a manner as possible and to interpret the results with caution.

2.5. DIASTEREOMERIC ENERGY DIFFERENCES

2.5.1 Background

Molecular recognition, particularly of chiral substrates, is an active area of chemical research (Lehn, 1988). One practical goal of this work is the development of efficient chiral separation techniques, both by chromatography (Pirkle, 1983; Schurig, 1983) and through diastereomeric salt crystallization (Jacques et al., 1981). Due to the complex interplay of a large number of weak interactions, the development of these techniques has largely been empirical. Nonetheless, there is hope that molecular mechanics can lead to the systematic development of specific chiral separation agents.

That the interaction of two chiral centers will potentially lead to energetic discrimination between enantiomers can be seen from the interaction of molecule **2.14** with both enantiomers of molecule **2.15** as shown in Figure 2.19(a) (**2.15** and **2.15′**) (Dalgliesh, 1952; Salem et al., 1987; Topiol et al., 1988).

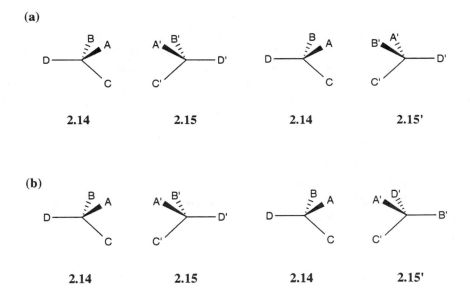

FIGURE 2.19.
Interactions between enantiomers to make diastereomeric pairs.

It should be apparent that at the pairwise interaction level it is not possible for all of the nonbonded distances between the substituents of molecules **2.14** and **2.15′** to be the same as the nonbonded distances between the substituents of molecules **2.14** and **2.15**. It is possible however, to have two of the three close approach interactions the same for both diastereomeric pairs, see Figure 2.19(b). Here the interactions between A and A′ and C and C′ are the same; the differential effect is that the interaction between B and B′ is replaced by an interaction between B and D′. Because the molecules will bind so as to retain the largest attractive substituent–substituent interactions, the magnitude of the diasteromeric energetic differentiation possible between discrete pairs of species is limited to being approximately the magnitude of the third largest pairwise interaction in the system. Additionally, as the binding between the two molecular species increases, the translational and rotational entropic penalty associated with binding will increase; this ΔH, ΔS compensation effect causes the diastereomeric $\Delta\Delta G$ achievable between two discrete moderately sized molecular species to be rather small. Experimentally, this is not much of a problem. Chromatographic techniques can exploit quite small energetic differences because the discriminating binding equilibria are reestablished many times as the solution flows through the column. Computationally, the small intrinsic energetic differentiation represents a significant challenge. In order to begin to explain a phenomena based on a computational methodology the phenomena must first be computationally reproduced. Chiral discriminations in solution are typically less than 0.5 kcal/mol—this is a smaller energy difference than any present day computational methodology can routinely achieve, particularly for molecules the size of pharmacologically active agents that would be of interest for enantiomeric separation. An additional complication is that there are likely several low-energy conformations available to each complex making a rather extensive conformational searching procedure necessary.

However, because enantiomers (**2.15** and **2.15′**) are of identical energy and because the reagent used for chiral discrimination **2.14** is the same in both calculations (and hence of constant energy) there is reason to believe that molecular mechanics, with a van der Waals potential and electrostatics model selected to reproduce intermolecular interactions, could accurately account for the effects responsible for diasteromeric differentiation.

2.5.2 Chiral Chromatography

Lipkowitz et al. (1989) of Indiana-Purdue University reported a molecular mechanics study on two of the classic Pirkle chiral stationary phases (CSP) (Pirkle et al., 1984, Pirkle et al., 1985, Pirkle and Pochapsky, 1986), **2.16** and **2.17**, the symbol Si-Surf indicates the point of attachment to the silica support.

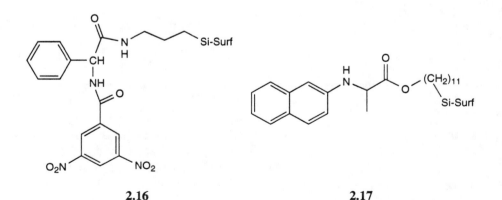

2.16   **2.17**

METHODS

The Pirkle CSPs **2.16** and **2.17** were modeled by replacing the alkyl chains used to attach the CSP to the silica support with methyl groups. The nitro groups of **2.16** were replaced by aldehyde functional groups because nitro groups were not parameterized in the MM2 force field used in this study. Simplified models **2.18** and **2.19** resulted.

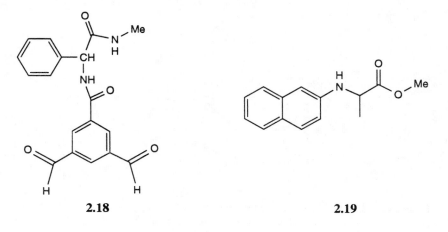

2.18   **2.19**

A set of four chiral molecules (analytes to be "separated") were chosen for this initial study, **2.20–2.23** of Figure 2.20. where Et = C_2H_5)

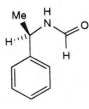

2.20   **2.21**

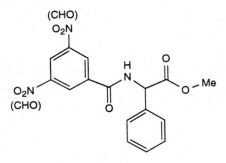   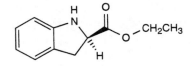

2.22   **2.23**

FIGURE 2.20.
Chromatographic analytes.

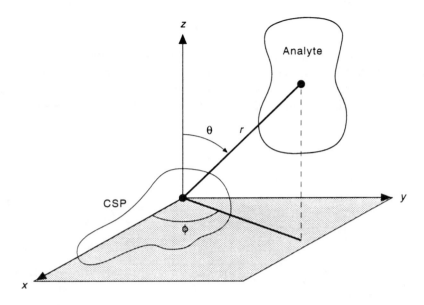

FIGURE 2.21.
Spherical polar coordinate system used to orient the analyte with respect to the CSP. [Reproduced with permission from Lipkowitz, K. B.; Demeter, D. A.; Zegarra, R.; Larter, R.; Darden, T. (1988), *J. Am. Chem. Soc.* **110**, 3446. Copyright © 1988 American Chemical Society.]

As there was no obvious "lock-and-key" binding interaction between the CSP and ana-lytes, Lipkowitz et al. (1989) developed a rather exhaustive search procedure using spheri-cal polar coordinates to obtain a set of low-energy conformations for the interaction between the CSP and the analytes. If the CSP is considered to be positioned at the origin, the center of mass of the analyte can be positioned in spherical polar coordinates in terms of r, θ, and ϕ, as shown in Figure 2.21. With the center of mass for the analyte determined by r, θ, and ϕ there are three degrees of freedom left for a rigid analyte. These degrees of freedom are described in terms of three Euler angles of orientation. Lipkowitz et al. (1989) established a grid of θ and ϕ angles (0°–180° and 0°–360°, respectively, each with 10° increments), varied r to minimize van der Waals contacts, and evaluated the intermolecular energy over a set of 1020 Euler angles. This set of 1020 configurations were Boltzmann weighted (see Appendix B) to obtain an energy for each choice of θ and ϕ angles. This analysis was carried out for the interaction of each of the low-lying conformations of each CSP with each of the low-lying conformations of each analyte. The resulting set of conformational energies was then used to obtain a $\Delta\Delta G$ for chromatographic separation (Lipkowitz et al., 1988). As discussed in Appendix B, the free energy of a system in general can be expressed as in Eq. (2.32) where Q is the partition function of the system, given by Eq. (2.33).

$$G = -kT \ln Q + PV \qquad (2.32)$$

$$Q = \sum_i f_i e^{-E_i/kT} \qquad (2.33)$$

TABLE 2.10. Interaction Energies (Averaged over Conformational Space)

Complex	Averaged Energy	
	(R)·(R) Diastereomer	(R)·(S) Diastereomer
CSP3(**2.18**)·(**2.20**)	−8.23	−8.35
CSP3(**2.81**)·(**2.21**)	−6.09	−6.29
CSP3(**2.18**)·(**2.22**)	−6.21	−6.57
CSP4(**2.19**)·(**2.23**)	−6.98	−7.55

The energies of the states of a system, E_i, have an arbitrary zero, and the i summation is over all the energy states of the system. In the present case the pressure, P, and volume, V, of the system were properly assumed constant.

RESULTS

Lipkowitz et al. (1989) carried out a molecular mechanics investigation of the interaction energies between two CSPs and four analytes to obtain a trend in α, the chromatographic separability factor ($\Delta\Delta G = -RT \ln \alpha$). The results are summarized in Table 2.10 and Figure 2.22. As seen in Figure 2.22, the trend in α is well reproduced by this computational procedure, despite the small energy differences involved.

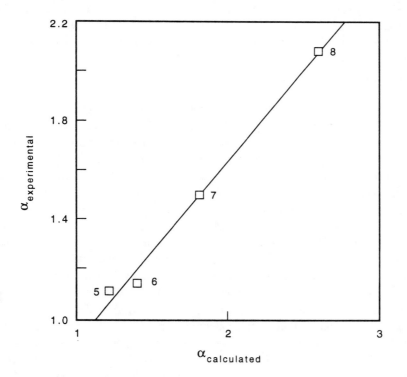

FIGURE 2.22.
Plot of computed chromatographic separability factors, α, versus experimental seperability factors. [Reproduced with permission from Lipkowitz, K. B.; Baker, B.; Zegarra, R. (1989), *J. Comp. Chem.* **10**, 718.]

Lipkowitz et al. (1989) attributed the deviation of the slope of the line in Figure 2.22 from a unit slope to the use of a dielectric constant of one in the calculations. Experimentally, the separations are carried out in a 3–10% 2-propanol in hexane solvent, which has a dielectric constant greater than 1 (~4). Either the dielectric constant in the calculations should be increased or, more properly, the simultions carried out in a solvent.

This type of study is a severe test of the intermolecular portion of a force field. Lipkowitz et al. (1989) used the MM2 force field (see Section 2.2) to obtain the geometries and energies of the conformations of the CSPs and analytes (which the force field is well documented to be able to reproduce). However, for obtaining intermolecular energetics they found MM2 to be lacking. This is not surprising because MM2 was not parameterized for intermolecular interactions. In particular, it was found that replacing the bond dipoles of MM2 [see Eq. (2.15)] with point charges significantly improved the results; the results reported in Figure 2.22 utilized a point charge model. They interpreted this improvement as an indication that, at the intermolecular distances associated with chiral recognition and complexation, bond dipoles do not properly describe the distance dependence of electrostatics (dipole–dipole interactions fall off as $\frac{1}{r^3}$, whereas point charge–point charge interactions fall off as $\frac{1}{r}$).

In conclusion, it appears that because enantiomers **2.18** and **2.18′** were of identical energy and the reagent used for chiral discrimination (e.g., **2.20**) was the same in both complexes (and hence of constant energy) there was a precise cancellation of errors. Furthermore, this study suggests that molecular mechanics should prove useful in the design of chiral chromatography columns, particularly if care is taken to sample conformational space well and nonbond potentials suited for intermolecular interactions are used.

2.5.3 Diastereomeric Salt Formation

The magnitude of energetic discrimination between enantiomers can be greatly enhanced by placing the diastereomeric pair (e.g. **2.14** and **2.15** or **2.15′**) in a crystalline environment, where nonbonded interactions are larger due to the close proximity of many molecules. Differences in heats of solution or heats of fusion (ΔH_{fus}) of up to 5 kcal/mol have been observed for diastereomeric salts, in contrast to the small energy differences found in the isolated complex example (Section 2.5.2). Experimentally, the strategy in diastereomeric salt separation is to react the racemic mixture of a chiral carboxylic acid with a chiral base such as an amine or phosphate. (Alternatively, a racemic mixture of a chiral base could be separated with a chiral acid.) One of the enantiomers of the chiral carboxylate will preferentially crystallize with the conjugate acid of the chiral amine forming a diastereomeric salt. If the crystallization event is thermodynamically controlled, the differences in heats of solution can be exploited to preferentially form the more stable crystalline species. Diastereomeric salt crystallization was, in fact, the earliest form of enantiomeric separation and continues to be a common procedure (Jacques et al., 1981).

Leusen, Bruins Slot, and Noordik of the University of Nijmegen, van der Haest and Wynberg of the University of Groningen, and Bruggink of Andeno BV (Leusen et al., 1992) reported a molecular mechanics study of the energetics of diastereomeric salt formation for the base ephedrine, **2.24** with a series of five halogen-substituted chiral phosphoric acids, **2.25–2.29**.

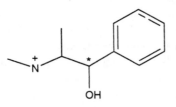

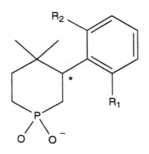

2.24

2.25 $R_1 = H$ $R_2 = H$

2.26 $R_1 = F$ $R_2 = H$

2.26 $R_1 = Br$ $R_2 = H$

2.28 $R_1 = Cl$ $R_2 = H$

2.29 $R_1 = Cl$ $R_2 = Cl$

The stars indicate stereogenic centers in **2.24–2.29**. Ephedrine is used in the treatment of asthma or nasal congestion (Hoffman and Lefkowitz, 1990). As illustrated in **2.24**, ephedrine has two stereogenic centers, leading to four stereoisomers (*SS, RR, RS,* and *SR*). The researchers focused on the enantiomeric pair (1*R*2*S*)-(−)-ephedrine, **2.24**(−), and (1*S*2*R*)-(+)-ephedrine, **2.24**(+). There were experimental differences between the salts formed by **2.24**(+) and **2.24**(−) with the various chiral phosphoric acids, see Table 2.11. For the halogen substituted acids there was a substantial difference in solubility for the diastereomeric salts ranging from a factor of 2 to a factor of 6, suggesting a significant range in differential heats of solution ($\Delta\Delta H_{sol}$). Substantial (> 2 kcal/mol) differential heats of fusion ($\Delta\Delta H_{fus}$) were found for **2.27** and **2.28**. Small $\Delta\Delta H_{fus}$ values were observed for the other three diastereomeric salts.

TABLE 2.11. Experimental Data on Diastereomeric Salts

Complex	Solubility	Resolution Efficiency	Melting Point (K)	ΔH_{fus} (kcal/mol)	$\Delta\Delta H_{fus}$
2.25·2.24(+)	8.8	−0.05	504.1	11.6±0.1	0.2±0.4
2.25·2.24(−)	8.4		508.3	11.4±0.3	
2.26·2.24(+)	22.1	0.52	483.0	9.71±0.02	0.8±0.1
2.26·2.24(−)	46.0		474.2	8.9±0.1	
2.27·2.24(+)	5.7	0.76	492.7	11.4±0.2	2.0±0.4
2.27·2.24(−)	23.6		491.3	9.4±0.2	
2.28·2.24(+)	9.3	0.43	491.6	10.2±0.1	0.5±0.3
2.28·2.24(−)	16.4		482.0	9.7±0.1	
2.29·2.24(+)	10.0	0.84	488.4	9.7±0.1	2.1±0.9
2.29·2.24(−)	62.6		476.3	7.5±0.1	

When Leusen et al. (1992) could not explain the trends in $\Delta\Delta H_{fus}$ and solubility in terms of hydrogen-bonding contacts, they attempted to use molecular mechanics to reproduce the trends. They reasoned that perhaps the $\Delta\Delta H_{fus}$ was a function of specific electrostatic or van der Waals interactions or due to a simple packing density differential. If the diastereomeric salt were completely dissociated in the melt, then trends in $\Delta\Delta H_{fus}$ could be directly related to the crystal energies of salts, since the energies of the enantiomers **2.24(−)** and **2.24(+)** are the same.

Placing the diastereomeric salts of **2.24(−)** or **2.24(+)** with **2.25–2.29** in a crystalline environment presents computational difficulties. Rather than having to calculate nonbonded interactions over the few angstroms in an isolated pair of molecules, the nonbonded interactions in a crystal span the entire infinite system. The infinite nature of crystalline nonbonded interactions is simulated by invoking the concept of periodic boundary conditions. The construct of periodic boundary conditions is reviewed below.

METHODS

To evaluate the energy of a crystalline material the energies of interaction between the atoms within the crystal unit cell and all other atoms in the entire crystal need to be calculated. The construct of periodic boundary conditions provides at least three different means of carrying out this infinite summation. These conditions are discussed in Box 2.4.

Of the several modeling programs available at the time their work was carried out by Leusen et al. (Polygraf, Insight, GROMOS, and CHARMM) none could evaluate the nonbonded interactions for a crystalline system using the Ewald summation technique discussed above. Only CHARMM was able to carry out a real space summation to more than 14 Å, the interlayer distances for this set of diastereomeric salts. Thus, CHARMM was used to carry out the condensed state simulations (see Section 3.2 for a discussion of the CHARMM force field).

RESULTS

For the series of diastereomeric salts formed from **2.24(±)** and **2.25–2.29**, Leusen et al. found no correlation between calculated diastereomeric energy differentiation as probed by crystal energy. The poor correlation was independent of the source of charges (see Table 2.12) and computational procedure. Neither using a large dielectric constant

TABLE 2.12. Calculated Lattice Energy Differences for Diastereomeric Salts

Complex	Source of Atomic Charges			Constraints		Experiment
	AM1	PM3 ($\varepsilon=5$)	Quanta	Harmonic $\varepsilon=5$	Dielectric $\varepsilon=12$	
2.25·2.24(+/−)	−3.0	−3.4	−3.3	2.1	3.2	0.2±0.4
2.26·2.24(+/−)	−3.3	−3.7	−5.8	1.7	3.2	0.8±0.1
2.28·2.24(+/−)	−5.8	−5.6	−0.9	0.6	3.8	2.0±0.4

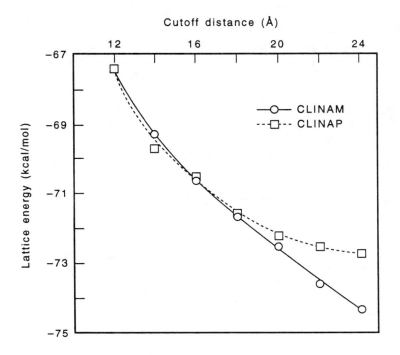

FIGURE 2.23.
Convergence of energy as a function of cutoff distance for **2.28** with **2.24(±)**. [Reproduced with permission from Leusen, F.J.J., Bruins Slot, H. J.; Noordik, J. H.; van der Haest, A. D.; Wynberg, H.; Bruggenk, A. (1992), *Recl. Trav. Chim. Rays-Bas* **111**, 111.]

nor applying harmonic constraints to force the calculated valence geometries to be the same as the experimental geometries helped matters. By using the X-ray coordinates for the diastereomeric salt formed from **2.24(±)** and the chloro-substituted acid, **2.28**, Leusen et al. carried out a real space summation as a function of cutoff radius (see Fig. 2.23) with a dielectric constant of 12 (it should be 1). They found that the lattice energy had not converged by 24 Å; in fact, the energy of one salt had nearly converged but the other was dropping off nearly linearly, suggesting that either a cutoff radius many times 24 Å should be used or that an infinite summation procedure such as the Ewald method, is needed.

Since this work was carried out, numerous commercial molecular modeling packages have added the Ewald summation technique: It would be interesting to repeat the study with more modern technology.

BOX 2.4   Periodic Boundary Conditions

Crystalline samples possess the useful property that they have translational symmetry. That is, there is a smallest unique entity within a crystalline sample (called the unit cell) that can be translated to identical images merely by adding a multiple of the lattice coordinates to the coordinates of each atom in the unit cell. For a general unit cell there are six lattice coordinates a, b, c, α, β, and γ, which are identified in Figure 2.24. For the simplest case, a cubic unit cell ($a = b = c = L$, $\alpha = $

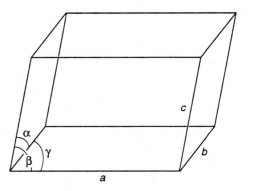

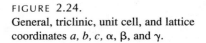

FIGURE 2.24.
General, triclinic, unit cell, and lattice coordinates a, b, c, α, β, and γ.

$\beta = \gamma = 90°$), translation in the x direction by one unit is given by Eq. (2.34) where x_t, y_t, and z_t are the translated x, y, and z coordinates. The coordinates of the nearest-neighbor image cells, in two dimensions, are given in Figure 2.25(a).

$$x_t = x + L$$
$$y_t = y \qquad\qquad (2.34)$$
$$z_t = z$$

Evaluation of the nonbonded interactions between the atoms in the unit cell and all the image cells consists of considering the interaction of the unit cell with each of the infinite number of image cells. In practice, three procedures have proven useful for carrying out this lattice summation. The first approach is to simply sum out over image cells until the magnitude of the interactions are small enough to ignore. The decision to stop adding in additional terms is either cell based (all interactions from a given cell are included) or atom based (a spherical cutoff could be used). There is significant difficulty with either of these techniques since the number of interactions increases rather dramatically ($\sim r^2$) as the cutoff distance increases and the magnitude of electrostatic interactions only dies off as $\frac{1}{r}$.

The second commonly used approach is called the minimum image convention. Here the unit cell is made large enough so that all important interactions are within the cell. (The distance range of interactions is less than the cell length.) Interactions are included between atoms within the cell or between an atom in the cell and the nearest image cell depending on whether the intracell distance or intercell is shorter (Metropolis et al., 1953). For the cell in Figure 2.25(b) the distance within the cell is shortest—the distance between a and b is shorter than the distance between a and the image of b, b'. For the cell in Figure 2.25(c) the distance between a and the image of b, b' is shorter than the distance between a and b—the image distance will be used.

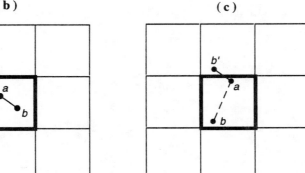

FIGURE 2.25.
(a) A two-dimensional lattice with nearest-neighbor image cells. (b) A configuration with shorter distance within unit cell. (c) A configuration with shorter image distance.

In the third approach, called Ewald summation, the full infinite summation is obtained by considering both a real space and a reciprocal space summation (Ewald, 1921). In order to minimize the number of terms used in the lattice summation of nonbonded interactions (particularly electrostatic interactions) each term in the summation is multiplied by a damping function, says ϕ, and separately by $(1 - \phi)$. The damping function is chosen so that the real space summation containing ϕ converges quickly. The complementary $(1 - \phi)$ summation will converge quite slowly in real space. That is, the $(1 - \phi)$ function will slowly decay in real space, but the $(1 - \phi)$ summation will converge rapidly in reciprocal space. The Ewald approach is substantially more computationally demanding than either of the straightforward real space techniques, but it does insure that all interactions are properly accounted for.

Homework

2.1. Sketch in and minimize methane, ethane, *trans-* and *gauche*-butane, tri-*tert*-butyl methane, and chair and twist–boat cyclohexane using the native force field of your modeling software. How do the structures and conformational energy differences you have obtained compare to experiment and those reported in the text using the MM2 force field?

2.2. Sketch in and minimize the structures for **2.1t** and **2.1c**. Which configuration is preferred? What terms in the energy expression are the source of the configurational preference?

2.3. Sketch in and minimize the three configurations of **2.8**, the hydrindanone model of androsterone. Which configuration is preferred? What terms in the energy expression are the source of the configurational preference?

2.4. Sketch in the closed-below conformation of **2.9**, R $=$ H, with the correct stereochemistry and minimize the structure. Next, rotate about the $C_{4'}$–C_9 and C_8–C_9 single bonds to form the open-below conformation. Measure the H–H distances associated with NOE close contacts. Are there other H–H pairs that might show a NOE effect? (The dihedral angles in Table 2.6 might help you construct reasonable models.)

2.5. Use the molecular mechanics energies in Table 2.6 and Eq. (1.29) to determine the 300 K Boltzmann population of each of the four reported conformations for **2.9**, R $=$ H.

2.6. Sketch in and minimize the structure of the cyclic amide, **2.30**, using your modeling software.

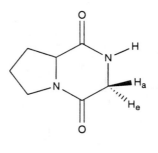

2.30

Measure the H−N−C−H_a and H−N−C−H_e dihedral angles. Use Eq. (2.20) with $A =$ 7.76, $B = -1.0$, and $C = 1.40$ to calculate $^3J_{HH}$. How do your results compare to the experimental values of $^3J_{HHe} = 4.5$ and $^3J_{HHa} = 1.5$?

2.7. Sketch in **2.10** with the correct stereochemistry, minimize the structure, and carry out a conformational search. For your best structure measure the distance pairs in Table 2.4 associated with the two aromatic rings (e.g., H-10^{pro-S} and P2,P6; T2, T6 and H-3; P3, P5 and H-10^{pro-S}; T2, T6 and H-1; P3, P5 and H4^{pro-R}, P3, P5, and H-3; and T2, T6, and H-4^{pro-S}, and compare the distances you calculate to the NMR distances.

2.8. With the use of a simple arithmetic average $[r_{ave} = \frac{1}{2}(r_a + r_b)]$ and the "NOE average" given by Eq. (2.28) determine the averages for the distances obtained in Problem 2.7. How do these average distances compare to the NMR distances?

2.9. Sketch in, minimize, and carry out a conformational search on the two enantiomers of lactic acid. Are the strain–steric energies, bond distances, and bond angles the same?

2.10. Carboxylic acids such as lactic acid can condense to form anhydrides, see Eq. (2.35).

$$(2.35)$$

Condense the two enantiomers of lactic acid from Problem 2.9 to form the four combinations of anhydrides (L + L, L + D, D + L, and D + D). Minimize and carry out a conformational search on each of the four structures. Do they each have the same strain energy? Is each energy unique? If your modeling package has a docking facility, generate superimposed images of the structures that are energetically close. Are any pairs of the structures superimposable?

2.11. In solution lactic acid tends to form a lactide, **2.31**, the result of undergoing two condensation reactions.

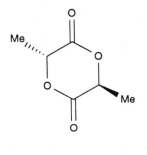

2.31

From a solution of racemic lactic acid there are four combinations of enantiomers possible (L with L, L with D, D with L, and D with D). Sketch in and minimize the four possibilities. Which conformation(s) is(are) lowest? If your modeling package has a docking facility, generate superimposed images of the structures that are energetically close. Are there actually four structures?

2.12. Lactic acid could react with alanine to generate a structure analogous to the lactide discussed in Problem 2.11, Compound **2.32**.

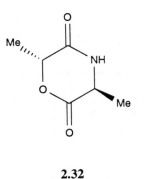

2.32

From a solution of racemic lactic acid and racemic alanine there are four mixed combinations of enantiomers possible (L lactic acid with L alanine, L lactic acid with D alanine, D lactic acid with L alanine, and D lactic acid with D alanine). Sketch in and minimize the four possibilities. Which combination(s) is(are) lowest? If your modeling package has a docking facility, generate superimposed images of the structures that are energetically close. Are there actually four structures?

2.13. Sketch in and minimize **2.19** and **2.20**. Bring the molecules together and convince yourself that, as Lipkowitz et al. found, there are not any obvious binding preferences.

2.14. A series of substituted biphenyls have been constructed to systematically investigate the biphenyl rotational barriers for systems where the barrier is small (see Bott, 1980), see **2.33**.

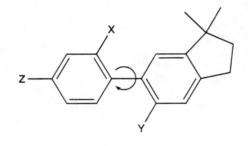

2.33

Use your molecular modeling program to estimate the activation energies for Y = Me, Z = H, and X = F, Cl, and Br. How do the results compare to the experimental values reported below?

X	Y	Z	$\Delta H^{\ddagger}$ (kcal/mol)
Br	Me	H	15.8 ± 0.4
Cl	Me	H	11.8 ± 0.2
F	Me	H	5.5 ± 0.5

Experimental data are also available for the series with Z = NO$_2$. Do you think your modeling software could reproduce the effect of substituting H by NO$_2$? Why or why not?

X	Y	Z	$\Delta H^{\ddagger}$ (kcal/mol)
Cl	Me	NO$_2$	13.4 ± 0.2
F	Me	NO$_2$	4.7 ± 0.5

2.15. The crystal structures of both members of the diastereomeric salt pair of **2.24($\pm$)** with **2.28** are present in the Cambridge Structural Database (FIMTUQ and FIMVAY). Read in and minimize these two structures. Compare the energy difference you obtain to the experimental $\Delta\Delta H_f$ of 2.0 kcal/mol.

References

Allinger, N. L. (1977), Conformational Analysis. 130. MM2. A hydrocarbon force field utilizing V_1 and V_2 torsional terms, *J. Am. Chem. Soc.* **99**, 8127.

Allinger, N. L.; Tribble, M. T. (1972), Conformational analysis—LXXX The hydrindanone ring system, *Tetrahedron* **28**, 1191.

Allinger, N. L.; Tribble, M. T.; Miller, M. A. (1972), Conformational analysis—LXXIX An improved force field for the calculation of the structures and energies of carbonyl compounds, *Tetrahedron* **28**, 1173.

Allinger, N. L.; Tribble, M. T.; Miller, M. A.; Wertz, D. H. (1971), Conformational analysis. LXIX. An improved force field for the calculation of the structures and energies of hydrocarbons, *J. Am. Chem. Soc.* **93**, 1637.

Allinger, N. L.; Yuh, Y. H.; Lii, J.-H. (1989), Molecular mechanics. The MM3 force field for hydrocarbons, *J. Am. Chem. Soc.* **111**, 8551.

Allinger, N. L.; Zhu, Z. S.; Chen, K. (1992), Molecular mechanics (MM3) studies of carboxylic acids and esters, *J. Am.Chem.Soc.* **114**, 6120.

Altona, C.; Ippel, J. H.; Westra Hoekzema, A. J. A.; Erkelens, C.; Groesbeek, M.; Donders, L. A. (1989), Relationship between proton–proton NMR coupling constants and substituent electronegativities. V. Empirical substituent constants deduced from ethanes and propanes, *Magn. Res. Chem.* **27**, 556.

Bartell, L. S.; Bürgi, H. B (1972), Structure of tri-*tert*-butylmethane. II. Inferences combining electron diffraction, spectroscopy, and molecular mechanics, *J. Am. Chem. Soc.* **94**, 5235.

Bigger, J. T.; Hoffman, B. F. (1990), Antiarrhythmic drugs. In Gilman, A. G.; Rall, T. W.; Nies, A. S.; Taylor, P., Eds., *Goodman and Gilman's The Pharmacological Basis of Therapeutics*, Mc-Graw-Hill, New York, pp. 848–857.

Bott, G.; Field, L. D.; Sternhill, S. (1980), Steric effects. A study of a rationally designed system *J. Am. Chem. Soc.* **102**, 5618.

Buckingham, R. A. (1938), The classical equation of state of gaseous helium neon and argon. *Proc. R. Soc. London, Ser. A* **168**, 264.

Burkert, U.; Allinger, N. L. (1982), *Molecular Mechanics*, American Chemical Society, Washington, DC.

Cherest, M.; Felkin, H. Prudent, N. (1968a), Torsional strain involving partial bonds. The stereochemistry of the lithium aluminum hydride reduction of some simple open-chain ketones, *Tetrahedron Lett.*, 2199.

Cherest, M.; Felkin, H. Prudent, N. (1968b), Torsional strain involving partial bonds. The steric course of the reaction between allyl magnesium bromide and 4-*t*-butyl cyclohexanone, *Tetrahedron Lett.* , 2205.

Clore, G. M.; Gronenborn, A. M.; Brunger, A. T.; Karplus, M. (1985), Solution conformation of a heptadecapeptide comprising the DNA binding helix F of the cycle AMP receptor protein of *Escherichia coli*; combined use of ^{1}H nuclear magnetic resonance and restrained molecular dynamics, *J. Mol. Bio.* **186**, 435.

Cram, D. J.; Abd Elhafez, F. A. (1952), Studies in stereochemistry. X. The rule of "Steric control of asymmetric induction" in the synthesis of acyclic systems, *J. Am. Chem. Soc.* **74**, 5828.

Dalgliesh, C. E. (1952), The optical resolution of aromatic amino-acids on paper chromatograms, *J. Chem. Soc.* **137**, 3940.

Dauben, W. G.; Fonken, G. S.; Noyce, D. S. (1956), The stereochemistry of hydride reductions, *J. Am. Chem. Soc.* **78**, 2579.

DeTar, D. F. (1974), Quantitative predictions of steric acceleration, *J. Am. Chem. Soc.* **96**, 1255.

Diez, E.; San-Fabian, J.; Guilleme, J.; Altona, C.; Donders, L. A. (1989), Vicinal proton–proton coupling constants. I. Formulation of an equation including interactions between substituents, *Mol. Phys.* **68**, 49.

Dijkstra, G. D.; Kellogg, R. M.; Wynberg, H.; Svendsen, J. S.; Marko, I.; Sharpless, K. B. (1989), Conformational Study of Cinchona Alkaloids. A combined NMR, molecular mechanics, and X-ray approach, *J. Am. Chem. Soc.* **111**, 8069.

Donders, L. A.; de Leeuw, F. A. A. M.; Altona, C. (1989), Relationship between proton–proton NMR coupling constants and substituent electronegativities. IV. An extended Karplus equation accounting for interactions between substituents and its application to coupling constant data calculated by the extended Hückel method, *Magn. Res. Chem.* **27**, 556.

Eurenius, K. P.; Houk, K. N. (1994), Rational vs random parameters in transition state modeling: MM2 transition state models for intramolecular hydride transfers, *J. Am. Chem. Soc.* **116**, 9943.

Ewald, P. (1921), Die Derechnung optischer und elektrostatischer Gitterpotentiale, *Annu. Phys.* **64**, 253.

Garbisch, E. W.; Griffith, M. G. (1968), Proton couplings in cyclohexane, *J. Am. Chem. Soc.* **90**, 6543.

Güntert, P.; Braun, W.; Billeter, M.; Wüthrich, K. (1989), Automatic stereospecific ^{1}H assignments and their impact on the precision of protein structure determinations in solution, *J. Am. Chem. Soc.* **111**, 3997.

Haasnoot, C. A. G.; de Leeuw, F. A. A. M.; Altona, C. (1980), The relationship between proton–proton NMR coupling constants and substituent electronegativities-I, *Tetrahedron* **36**, 2783.

Hehre, W. J.; Radom, L.; v.R. Schleyer, P.; Pople, J. A. (1986), Applications of the theory. In *Ab Initio Molecular Orbital Theory*, Wiley, New York, pp. 466–504.

Hill, T. L (1946), On Steric Effects, *J. Chem. Phys.* **14**, 465.

Hill, T. L. (1948), Steric effects. I. Van der Waals Potential Energy curves, *J. Chem. Phys.* **16**, 399.

Hoffman, B. F.; Lefkowitz, R. J. (1990), Catecholamines and sympathomimetic drugs. In Gilman, A. G.; Rall, T. W.; Nies, A. S.; Taylor, P., eds., *Goodman and Gilman's The Pharmacological Basis of Therapeutics*, McGraw-Hill, New York, pp 214–215.

Hoppe, I.; Hoffmann, H.; Gartner, I.; Krettek, T.; Hoppe, D. (1991), Diastereoselective synthesis of enantiomerically pure 3-organosulfonyl-2-(2-oxocycloalkyl)-1,3-oxazolidines from 2-formyl-cycloalkanones and β-aminoalkanols, *Synthesis* 1157.

Houk, K. N.; Tucker, J. A.; Dorigo, A. E. (1990), Quantitative modeling of proximity effects on organic reactivity, *Acc. Chem. Res.* **23**, 107.

House, H. O. (1972), Metal hydride reductions and related reactions. In *Modern Synthetic Reactions*, Benjamin/Cummings, Menlo Park, CA, pp. 45–144.

Huggins, M. L. (1953), Bond energies and polarities, *J. Am. Chem. Soc.* **75**, 4123.

Jacques, J.; Collet, A.; Wilen, S. H. (1981), Formation and separation of diastereomers. In *Enantiomers, Racemates and Resolution*, Wiley, New York, pp. 251–328.

Karplus, M. (1959), Contact electron-spin coupling of nuclear magnetic moments, *J. Chem. Phys.* **30** 11.

Karplus, M. (1963), Vicinal proton coupling in nuclear magnetic resonance, *J. Am. Chem. Soc.* **85**, 2870.

Lehn, J.-M. (1988), Supramolecular chemistry-scope and perspectives molecules, supermolecules, and molecular devices, *Angew. Chem. Int. Ed. Engl.* **27**, 89.

Leusen, F. J. J.; Bruins Slot, H. J.; Noordik, J. H.; van der Haest, A. D.; Wynberg, H.; Bruggink, A. (1992), Towards a rational design of resolving agents. Part IV. Crystal packing analyses and molecular mechanics calculations of five pairs of diastereomeric salts of ephedrine and a cyclic phosphoric acid, *Recl. Trav. Chim. Rays-Bas* **111**, 111.

Lii, J.- H.; Allinger, N. L. (1989a), Molecular mechanics. The MM3 force field for hydrocarbons. 2. Vibrational frequencies and thermodynamics, *J. Am. Chem. Soc.* **111**, 8566.

Lii, J.- H.; Allinger, N. L. (1989b), Molecular mechanics. The MM3 force field for hydrocarbons. 3. The van der Waals' potentials and crystal data for aliphatic and aromatic hydrocarbons, *J. Am. Chem. Soc.* **111**, 8576.

Lipkowitz, K. B.; Baker, B.; Zegarra, R. (1989), Theoretical studies in molecular recognition: enantioselectivity in chiral chromatography, *J. Comp. Chem.* **10**, 718.

Lipkowitz, K. B.; Demeter, D. A.; Zegarra, R.; Larter, R.; Darden, T. (1988), A protocol for determining enantioselective binding of chiral analytes on chiral chromatographic surfaces, *J. Am. Chem. Soc.* **110**, 3446.

Mayo, S. L.; Olafson, B. D.; Goddard III, W. A. (1990), DREIDING: A generic force field for molecular simulations, *J. Phys. Chem.* **94**, 8897.

Menger, F. M.; Sherrod, M. J. (1990), Origin of high predictive capabilities in transition-state modeling, *J. Am. Chem. Soc.* **112**, 8071.

Metropolis, N.; Rosenbluth, A. W.; Rosenbluth, M. N.; Teller, A. H.; Teller, E. (1953), Equation of state calculations by fast computing machines, *J. Chem. Phys.* **21**, 1087.

Meyers, A. I.; Sielecki, T. M.; Crans, D. C.; Marshman, R. W.; Nguyen, T. H. (1992), (−)-Cryptaustoline: Its synthesis, revision of absolute stereochemistry, and mechanism of inversion of stereochemistry, *J. Am. Chem. Soc.* **114**, 8483.

Murphy, W. F.; Fernandez-Sanchez, J. M.; Raghavachari, K. (1991), Harmonic force field and raman scattering intensity parameters of *n*-butane, *J. Phys. Chem.* **95**, 1124.

Neuhaus, D.; Williamson M. P. (1989), *The Nuclear Overhauser Effect in Structural and Conformational Analysis,* VCH, New York.

Pirkle, W. H. (1983), Separation of enantiomers by liquid chromatographic methods. In Morrison, J. D., Ed., *Asymmetric Synthesis*, Vol. 1 Academic, New York, pp. 87–124.

Pirkle, W. H.; Hyun, M. H.; Bank, B. (1984), A rational approach to the design of highly-effective chiral stationary phases, *J. Chromatogr.* **316**, 585.

Pirkle, W. H.; Hyun, M. H.; Bank, B. (1985), Chromatographic separation of the enantiomers of 2-carboalkoxyindolines and *N*-aryl-α-amino esters on chiral stationary phases derived from *N*-(3,5-dinitrobenzoyl)-α-amino acids, *J. Chromatogr.* **348**, 89.

Pirkle, W. H.; Pochapsky, T. C. (1986), A new, easily accessible reciprocal chiral stationary phase for the chromatographic separation of enantiomers, *J. Am. Chem. Soc.* **108**, 352.

Pluim, H.; Wynberg, H. (1980), Catalytic asymmetric induction in oxidation reactions. Synthesis of optically active epoxynaphthoquinones, *J. Org. Chem.* **45**, 2498.

Reggelin, M.; Hoffmann, H.; Kock, M.; Mierke, D. F. (1992), Determination of conformation and relative configuration of a small, rapidly tumbling molecule in solution by combined application of NOESY and restrained MD calculations, *J. Am. Chem. Soc.* **114**, 3272.

Salem, L.; Chapuisat, X.; Segal, G.; HIberty, P. C.; Minot, C.; Leforestier, C.; Sautet, P. (1987), Chirality forces, *J. Am. Chem. Soc.* **109**, 2887.

Salvadori, P.; Rosini, C.; Pini, D.; Bertucci, C.; Altemura, P.; Uccello-Barretta, G.; Raffaelli, A. (1987), A novel application of cinchona alkaloids as chiral auxiliaries: preparation and use of a new family of chiral stationary phases for the chromatographic resolution of racemates, *Tetrahedron* **43**, 4969.

Schmitz, L. R.; Allinger, N. L (1990), Molecular mechanics calculations (MM3) on aliphatic amines, *J. Am. Chem. Soc.* **112**, 8307.

Schurig, V. (1983), Gas chromatographic methods. In Morrison, J. D., Ed., *Asymmetric Synthesis*, Vol. 1, Academic, New York, pp. 59–86.

Sharpless, K. B.; Amberg, W.; Bennani, Y. L.; Crispino, G. A.; Hartung, J.; Jeong, K.-S.; Kwong, H.-L.; Morikawa, K.; Wang, Z.-M.; Xu, D.; Zhang, X.-L. (1992), The osmium-catalyzed, asymmetric dihydroxylation: a new ligand class and a process improvement, *J. Org. Chem.*, **57**, 2768.

Squillacote, M. S; Sheridan, R. S.; Chapman, O. L,; Anet, F. A. L (1975), Spectroscopic detection of the twist–boat conformation of cyclohexane. A direct measurement of the free energy difference between the chair and the twist–boat, *J. Am. Chem. Soc.* **97**, 3244.

Topiol, S.; Sabio, M.; Moroz, J.; Caldwell, W. B. (1988), Computational studies of the interaction of chiral molecules: complexes of methyl *N*-(2-Naphthyl)alaninate with *N*-(3,5-dinitrobenzoyl)leucine *n*-propylamide as a model for chiral stationary-phase interactions, *J. Am. Chem. Soc.* **110**, 8367.

Webster, L. T. (1990), Drugs used in the chemotherapy of protozoal infections. In Gilman, A. G.; Rall, T. W.; Nies, A. S.; Taylor, P., Eds., *Goodman and Gilman's The Pharmacological Basis of Therapeutics*, McGraw-Hill, New York, pp. 991–994.

Westheimer, F. H. (1956), Calculation of the magnitude of steric effects, In Newman, M. S., Ed. *Steric Effects in Organic Chemistry*, Chapter 12 Wiley, New York, p. 523.

Westheimer, F. H.; Mayer, J. E. (1946), The theory of the racemization of optically active derivatives of diphenyl, *J. Chem. Phys.* **14**, 733.

Wu, Y.-D.; Houk, K. N. (1987), Electronic and Conformational Effects on π-Facial Stereoselectivity in Nucleophilic Additions to Carbonyl Compounds, *J. Am. Chem. Soc.* **109**, 908.

Wynberg, H.; Helder, R. (1975), Asymmetric Induction in the alkaloid-catalyzed michael reaction. *Tetrahedron Lett.*, **1975** 4057.

Wynberg, H.; Staring, E. G. J. (1984), The absolute configuration of 4-(trichloromethyl)oxetan-2-one; a case of double anchimeric assistance with inversion, *J. Chem. Soc., Chem. Commun.* **1984**, 1181.

Further Reading

Asymmetric Induction

Morrison, J. D. (1983), A summary of ways to obtain optically active compounds. In Morrison, J. D., Ed., *Asymmetric Synthesis*, Vol. 1, Academic, New York, pp. 1–12.

Molecular Mechanics of Organic Molecules

Burkert, U.; Allinger, N. L (1982), *Molecular Mechanics* ACS Monograph 177, American Chemical Society, Washington, DC.

Eksterowicz, J. E.; Houk, K. N. (1993), Transition-state modeling with empirical force fields, *Chem. Rev.* **93**, 2439.

Lipkowitz, K. B.; Peterson M. A. (1993), Molecular mechanics in organic synthesis, *Chem. Rev.* **93**, 2463.

NOE Spectroscopy

Neuhaus, D.; Williamson M. P. (1989), *The Nuclear Overhauser Effect in structural and conformational analysis*, VCH, New York.

Organic Crystals

Karfunkel, H. R.; Gdanitz, R. J. (1992), *Ab Initio* prediction of possible crystal structures for general organic molecules, *J. Comp. Chem.* **13**, 1171.

Supplemental Case Studies

Batta, G.; Gunda, T. E.; Szabó Berényi, Gulyás, G.; Makleit, S. (1992), C-19 configurational assignments in some morphine derivatives by homonuclear NOE, *Magn. Res. Chem.* **30**, S96.

Ejchart, A.; Dabrowski, J. (1992), Solution conformation of mono- and difucosyllactoses as revealed by rotating-frame NOE-based distance mapping and molecular mechanics and molecular dynamics calculations, *Magn. Res. Chem.* **30**, S105.

Hofer, O.; Kählig, H.; Reischl, W. (1993), On the conformational flexibility of vitamin D, *Monatsh. Che.* **124**, 185.

Kulishov, V. I.; Kutulya, L. A.; Tolochko, A. S.; Vashchenko, V. V.; Yarmolenko, S. N.; Mitkevich, V. V.; Tret'yak, S. M. (1991), Molecular and crystal structures of diastereomeric (−)-2[O-(*p*-chlorbenzoyl)oxymethylene]-*p*-menthane-3-ones effective chiral components of induced cholesteric systems. 1. *cis*-diastereomer, *Sov. Phys. Crystallogr.* **36(5)**, 669.

Meyers, A. I.; Sielecki, T. M.; Crans, D. C.; Marshman, R. W.; Nguyen, T. H. (1992), (−)-cryptaustoline: Its synthesis, revision of absolute stereochemistry, and mechanism of inversion of stereochemistry, *J. Am. Chem. Soc.* **114**, 8483.

Shea, K. J.; Stoddard, G. J.; England, W. P.; Haffner, C. D. (1992), Origin of the preference for the chair conformation in the cope rearrangement. Effect of phenyl substituents on the chair and boat transition states, *J. Am. Chem. Soc.* **114**, 2635.

CHAPTER 3

Peptides and Proteins

3.1. INTRODUCTION

Peptides and proteins are the most abundant macromolecules in cells. They carry out a variety of essential functions including catalysis of cellular reactions, storage and transport, cellular movement, defense, regulation of cellular activity, and serving as structural materials. Proteins can be classified into three main categories: fibrous, globular, and membrane. Fibrous proteins provide mechanical and structural support to the cell, and are built upon a single repetitive structure. Membrane proteins float in, or are attached to, the cell membrane, and are involved in transport of molecules and transduction of signals. Globular proteins are the primary agents of biological action in the cell, and are compact. In this chapter we will be concerned primarily with peptides and globular proteins.

3.1.1. Amino Acids

Peptides and proteins are chains of α-amino acids. The distinction between peptides and proteins is not always sharp, but a reasonable rule of thumb might be to call macromolecules with more than 50 amino acids proteins or polypeptides, and those with less than 50 amino acids peptides. The peptides and proteins in all plants and animals are constructed from the same basic set of 20 amino acids. The general formula of an amino acid in un-ionized form (**3.1**) is given below.

$$H_2NCH(R)CO_2H$$

3.1

The R group represents the side chain, which is different in each amino acid. In all amino acids except for glycine (where R = H), the α carbon (the first carbon atom on a chain from a carbonyl group) has four different groups and is thus a stereogenic center or site

87

of chirality. Naturally occurring amino acids are only of the L absolute configuration; the stereochemistry of an amino acid (**3.2**) is illustrated below.

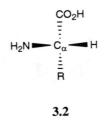

3.2

The structures of the 20 naturally occurring amino acids are illustrated in Figure 3.1. The R groups vary widely with respect to shape, size, charge, and polarity.

3.1.2. Protein and Peptide Structure

Chains of amino acids are covalently linked through an amide bond in proteins or peptides. The linkage is formed by removal of H_2O from the carboxyl group of one amino acid and the α-amino group of the other. The formation of a dipeptide is given by Eq. (3.1).

$$H_2NCH_2\overset{O}{\overset{||}{C}}OH \quad + \quad H_2NC(CH_3)HC\overset{O}{\overset{||}{}}OH \quad \xrightarrow{-H_2O} \quad H_2NCH_2\overset{O}{\overset{||}{C}}-NHC(CH_3)HC\overset{O}{\overset{||}{}}OH \qquad (3.1)$$

Once joined together, the amino acid units in a peptide are called residues. The amino acid residue at the end of the peptide having a free α-amino group is called the N-terminal or amino-terminal residue; the residue at the opposite end with the free carboxyl group is called the carboxy-terminal or C-terminal residue. The convention for naming peptides is to begin with the N-terminal residue, and then list the sequence of constituent amino acids. The dipeptide formed in Eq. (3.1) is called Gly-Ala, rather than Ala-Gly.

The amino acid sequence determines the physical and chemical properties of a protein or peptide, and hence determines its shape and biological function. Short-to-medium sized peptides tend to have little regular structure in dilute aqueous solution. These peptides adopt many different conformations in dynamic equilibrium. In contrast, globular proteins tend to adopt fairly well defined coiled and folded 3-D structures. The structure of a protein can be classified according to its primary, secondary, tertiary, and quaternary structure.

PRIMARY STRUCTURE

The primary structure of a peptide is the sequence of the amino acids in the chain. The primary structure dictates the secondary, tertiary, and quaternary structures. The peptide backbone (the repeating series of α-carbon, carboxyl carbon, and amino nitrogen atoms

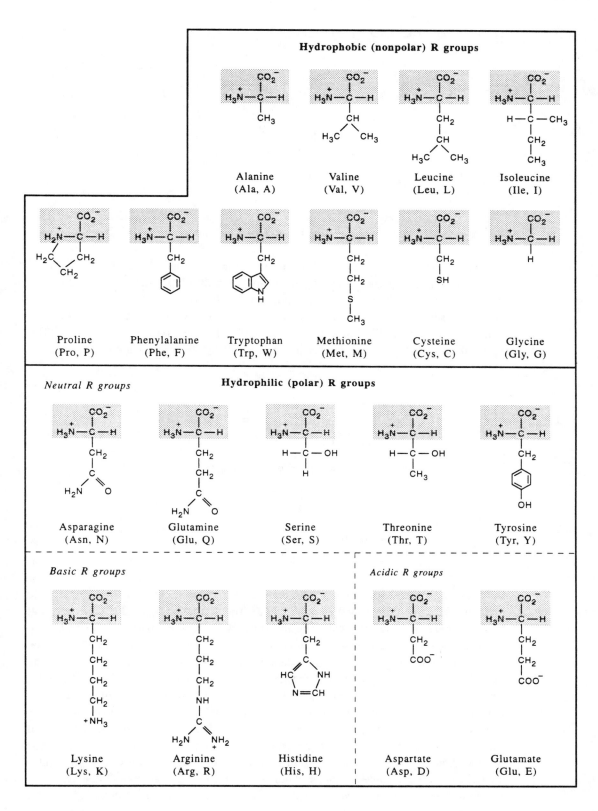

FIGURE 3.1.
Structure, abbreviations, and classifications of the amino acids at pH 7.0. The shaded portions are those common to all amino acids.

89

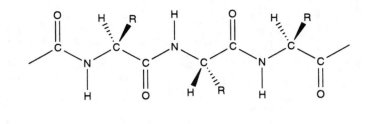

3.3

in a peptide chain) consists of a series of planar trans amide linkages (**3.3**). The cis link-ages are observed only when proline is one of the residues.

The α-carbon atom is a pivot that links two adjacent amide planes of **3.3**. To a rough approximation, two angles specify the conformation of a polypeptide: the angle φ defines the torsional angle about the N–C$_\alpha$ bond, and ψ defines the angle about the C$_\alpha$–C=O bond. The angle ω defines the amide dihedral, and χ defines the side-chain dihedral. The nomenclature for the dihedral angles in a dipeptide linkage is given in Figure 3.2. This figure shows a polypeptide fragment in the fully extended conformation φ = ψ = ω = 180°.

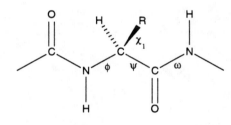

FIGURE 3.2.
Nomenclature of dihedral angles in peptides.

Assuming the amide bond is rigorously planar, there are certain sterically allowed values of φ and ψ (Ramachandran et al., 1963). The normally observed conformations of the polypeptide backbone, calculated using simple hard-sphere repulsions, can be graphi-cally represented by a 2-D Ramachandran plot (Fig. 3.3). The sterically allowed conforma-tions of a polypeptide chain are represented as areas within solid lines. Dashed lines are outer-limit torsional angle ranges. Forbidden conformational space due to unfavorable steric interactions are white spaces in the plot. Glycine frequently can be in the forbidden region because it has considerable conformational mobility due to its small side chain. Proline is also an interesting special case. The pyrrolidone ring restricts the geometry of the backbone and can introduce abrupt changes in the direction of the chain. Proline φ is fixed at about −60° due to the constraints of the pyrrolidone ring; the values allowed for the corresponding ψ angle are either about −55° or 130°. The large solid dots in Figure 3.3 correspond to φ and ψ values that produce regular repeating structures, called the secondary structure, discussed in the following sections.

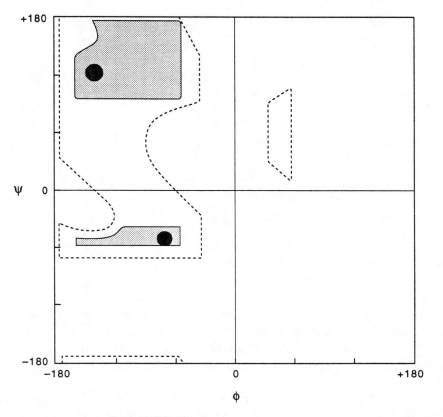

FIGURE 3.3.
Ramachandran plot of φ angles versus ψ angles.

SECONDARY STRUCTURE

The configuration of continuous sections of protein chain is called its secondary structure. It results largely from short-range hydrogen-bonding interactions between amino acids that are close together in the primary structure. Three kinds of secondary structure commonly observed in proteins are the α helix, the β-pleated sheet, and turns.

The most stable type of helix is the right-handed α helix, with about 3.7 residues per turn, involving hydrogen-bonding interactions between the i residue and $i + 4$ residue (Fig. 3.4). The side chains of the amino acids lie outside the coil of the helix, and are in close proximity to the side chains three or four amino acid units apart.

In β sheets, two or more extended, or stretched out, β strands are arranged in rows (see Fig. 3.5). Pleated sheet structures may involve peptide chains having all N termini at one end or with every other N terminus at one end. These are parallel and antiparallel forms, respectively.

Secondary structural elements are often connected by sharp turns, particularly at protein surfaces. The β turns (**3.4**) describe four residue turns, frequently stabilized by a hydrogen bond between the carbonyl of the first residue and amide of the fourth residue.

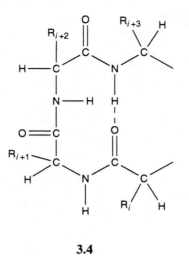

3.4

A β hairpin is a common structural element in proteins. In a β hairpin antiparallel β strands that are sequential in the primary structure are adjacent in the β sheet and linked by a β turn. The β turns have been subclassified into conformational types according to the torsional angles of the second $(i + 1)$ and third $(i + 2)$ "corner" residues (Rose et al., 1985).

The γ turns (**3.5**) describe turns of three residues and are often stabilized by an hydrogen bond between the carbonyl of the first residue and amide of the third residue.

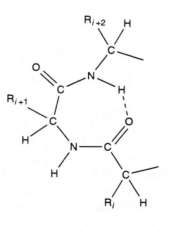

3.5

The γ turns have been subclassified into two conformational types according to the torsional angles of the second $(i + 1)$ residue.

Secondary structure can be defined by the φ and ψ torsional angles as illustrated in the Ramachandran plot of Figure 3.3. Ideal or canonical values for the angles of some secondary structures are collected in Table 3.1.

TABLE 3.1. Torsion Angles for Various Secondary Structures

Secondary Structure	$i + 1$		$i + 2$	
	ϕ	ψ	ϕ	ψ
Fully extended chain	180°	180°		
α helix (right handed)	−57°	−47°		
Antiparallel β sheet	−119°	+113°		
Parallel β sheet	−139°	+135°		
β Turns				
Type I	−60°	−30°	−90°	0
Type I′	+60°	+30°	+90°	0
Type II	−60°	+120°	+80°	0
Type II′	+60°	−120°	−80°	0
Type III	−60°	−30°	−60°	−30°
Type III′	+60°	+30°	+60°	+30°
γ Turns				
Turn	+70° to +85°	−60° to −70°		
Inverse turn	−70° to −85°	+60° to +70°		

TERTIARY STRUCTURE

The 3-D protein architecture resulting from the further folding, bending, twisting, and packing of secondary structures is called its tertiary structure. It is the biologically active or native conformation of a protein. Tertiary structure is a consequence of long-range interactions, that is, interactions between residues that are far apart in the primary structure. These long-range interactions include hydrophobic interactions (van der Waals interactions between nonpolar, ''water-hating'' groups such as aromatic residues), S−S bonds between cystine residues, hydrogen bonding, and electrostatic interactions. In general, proteins fold to minimize the contact between water and hydrophobic amino acids. The protein interior is crystalline, with densities typical of organic solids.

The tertiary structure of larger globular proteins frequently consists of several domains, which are compact, spatially distinct secondary structural units, such as helices or sheets. Proteins can also contain nonpeptide, or prosthetic, groups, which are essential to the biological activity and contribute to the 3-D structure; these are molecules such as heme groups, metal ions, lipids, and sugars.

The tertiary structure of a protein can be drawn schematically in several ways. The simplest is a stick-figurelike set of line segments linking successive α-carbon atoms. The smooth and kinked ribbon representations are also commonly used. In the kinked ribbon representation, the main chain of each residue is represented by a small polygon in the plane of the peptide group. Proteins can also be represented by combining the cylinder and arrow icons illustrated in Figures 3.4 and 3.5. These drawing methods will be used throughout this chapter and Chapter 4.

QUATERNARY STRUCTURE

Quaternary structure is the aggregation of two or more separate polypeptide chains held together by covalent cross-links or noncovalent interactions. The constituent peptide chains are called subunits. Not all proteins have quarternary structure.

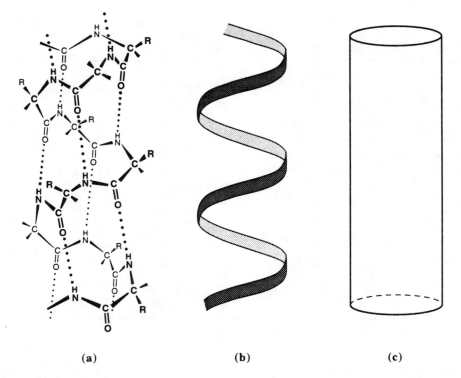

FIGURE 3.4.
(a) The right-handed α helix. Hydrogen bonds are shown as dashed lines. The spiral ribbon (b) and cylinder (c) are schematic representations of helices.

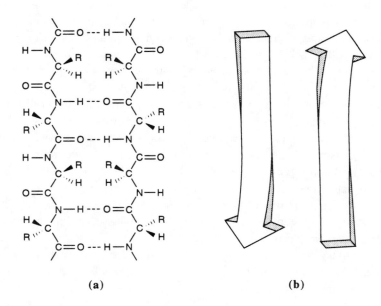

FIGURE 3.5.
(a) The antiparallel β-pleated sheet. Hydrogen bonds are shown as dashed lines. (b) The β-pleated sheet is schematically represented by two large thick arrows with an arrowhead at the C terminus indicating chain direction. Note the twist in the sheet.

DETERMINING STRUCTURES OF PEPTIDES AND PROTEINS

X-ray crystallography is currently the most powerful tool for the determination of the structures of biological macromolecules. Resolution (not to be confused with R factor) is an important criterion for the accuracy of the X-ray structural determination. Resolution is a measure of how much data was collected. The more data collected, the more detailed the features that can be distinguished. Resolution is expressed in angstroms: the lower the number the higher the resolution. Protein structures are usually determined between 1.7 and 3.5 Å. A resolution of 3 Å allows a fair determination of only the secondary structure. Protein structures determined at 2.0 Å or better are considered high resolution; the positions of the side chains and orientation of peptide planes are distinct. At 1-Å resolution the individual atoms of the protein can be resolved.

In recent years ^{1}H NMR spectroscopy, in conjunction with modeling, has been used to determine the structure of proteins and peptides. The technique is limited to relatively small proteins but it has the advantage of characterizing the structure of proteins in its native aqueous environment. The resolution of an NMR structure is generally less than an X-ray diffraction structure. Finding one unique structure is generally not possible; the calculated structure is actually a collection of related structures all of which satisfy the NMR observations. The spread between these structures gives an indication of the precision of the structure. The structures obtained from NMR and X-ray crystallography are usually in general agreement, with the least structural correspondence at the protein surface.

3.1.3. Dynamics of Proteins and Peptides

There is much experimental evidence to suggest that proteins and peptides have fluctuating dynamic structures at ordinary temperatures; and that this internal mobility often plays an important role in their biological function. Typical motions of proteins and peptides include the fast vibrations of individual atoms, and the slower short-range, or local, motions of groups of atoms, such as torsional oscillations. Long-range, or collective, movement of large regions in the molecule is rarer.

3.1.4. Prediction of Protein Structure

A small protein folds to its native, biologically active form spontaneously and quickly under physiological conditions. Although it is well known that the tertiary structure of a small protein is a function only of its amino acid sequence, how and why it folds are two parts of the unsolved "protein folding problem." Interest in predicting protein structure based on the amino acid sequence stems largely from the fact that hundreds of thousands of protein sequences are known, yet only hundreds of complete structures have been determined. The desire to design and construct new proteins with specific properties by genetic engineering has also renewed research interest in the folding problem.

Molecular mechanics and dynamics algorithms cannot take the linear amino acid sequence of a protein and predict a meaningful tertiary structure. Even if proteins did fold by random search, sampling every possible structure until it found the energy minimum, a folding simulation would be computationally intractable because of the large number of conformational states available; the multiple minima; the small steps [1–2 fs (fs =

10^{-15} s)] required to traverse detailed energy surfaces; and the time scale of global folding (less than a second to a few minutes real time). An optimistic estimate for the amount of computer time it would take to fold a small 58 residue protein is 10^8 years.

There are additional problems with predicting protein folding by molecular mechanics. First, it is not known whether the native, or biologically relevant, folded protein conformation is at the global energy minimum, or in a local energy minimum as a result of kinetic trapping. Second, the methods described in this text generally calculate enthapy rather than free energy minima. Entropic contributions are largely ignored (see Appendix B for a discussion of thermodynamics).

Protein folding investigators have therefore searched for basic principles important in the folding and structural stability of proteins. Luckily, there are hundreds of experimental solutions to the protein folding problem: the protein structures in the Brookhaven Protein Data Bank. By a statistical analysis of the known protein structures, many empirical rules that determine secondary structure have been discovered. The simplest statistical approach involves determining the frequencies of occurrence of a given amino acid in the basic secondary structure types; residues that occur frequently in a given secondary structure type are assumed to favor that structure type (see, e.g., Richardson and Richardson, 1988).

There are two broad approaches to the prediction of tertiary structure: a probabilistic approach using the data base of experimental structures, and an empirical energy approach using highly simplified physical models of proteins. We have listed several reviews on the subject of predicting tertiary protein structure in Further Reading at the end of the chapter. Briefly, empirical energy approaches reduce the many degrees of freedom in a protein by representing each residue by a single point. Lattice models are one example of this method (see, e.g., Lau and Dill, 1989). The statistical methods include assembling secondary structures into tertiary structures using empirically derived folding rules, and homology modeling.

Homology modeling has been the most successful method for predicting the tertiary structure of a protein. The procedure uses a known structure as a scaffold to build the structure of an evolutionarily related or homologous protein. It works because the secondary structure framework tends to be relatively well conserved between homologous structures, even if the amino acid sequence is not. Residues change during evolution as a result of mutation; if the mutant protein functions satisfactorily the mutation is accepted. The structural basis of the acceptance or rejection of a mutation is part of the folding problem. Mutated proteins can also be "engineered" using biotechnology. This procedure is called site-directed mutagenesis. During the short time site-directed mutagenesis has been used to explore protein structure and function, it has been discovered that, in general, protein function is not dependent on one critical interaction or one critical residue (Reidhaar-Olson and Sauer, 1988). Changes in protein function occur incrementally. Amino acid substitutions usually lead to highly localized changes in protein structure, and substitutions of residues at the surface are especially well tolerated in the mutant.

COMPARING STRUCTURES

When we compare tertiary structures (e.g., compare a modeled structure with an experimental structure, or compare two homologous structures, or compare a set of conformers derived from NMR experiments and modeling), what do we mean by the same or different? The most commonly used comparison is the main-chain atom rms, or root mean

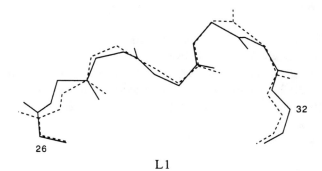

L1

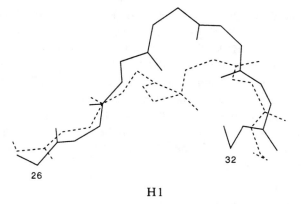

H1

FIGURE 3.6.
Root-mean-square deviations of the main-chain atoms (L1) 0.85 Å ; (H1) 2.07 Å. [Reproduced with permission from Chothia, C.; Lesk, A. M.; Levitt, M.; Amit, A. G.; Mariuzza, R. A.; Phillips, S. E. V.; Poljak, R. J. (1986), The predicted structure of immunoglobulin D1.3 and its comparison with the crystal structure, *Science* **233**, 755. Copyright © 1986 by AAAS.]

square, as defined in Equation (2.19). The rms fit of the C_α atoms is also frequently used. An rms fit of about 0.5 Å is the same as experimental error. Two examples taken from loops of immunoglobulin structures (Chothia et al., 1986) are illustrated in Figure 3.6. The L1 structures, with an rms deviation of 0.85 Å, are nearly the same. The H1 structures, with an rms deviation of 2.07 Å, are quite different.

APPLICATIONS

In Section 3.2, the force fields used to study biomolecules are described and compared. In Sections 3.3–3.8, applications of molecular modeling to peptides and proteins are discussed. Section 3.3 describes molecular mechanics studies of the solvated protein crambin and the solvated metalloenzyme carbonic anhydrase I. Section 3.4 discusses the molecular graphics modeling of the site-directed mutagenesis of insulin, resulting in the development of a useful drug. Molecular dynamics (MD) investigations are described in Sections 3.5 and 3.6. (A review of the MD method is given in Appendix C). In Section

3.5, high-temperature quenched dynamics is used to search configuration space for the tetrapeptide tuftsin, and in Section 3.6, dynamics is used to study large-scale movement in the solvated protein HIV-1 protease. Section 3.7 describes the combination of molecular modeling and 2-D NMR to derive the solution structures of tendamistat and the *lac* repressor headpiece. Homology modeling is used to predict the structures of the hypervariable regions of immunoglobulins in Section 3.8.

3.2. BIOLOGICAL FORCE FIELDS

A number of empirical potential functions have been proposed for peptides and proteins. The common biological force fields in use today have retained the functional forms originating with Hill and Pitzer (harmonic stretch and bend, cosine torsional potential, and Lennard-Jones 6-12 van der Waals potential). These specialized biological force fields include AMBER (Assisted Model Building and Energy Refinement) (Weiner and Kollman, 1981; Weiner et al., 1984, 1986), CHARMM (Chemistry at Harvard Macromolecular Mechanics) (Brooks et al., 1983 and Nilsson and Karplus, 1986), and GROMOS (Groningen Molecular Simulation System) (van Gunsteren and Berendsen, 1987, 1990). These force fields all originated as extensions to the Gelin–Karplus protein force field and program (Gelin and Karplus, 1979), which in turn can be traced back to the early work by Hill and Pitzer. Other harmonic force fields include DREIDING (Mayo et al., 1990) and Tripos 5.2 (Clark et al., 1989).

3.2.1. Comparison with MM2 and MM3

As originally formulated, AMBER, CHARMM, and GROMOS were specifically designed to study large nucleic acids and proteins; and as such, a number of simplifications have been made relative to the MM2 and MM3 force fields discussed in Section 2.2. In the interest of computational time, parameter development effort, and numerical stability during MD simulation, harmonic stretch and bend terms replaced the more complex functions of MM2 and MM3. In addition, the interaction cross-terms of MM2 and MM3 (stretch–bend, stretch–torsion, and bend–bend in MM3) were left out.

Each of the biological harmonic force fields discussed here uses the Lennard-Jones 6-12 van der Waals potential rather than the exponential-6 form used in MM2 and MM3. A 6-12 potential requires fewer parameters, and does not have a singularity at short distances. It also takes less computer time to evaluate: $\frac{1}{r^{12}}$ is related to $\frac{1}{r^6}$ by a simple multiplication, whereas the exponential-6 potential requires a more time consuming exponential evaluation.

In addition to the conventional force field terms discussed in Chapter 1, all of the harmonic force fields discussed in this section utilize an improper torsion or inversion term, electrostatics, and sometimes a hydrogen-bond term.

3.2.2. Inversion

For an atom I, bonded exactly to three other atoms J, K, L, such as the amide nitrogen and carbon atoms in a peptide linkage, it is usually necessary to include an energy term

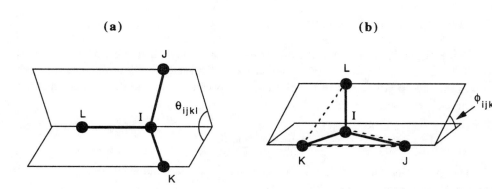

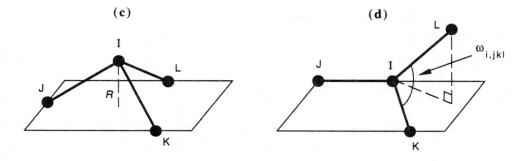

FIGURE 3.7.
Geometric definitions of (a) the AMBER inversion term, (b) the CHARMM and GROMOS inversion terms, (c) the Tripos 5.2 inversion term, and (d) the DREIDING inversion term.

in the force field that describes the energy associated with distorting the carbon or nitrogen in an amide away from planarity. Angle bend terms of appropriate magnitude to describe the angle bend vibrational motion significantly underestimate the nitrogen or carbon puckering motions in an amide linkage. There is little agreement as to how to treat the inversion term, and so each approach is discussed below.

For the inversion (INV) term involving an atom I, bonded exactly to three other atoms J, K, L, AMBER considers one of the bonds as being special (say, IL) and uses Eq. (3.2)

$$V_\theta = \frac{1}{2}K_{INV}\{1 - \cos[n(\theta - \theta_0)]\} \qquad (3.2)$$

where θ is the angle between the JIL and KIL planes, see Figure 3.7(a). The parameter n is 2 for planar centers and 3 for tetrahedral centers.

Both CHARMM and GROMOS use Eq. (3.3) for the inversion term involving an atom I, bonded exactly to three other atoms J, K, L:

$$V_\phi = \frac{1}{2}K_{INV}(\phi - \phi_0)^2 \qquad (3.3)$$

where ϕ is the torsion angle of the IJ bond with respect to the KL ''bond'' using the JK ''bond'' dihedral (the angle between the IJK and JKL planes), see Figure 3.7(b).

In the Tripos 5.2 force field, the inversion or out-of-plane bending term is only used for planar systems. For a central atom I surrounded by atoms J, K, and L, Tripos 5.2 defines the inversion in terms of the distance R between atom I and the plane containing J, K, and L, see Figure 3.7(c) and Eq. (3.4).

$$V_{INV} = K_{IJKL}R^2 \tag{3.4}$$

The DREIDING force field uses a spectroscopic or umbrella inversion, Eq. (3.5) for nonplanar systems

$$V_{INV} = \frac{1}{2}\frac{K_I}{(\sin \omega_I^0)^2}(\cos \omega - \cos \omega_I^0)^2 \tag{3.5}$$

and Eq. (3.6)

$$V_{INV} = K_{INV}(1 - \cos \omega) \tag{3.6}$$

for planar systems. The parameter ω is defined in terms of the angle between one bond, say between I and L, and the plane containing I, J, and K, see Figure 3.13(d).

3.2.3. van der Waals

In each of these harmonic force fields, nonbonded interactions are calculated using a Lennard-Jones 6-12 function (Lennard-Jones, 1924) Eq. (3.7):

$$V_{vdW} = D_{IJ}\left\{-2\left[\frac{\rho_{IJ}}{\rho}\right]^6 + \left[\frac{\rho_{IJ}}{\rho}\right]^{12}\right\} \tag{3.7}$$

where ρ is the nonbonded distance in angstroms, D_{IJ} and ρ_{IJ} are the nonbond parameters, D_{IJ} is the well depth in kilocalories per mole, and ρ_{IJ} is the natural van der Waals (vdW) distance in angstroms. Conventionally, general ρ_{IJ} and D_{IJ} parameters are obtained from homonuclear parameters, D_{II} and D_{JJ}, through the use of combination rules.

As published, AMBER uses geometric combination rules for both distance and energy [Eq. (3.8) and (3.9)].

$$\rho_{IJ} = \sqrt{\rho_I \times \rho_J} \tag{3.8}$$

$$D_{IJ} = \sqrt{D_I \times D_J} \tag{3.9}$$

As published, CHARMM expresses the Lennard-Jones 6-12 function in terms of A_{IJ} and B_{IJ} rather than D_{IJ} and ρ_{IJ}; $B_{IJ} = 2D_{IJ}\rho_{IJ}^6$ and $A_{IJ} = D_{IJ}\rho_{IJ}^{12}$ [Eq. (3.10)]:

$$V_{vdW} = \left\{\frac{A_{IJ}}{\rho^{12}} - \frac{B_{IJ}}{\rho^6}\right\} \tag{3.10}$$

where A_{IJ} and B_{IJ} are defined with the Slater–Kirkwood formulas [Eqs. (3.11) and (3.12)] rather than either arithmetic or geometric combination rules:

$$B_{IJ} = 361.67 \frac{\alpha_I \alpha_J}{\left(\dfrac{\alpha_I}{N_I}\right)^{1/2} + \left(\dfrac{\alpha_J}{N_J}\right)^{1/2}} \tag{3.11}$$

$$A_{IJ} = \frac{1}{2} B_{IJ}(\rho_I + \rho_J)^6 \tag{3.12}$$

where α_I is the polarizability for center I (in cubic angstroms, $\mathring{A}^3$), N_I is the effective number of outer-shell electrons on center I, ρ_I is the van der Waals radius for center I (in angstroms), and B_{IJ} has the units kilocalorie angstrom^6 per mole.

The GROMOS force field also expresses the Lennard-Jones 6-12 function in terms of A_{IJ} and B_{IJ} rather than D_{IJ} and ρ_{IJ}. Geometric combination rules are used [see Eqs. (3.13) and (3.14)].

$$A_{IJ} = \sqrt{A_I \times A_J} \tag{3.13}$$

$$B_{IJ} = \sqrt{B_I \times B_J} \tag{3.14}$$

Tripos 5.2 and DREIDING use geometric combination rules for the energy term [Eq. (3.9)] and arithmetic combination rules for the distance term [see Eq. (3.15)].

$$\rho_{IJ} = \frac{1}{2}(\rho_I + \rho_J) \tag{3.15}$$

3.2.4. Electrostatic Interactions

The force fields discussed here normally express electrostatic energies in terms of partial point charges, using the classical Coulomb potential Eq. (3.16), rather than the bond dipole model used in MM2 (see Section 2.2.7).

$$V_{el} = C \frac{q_I q_J}{\varepsilon \rho_{IJ}} \tag{3.16}$$

Charges q_I and q_J are separated by a distance ρ_{IJ}, ε is the dielectric constant, and C converts to the energy units of interest (332.06 for energies in kilocalories per mole).

3.2.5. Nonbonded Approximations

The number of nonbonded interactions in a molecule grows as $\dfrac{n \times (n-1)}{2}$, where n is the number of atoms in the molecule. For proteins, this is a large number. To limit the number of multiplications, divisions, and square root evaluations, two approximations are commonly used.

The first approximation is to assume electrostatic interactions die off as $\dfrac{1}{\rho^2}$ rather than $\dfrac{1}{\rho}$. This is equivalent to setting ε in Eq. (3.16) equal to ρ_{IJ}. This approximation saves a square root calculation for each distance pair because the distance ρ_{IJ} is calculated from $\rho_{IJ}{}^2$. The distance square $\rho_{IJ}{}^2$ is calculated using Eq. (3.17).

$$\rho_{IJ}{}^2 = (x_i - x_j)^2 + (y_i - y_j)^2 + (z_i - z_j)^2 \tag{3.17}$$

The second approximation is to only calculate interactions out to a particular distance, say 9 Å. For a protein such as carbonic anhydrase with 260 residues, using a 9-Å cutoff decreases the number of nonbond terms from 2,400,000 to 126,000! Simply using a cutoff does not actually save much time because the distance ρ_{ij} has to be evaluated to know which distances are longer than the cutoff. To save time, a list of distance pairs less than the cutoff distance is generated and then used for a number of energy evaluations. A 9-Å distance cutoff is appropriate for van der Waals interactions because they decay as $\dfrac{1}{\rho^6}$; use of $\dfrac{1}{\rho^2}$ for electrostatics also facilitates a short nonbond cutoff radius. To prevent a discontinuity in the energy at the cutoff distance, switching functions (sw) at intermediate distances are used [see Eqs. (3.18) and (3.19)]:

$$V_{el} = \frac{q_I q_J}{\varepsilon \rho_{IJ}} sw(\rho_{IJ}{}^2, \rho_{on}{}^2, \rho_{off}{}^2) \tag{3.18}$$

where

$$sw(x, x_{on}, x_{off}) = 1 \qquad x \le x_{on}$$

$$sw(x, x_{on}, x_{off}) = \frac{(x_{off} - x)^2(x_{off} + 2x - 3x_{on})}{(x_{off} - x_{on})^3} \qquad x_{on} < x \le x_{off} \tag{3.19}$$

$$sw(x, x_{on}, x_{off}) = 0 \qquad x > x_{off}$$

Another method to increase computational efficiency is the united atom approximation. The hydrogen atoms to carbon are not explicitly calculated in the united, or extended atom, representation. Instead the mass and van der Waals radius of carbon are expanded to approximately account for the hydrogen atoms in a spherically average way. This simplification reduces the number of atoms in the calculation and is applied only to hydrogen atoms not capable of hydrogen-bonding. As originally formulated, AMBER, CHARMM, and GROMOS used a united atom representation of CH_3, CH_2, and CH groups. In 1986 an "all-atom" extension to AMBER was reported, which explicitly included hydrogen atoms bonded to carbon (Weiner et al., 1986). Recently, CHARMM has also been extended to include an all-atom representation (Smith and Karplus, 1992; Momany and Rose, 1992). The DREIDING force field has both united atom and all-atom representations. As reported, Tripos 5.2 supports all-atom representations.

In addition to the usual convention in molecular mechanics of excluding van der Waals and electrostatic interactions for atoms that are bonded to each other (1, 2 interactions) or bonded to a common atom (1, 3 interactions), AMBER scales the 1,4 nonbonded interactions $\left(\times \frac{1}{2} \right)$. The CHARMM force field sometimes scales the 1,4 nonbonded interactions $\left(\times \frac{1}{2} \right)$. Because the van der Waals radii for united atoms cause excessive 1,4 interactions, GROMOS uses special van der Waals parameters for 1,4 interactions.

3.2.6. Hydrogen Bonding

The remaining term in each of the force fields is a hydrogen-bond potential. The AMBER force field uses a 10-12 nonbond potential between the hydrogen atom and the hydrogen-bond acceptor center to calculate hydrogen bonding [see Eq. (3.20)]:

$$V_{HB} = D_{IJ} \left\{ -6 \left[\frac{\rho_{IJ}}{\rho} \right]^{10} + 5 \left[\frac{\rho_{IJ}}{\rho} \right]^{12} \right\} \tag{3.20}$$

where D_{IJ} is the hydrogen-bond bond energy, ρ_{IJ} is the natural hydrogen-bond distance, and ρ is the distance between the hydrogen atom and the hydrogen-bond acceptor center.

The all-atom CHARMM force field paper (Smith and Karplus, 1992) states that no explicit hydrogen-bonding potential is required; hydrogen bonding can be described in terms of electrostatic and van der Waals interactions. Earlier versions of CHARMM used an angularly dependent 10-12 potential (Brooks et al., 1983). There is no special hydrogen-bond potential in the current version of CHARMM, although van der Waals parameters are adjusted for hydrogen-bonding interactions. Tripos 5.2 does not use a explicit hydrogen-bond potential, although the van der Waals radius x_I for hydrogen atoms attached to hydrogen donors is set to zero. The DREIDING force field uses what is described as a "CHARMM-like" hydrogen-bonding potential [see Eq. (3.21)]:

$$V_{HB} = D_{IJ} \left\{ -6 \left[\frac{\rho_{IJ}}{\rho} \right]^{10} + 5 \left[\frac{\rho_{IJ}}{\rho} \right]^{12} \right\} \cos^4 \theta_{DHA} \tag{3.21}$$

where θ_{DHA} is the angle between the hydrogen donor, the hydrogen, and the hydrogen acceptor, and ρ_{IJ} and ρ are the natural and current distances between the hydrogen donor and acceptor.

In GROMOS, hydrogen bonding is described in terms of electrostatic and van der Waals interactions; the van der Waals parameters are modified from their conventional values.

3.2.7. Force Field Parameters

In each of these force fields, the parameterization (choice of bond distances, bond angles, etc.) was based primarily on reproduction of amino acid and nucleic acid structure and secondarily on reproduction of experimental vibrational frequencies. Despite the similarities in force field functional forms, and the data base for parameterization, there are differences in the final parameters used in each of the force fields. A comparison of the published force field parameters for the amino acid alanine internal to a peptide sequence

TABLE 3.2. Bond Stretch Parameters[a]

Atom Pair	AMBER[b]		CHARMM[c]		DREIDING[d]		Tripos 5.2[e]	
	r_0	k	r_0	k	r_0	k	r_0	k
$N-C_{amide}$	1.334	980	1.33[c]	942[c]	1.34	700	1.345	870.1
$C_{amide}-O$	1.229	1140	1.215[f]	1190[f]	1.25	1400	1.22	1555.2
$C_{amide}-C_{sp^3}$	1.522	670	1.524[f]	374[f]	1.46	700	1.501	639.
$C_{sp^3}-C_{sp^3}$	1.526	620	1.530[f]	471[f]	1.53	700	1.54	633.6
$N-C_{sp^3}$	1.449	710	1.45[c]	844[c]	1.41	700	1.45	677.6
$C_{sp^3}-H$	1.090	622	1.100[f]	660[f]	1.09	700	1.1	662.4
$N-H$	1.010	868	0.98[c]	810[c]	0.97	700	1.0	700.
Average single-bond force constant		680		632		700		667

[a] Distances are in angstroms and force constants are in kilocalories per mole per square angstrom.
[b] From Weiner et al., 1986.
[c] From Brooks et al., 1983.
[d] From Mayo et al., 1990.
[e] From Clark et al., 1989.
[f] From Smith and Karplus, 1992.

TABLE 3.3. Angle Bend Parameters[a]

Atom Triple	AMBER[b]		CHARMM[c]		DREIDING[d]		Tripos 5.2[e]	
	θ_0	k	θ_0	k	θ_0	k	θ_0	k
$C_{amide}-N-C_{sp^3}$	121.9	100	120.[c]	160.[c]	120.	100	118.	288.
$C_{amide}-N-H$	119.8	70	120.[c]	60.[c]	120.	100	119.	105.
$N-C_{sp^3}-C_{amide}$	110.1	126	111.6[c]	140.[c]	109.5	100	109.5	144.
$N-C_{sp^3}-C_{sp^3}$	109.7	160	110.[c]	130.[c]	109.5	100	109.5	118.
$N-C_{sp^3}-H$	109.5	76	109.5[c]	100.[c]	109.5	100	110.	131.
$C_{sp^3}-C_{amide}-N$	116.6	140	117.5[c]	40.[c]	120.	100	117.	131.
$C_{sp^3}-C_{amide}-O$	120.4	174	122.3[f]	129.2[f]	120.	100	120.	171.
$C_{sp^3}-C_{sp^3}-H$	109.5	70	109.5[f]	89.8[f]	109.5	100	109.5	105.
$C_{sp^3}-C_{sp^3}-C_{amide}$	111.1	126	109.5[c]	140.[c]	109.5	100	109.5	118.
$H-C_{sp^3}-H$	109.5	70	108.5[f]	75.6[f]	109.5	100	109.5	158.
Average force constant		111		106		100		147

[a] Angles are in degrees and force constants are in kilocalories per mole per square radian.
[b] From Weiner et al., 1986.
[c] From Brooks et al., 1983.
[d] From Mayo et al., 1990.
[e] From Clark et al., 1989.
[f] From Smith and Karplus, 1992.

TABLE 3.4. van der Waals Parameters[a]

Atom Type	AMBER[b]		CHARMM[c]		DREIDING[d]		Tripos 5.2[e]	
	r_0	D_0	r_0	D_0	r_0	D_0	r_0	D_0
N	3.50	0.16	3.20[c]	0.238[c]	3.6621	0.0774	3.1	0.095
C_{amide}	3.70	0.12	3.60[f]	0.0903[f]	3.8983	0.0951	3.4	0.107
O	3.20	0.20	3.20[f]	0.1591[f]	3.4046	0.0957	3.04	0.116
C_{sp^3}	3.60	0.06	3.60[f]	0.0903[f]	3.8983	0.0951	3.4	0.107
H_C	3.08	0.01	2.936[f]	0.0045[f]	3.195	0.0152	3.0	0.042
H_N (hydrogen-bonded)	2.00	0.02	1.6[f]	0.0498[f]	3.195	0.0001	0.0	

[a] Distances are in angstroms and energies are in kilocalories per mole.
[b] From Weiner et al., 1986.
[c] From Brooks et al., 1983.
[d] From Mayo et al., 1990.
[e] From Clark et al., 1989.
[f] From Smith and Karplus, 1992.

for AMBER, CHARMM, DREIDING, and Tripos 5.2 is provided in Tables 3.2–3.4 (stretch, bend, and van der Waals, respectively). The molecular structure and the atom types used by each of the force fields to describe alanine are given in Figure 3.8. While the natural structural parameters are quite similar, greater variation is found in the force constants. Both AMBER and CHARMM, with a common source (Gelin and Karplus, 1979), use virtually identical natural distances and angles. The largest difference in r_0 is 0.03 Å, for the N$-$H bond, the second largest difference is 0.014 Å, for the $C_{amide}-$O bond. This difference is comparable to the experimental range, 0.012 Å, in amide C$-$O bond distances (Allen et al., 1987). The largest variation in angle is 1.9° for the $C_{amide}-$N$-$$C_{sp^3}$ and $C_{sp^3}-C_{amide}-$O angles. The Tripos 5.2 force field also uses quite similar distances and angles. The largest distance difference is 0.023 Å between Tripos 5.2 and CHARMM for the $C_{amide}-C_{sp^3}$ bond pair. The largest angle difference is 3.9° between Tripos 5.2 and AMBER for the $C_{amide}-$N$-C_{sp^3}$ angle. The DREIDING force field, rather than fitting any particular structures, generates bond distances from atomic radii. Logically, since the DREIDING C_{amide} radius was chosen to reproduce C$-$O double-bond distances, the largest distance difference between DREIDING and the other force fields is for the $C_{amide}-C_{sp^3}$ distance, where CHARMM and DREIDING differ by 0.066 Å. The idealized angles used in DREIDING are quite similar to the optimized angles used by AMBER, CHARMM, and Tripos 5.2. The largest difference is for the $C_{sp^3}-C_{amide}-$N angle where DREIDING and AMBER differ by 3.4°. The contrast between DREIDING and the other three harmonic force fields is greatest for the van der Waals parameters. For the DREIDING force field the van der Waals parameters were obtained by fitting heats of sublimation and crystal lattice coordinates. The remaining force fields used either Slater–Kirkwood rules, Eq. (3.11), or other qualitative data. The underlying philosophical reasons for the emphasis on van der Waals parameter development in DREIDING versus valence interaction parameter development in the other force fields are discussed in Chapter 8.

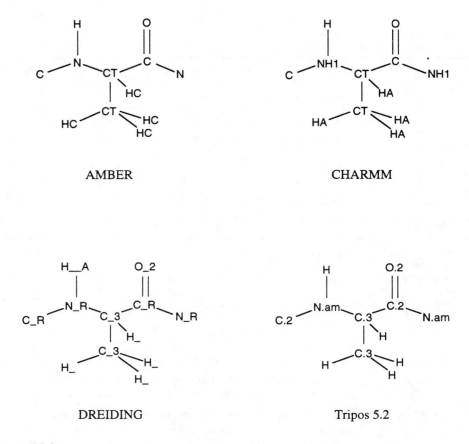

FIGURE 3.8.
Molecular structure and atom typing of alanine embedded in a peptide chain for AMBER, CHARMM, DREIDING, and Tripos 5.2.

The remaining parameters for each of these force fields are the partial charges used in Eq. (3.16). In AMBER, partial charges for amino and nucleic acid residues are obtained by fitting the ab initio molecular electrostatic potential (Singh and Kollman, 1984). The partial charges in CHARMM are set by an unpublished charge template scheme. The partial charges in GROMOS are set by amino acid or nucleic acid. The art or science of partial charge determination is still an active area of research and is discussed in more detail in Section 8.2.

3.2.8. Solvation

Because of the importance of solvent (usually water) to biopolymer modeling we discuss protein solvation and the ways in which solvent is simulated or approximated by protein force fields here. This discussion is also of relevance to nucleic acid simulations.

Water covers the surface of proteins and other biomolecules, strongly influencing their conformational structure and dynamics (Rupley and Careri, 1991; Teeter, 1991). Because water gives rise to the hydrophobic interactions that stabilize the core of the protein it is a critical factor in protein folding. Waters can also be found in small interior pockets, and these internal waters can play an integral role in the structure and function of proteins. When proteins are crystallized, water molecules are crystallized along with the protein, making the crystal more like a highly concentrated solution than a solid (Saenger, 1987). Protein surface waters are very mobile, so X-ray diffraction experiments do not always reveal all of the positions of the protein bound water molecules. Interior waters can be more easily located by X-ray diffraction, and it has been discovered that the number and positions of these waters are often evolutionarily conserved in related proteins.

The special behavior of water, arising from its ability to be both a hydrogen-bond donor and acceptor, and its property of diminishing charge–charge interactions, make simulating a solvated protein, or protein-bound water molecules, an important part of any biopolymer simulation. A variety of potential functions have been developed for explicitly including water in dynamics and mechanics calculations. Frequently used functions involve a rigid water molecule represented by three, four, or five interaction sites (Jorgensen et al., 1983). The functions include Lennard-Jones potentials between water, and Coloumb interaction energies. The three interaction site model has charges only on the oxygen and two hydrogen atoms (TIP3, SPC). In the four interaction site model, the negative charge is moved off the oxygen (TIPS2); the five site model has two charges on two lone-pair positions on the oxygen atoms (ST2). The GROMOS force field was developed for use with explicit solvent.

It is not always possible to explicitly treat water in a biopolymer simulation because of the considerable extra computational effort and expense. Various methods have been devised in order to approximate the effects of bulk water (Davis and McCammon, 1990). One of the most important properties of water is its high dielectric constant ε ($\varepsilon = 1$ for vacuum and 80 for water). The dielectric constant is a function of the polarizability of the medium; a high dielectric means the solvent is able to dampen out charge–charge interactions. Solvent effects have been modeled by adopting a constant value greater than 1 for the ε term. This is an artifice because the dielectric constant is a bulk, or macroscopic, property, being a function of the average behavior of a collection of molecules, rather than arising from a molecular level interaction. Solvent effects have also been approximated by using a relative dielectric constant proportional to r. Setting $\varepsilon = \varepsilon^* r$, Eq. (3.16) becomes Eq. (3.22), where ε^* is simply a scaling factor and has no physical basis.

$$V_{el} = \frac{q_i q_j}{\varepsilon^* \rho_{ij}{}^2} \qquad (3.22)$$

The distance dependent dielectric electrostatic potential of Eq. (3.22) was first used in order to save computer time, but it can be thought of as an attempt to mimic the polarization effect of a solvent by weighing closer interactions more heavily. This formalism has no sound physical foundation (van Gunsteren et al., 1990), although it seems to work. A distance dependent dielectric is used by AMBER and CHARMM.

A range of dielectric continuum models have also been used to simulate solvation (Davis and McCammon, 1990). The simplest of this set of nonexplicit solvation approaches involves placing the peptide or protein in an empty cavity surrounded by a polarizable continuous dielectric medium. Mathematically, the solute molecules are treated as irregularly shaped objects whose interior has a uniform dielectric constant ε, with point charges q at positions corresponding to atomic nuclei (Lim et al., 1991).

3.3. MINIMIZATIONS OF PROTEINS

3.3.1. Crambin

Crambin is a small hydrophobic protein derived from the seeds of *Crambe abyssinica*; its biological role is unknown. A schematic drawing of crambin is illustrated in Figure 3.9. The 46 residue protein has frequently been used as a benchmark for testing protein force fields (Whitlow and Teeter, 1986; Lii et al., 1989; Clark et al., 1989; Lii and Allinger, 1991). Crambin is a convenient benchmark because it is small, yet has a variety of secondary structure types: two α helixes, a small antiparallel β sheet, an extended chain region, and five turns. Crambin has six charged groups but is electrically neutral allowing counterions to be omitted from the calculation. Most importantly, the X-ray structure has been resolved to high resolution (0.945 Å) by Hendrickson and Teeter (1981; 1986).

We will examine a paper by Jorgensen and Tirado-Rives (1988) of Purdue University, who minimized the complete crambin crystal using two different force fields and compared the results to the crystal structure.

METHOD

Experimentally (Teeter, 1984), the crambin unit cell contains 2 protein, 4 ethanols (near the protein surface), and an estimated 182 water molecules. The calculations were carried out for the entire unit cell, as illustrated in Figure 3.10. The unit cell was generated

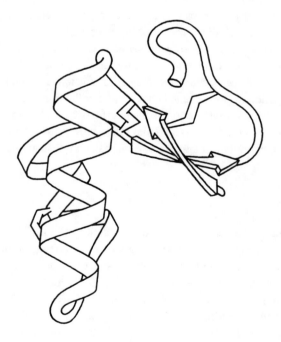

FIGURE 3.9.
A schematic drawing of the backbone of crambin. The β sheet is represented by arrows and the α helices are drawn as spiral ribbons. The disulfides are drawn as "lightning flashes." [Reproduced with permission from Richardson, J. S. (1981), *Adv. Protein Chem.* **34,** 276.]

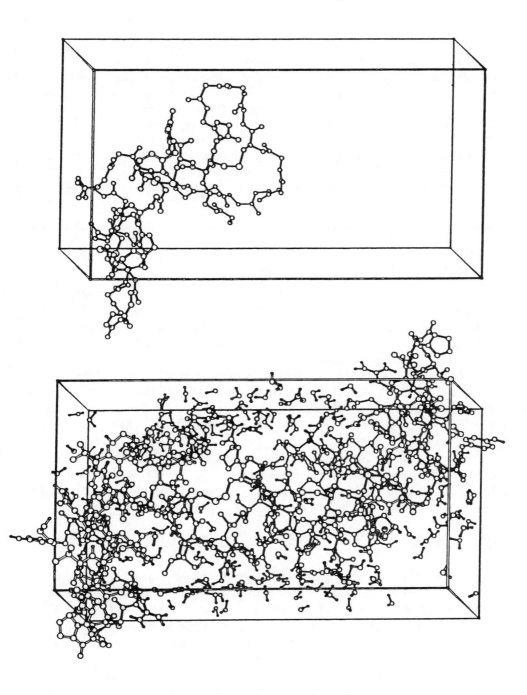

FIGURE 3.10.
(a) The backbone of a crambin molecule in the unit cell. (b) The complete unit cell with 2 protein, 4 ethanol, and 182 water molecules. The seemingly vacant regions in the corners are filled in by atoms from adjacent unit cells. [Reproduced with permission from Jorgensen, W. L.; Tirado-Rives, J. (1988), *J. Am. Chem. Soc.* **110**, 1657. Copyright © (1988) American Chemical Society.]

BOX 3.1.   OPLS

The OPLS model uses a simple Coloumb plus Lennard-Jones 6-12 function to describe the electro-static and van der Waals interactions. There are no explicit hydrogen-bonding terms in OPLS. The OPLS parameters for neutral amino acid residues are optimized by reproducing experimental densi-ties and heats of vaporization of organic liquids using Monte Carlo simulations; the parameteriza-tion of the charged protein residues are developed by reproducing ab initio geometries and experimental heats of hydration. The united atom approximation is used for the CH_n groups. The OPLS nonbonded terms are merged with the bond stretch, angle bend, and torsional terms of AMBER; this modified AMBER is called the AMBER/OLPS force field. A dielectric constant of 1 is used for evaluating the electrostatic energy, as OPLS parameters are derived in this way, and are intended for use in condensed-phase systems.

---

in steps. The positions of the non-hydrogen atoms of the protein and the ethanol were taken from the X-ray structure (Teeter, 1984), and hydrogen atoms on the heteroatoms added by standard geometrical parameters. The unit cell was filled with water taken from a TIP3P simulation of water; and all water molecules within 2.15 Å of non-hydrogen atoms of the protein were removed, giving 182 water molecules. A Monte Carlo simula-tion of the waters was carried out allowing only the waters to move (see Appendix D for a discussion of Monte Carlo configuration selection). After equilibration, energy minimi-zation was carried out on all 1356 explicit atoms with the united atom AMBER and AMBER/OPLS force fields. The calculations were executed using a Microvax II, and took several months!

The most difficult part of a force field to derive or develop are the electrostatic, van der Waals, and hydrogen-bonding interactions. One approach to describing these non-bonding interactions is the optimized potentials for liquid simulations (OPLS) parameter-ization, which is discussed in Box 3.1.

RESULTS

The results of the energy minimizations for the crambin crystal are given in Table 3.5. The resultant rms deviation for the non-hydrogen protein and ethanol atoms was 0.17 Å for the AMBER/OPLS force field, and 0.22 Å for the AMBER force field. The simu-lated and X-ray crystallographic structures were the same to within experimental error. The rms deviations were lower with the AMBER/OPLS force field for the backbone and side-chain atoms when considered separately, as well as for the main-chain dihedrals ϕ and ψ. The rms deviations for the atomic positions were decomposed as to residue type, and the variation in the errors for charged, polar, and nonpolar residues was small.

The paper by Jorgensen and Tirado-Rives (1988) shows the molecular mechanics minimization of a crystalline protein can give a structure nearly identical to the X-ray structure. The experimental structure, however, must be used as a starting point. The two different force fields (AMBER/OPLS and AMBER) resulted in slightly different struc-tures and rms deviations from the X-ray structure.

TABLE 3.5. Results of the Energy Minimizations for Crambin

rms	Force Field	
	AMBER	AMBER/OPLS
Protein (Å)	0.22	0.17
Backbone (Å)	0.19	0.14
Side chains (Å)	0.25	0.20
Φ (deg)	7.2	6.1
Ψ (deg)	7.9	5.6
ω (deg)	4.1	4.6
χ (deg)	10.9	11.5

3.3.2. Carbonic Anhydrase

Carbonic anhydrase (CA) is a zinc metalloenzyme found in plants, animals, and some bacteria. It catalyzes the reversible hydration of carbon dioxide to bicarbonate ion, Eq. (3.23).

$$CO_2 + H_2O \rightleftarrows HCO_3^- + H^+ \qquad (3.23)$$

The most important function of mammalian carbonic anhydrases is to facilitate the absorption of CO_2 into the blood and subsequent discharge into the lungs. There are several forms of the mammalian enzyme, differing in amino acid sequences and activity. The highest activity form, human carbonic anhydrase II (HCA II), is exceptionally efficient, and can be considered an "evolutionarily perfect" enzyme because the rate approaches the diffusion controlled limit of evolutionary perfection (Knowles and Albery, 1977). Carbonic anhydrase I, which occurs together with CA II in red blood cells, is less efficient, while CA III is specific to muscle and is even less active. Although CA has been known for 60 years, there is renewed interest in this enzyme partly because inhibitors of this enzyme are potentially useful for the treatment of the eye disease glaucoma (Hartman et al., 1992; Antonaroli et al., 1992).

The X-ray structures of three forms of native mammalian CA have been determined (Lindahl et al., 1991). The structures of forms I, II, and III are very similar. The active site is a deep conical cavity in the core of the protein. An essential zinc dication is located at the bottom, and is coordinated to the imidazole rings of three histidines. The zinc is also coordinated to a water or hydroxide, depending on the pH. The zinc coordination is almost tetrahedral. The active site cavity is divided into a hydrophobic region and a hydrophilic region with a large number of water molecules interconnecting the active site and external solvent.

An enormous number of studies have been devoted to the mechanism of the hydration of carbon dioxide catalyzed by carbonic anhydrase (Botre et al., 1991); the reactive spe-

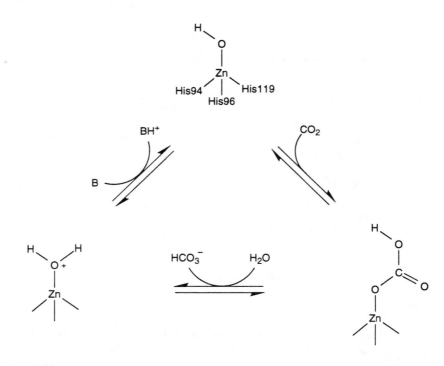

FIGURE 3.11.
Possible sequence for the catalysis of hydration of carbon dioxide by carbonic anhydrase.

cies is usually assumed to be a zinc-bound hydroxide (Fig. 3.11). Attack of the zinc hydroxide on carbon dioxide forms a metal-bound carbonate followed by bicarbonate dissociation. This reaction results in a zinc-bound water, which, in the second step, releases a proton to the solvent to regenerate the active zinc hydroxide. Proton transfer between the metal center and the external medium is the rate-limiting step for CO_2 hydration for CA II and III, but probably not for I. Experimental studies indicate there are several proton-transfer pathways (Tu et al., 1989). The fastest proton transfer is thought to involve His 64 acting as a "proton-shuttle group." That is, the solvent-exposed imidazole ring of His 64 transfers a proton from the zinc-bound water molecule to buffer molecules in solution. In this section, we will examine a paper by Vedani et al., (1989) of the University of Kansas who studied solvated human carbonic anhydrase I (HCA I) by molecular mechanics. A new proton relay network was discovered that provided a direct connection between a water weakly bound to zinc and the surrounding solvent.

Modeling metalloproteins can be difficult because of the variety of possible metal coordination environments, coordination numbers, and coordination geometries (Hoops et al., 1991). Approaches to model coordinated metals are discussed briefly in Box 3.2, and in more detail in Chapter 7 as it applies to nonbiological molecules. In this paper, a metal–ligand potential function that is a hybrid between the pure bonding and nonbonding approaches is used.

BOX 3.2   MODELING METAL–LIGAND BONDS

Correctly describing the easily distorted metal coordination, where electronic and noncovalent effects may be paramount, is a general problem in modeling metalloproteins. Given the importance of the metal to catalysis by metalloenzymes, modeling the active site correctly requires a reasonable simulation of both the metal and its coordination sphere. Two fundamental approaches are currently used to model the metalloenzyme metal–ligand bond. These are the bonded and nonbonded models. The bonded approach includes explicit covalent bonding between metal and ligand using bond length stretching and bond angle bending functions; there is no basic difference between the biological and inorganic parts of a metalloenzyme. Conversely, in the nonbonded formulation, the bonding between ligand and metal is assumed to involve only electrostatic and van der Waals interactions.

Both formulations have their limitations. Bonded approaches are limited in that only small changes in coordination can occur because the geometry of the metal must be defined beforehand; a distortion from square planar to tetrahedral coordination, for example, would not be possible. Nonbonded approaches are limited because the electronic origins of a deviation from valence shell electron-pair repulsion (VSEPR) geometries are not included in the force field. For example, square planar coordination is not possible in this model as the molecule would collapse to tetrahedral coordination.

METHOD

The coordinates of the HCA I and five active site waters were taken from the published crystal structure (Kannan et al., 1984), which was determined to a resolution of 2 Å. The metalloenzyme was solvated using 501 solvent molecules generated with the program SOLVGEN. The SOLVGEN algorithm, developed by Jacober and Vedani (Jacober, 1988), is based on length, linearity, and directionality of hydrogen bonds. The solvated structure is given by Figure 3.12.

Energy calculations were performed on the solvated HCA I using the molecular mechanics programs YETI (Vedani, 1988) and united atom AMBER (Singh et al., 1986). The force field YETI includes a special set of directional potential functions for hydrogen bonds, salt linkages (a bridge formed by a hydrogen bond between anionic and cationic amino acid residues), and metal–ligand interactions. The metal center potential function in YETI includes the metal–ligand bond distances and angles as coordination and metal-type variables, and by using a product of angle terms, allows specifically for distortions between types of coordination geometries such as square pyramidal and trigonal bipyramidal.

Atomic partial charges were taken from AMBER (Weiner et al., 1984); electrostatic energies were evaluated using a distance-dependent dielectric, $\varepsilon = 2r$.

RESULTS

Molecular mechanics refinement of the solvated native enzyme yielded a slightly distorted tetrahedron at the Zn; Zn$-$N bonds to His94, His96, His119, and a Zn to H_2O bond (1.923 Å). A second H_2O was also found, only weakly coordinated to the Zn (3.252

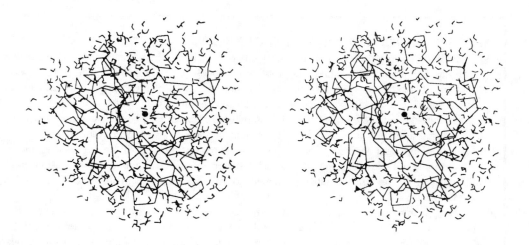

FIGURE 3.12.
Stereoscopic view of the solvated structure of native human carbonic anyhydrase after protein side-chain relaxation. Only the α-carbon atoms of the protein are shown; the zinc is represented by a sphere. [Reproduced with permission from Vedani, A.; Huhta, D. W.; Jacober, S. P. (1989), *J. Am. Chem. Soc.* **111**, 4075. Copyright © (1989) American Chemical Society.]

Å), opposite His96. The arrangement at zinc was described as a 4 + 1 distorted tetrahedron with four proximal ligands and one distal water. A close-up of the active site region is illustrated in Figure 3.13.

In view of the importance of water to the catalysis of hydrolysis of carbon dioxide, the structure of the internal water molecules was of special interest to the researchers.

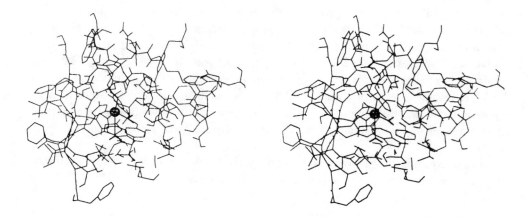

FIGURE 3.13.
Stereoscopic view of the active site cleft of native human carbonic anhydrase I. The Zn is represented by a sphere; the two zinc-bound water molecules have been drawn enhanced. [Reproduced with permission from Vedani, A.; Huhta, D. W.; Jacober, S. P. (1989), *J. Am. Chem. Soc.* **111**, 4075. Copyright © (1989) American Chemical Society.]

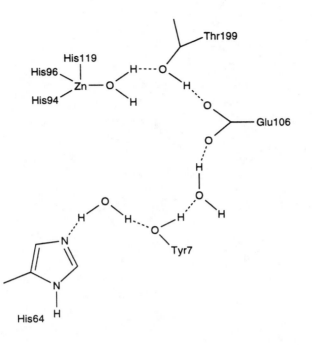

FIGURE 3.14.
Schematic representation of the hydrogen-bond network associated with the proximal zinc-bound water molecule.

Sixteen internal water molecules were identified. All five of the internal water molecules observed in the experimental X-ray structure were found by SOLVGEN. The hydrogen-bond network associated with three of these internal waters, and originating with the prox-imal zinc-bound water molecule is illustrated in Figure 3.14. The proximal zinc-bound water engaged in a hydrogen bond with the side-chain hydroxyl O atom of Thr199; Thr199 was hydrogen bonded to the side-chain carboxyl O atom of Glu106. The other O atom of Glu106 formed a H_2O mediated hydrogen bond with the side-chain hydroxyl O atom of Tyr7. Another water linked the O atom of Tyr7 with the imidazole N atom of His64, which was located at the entrance of the active site and provided the contact with the surrounding solvent. This hydrogen-bond network involving the ligated solvent, side-chain groups of Thr199, Glu106, Tyr7 and His64, and two bridging waters is thought to provide a relay mechanism in the deprotonation process that converts the Zn-bound water into a Zn-bound hydroxide.

An alternative proton relay network, originating with the distal Zn-bound water, was discovered by Vedani and co-workers (1989). This network was formed exclusively from water molecules, and is illustrated in Figure 3.15. An interesting feature was the formation of a water pentagon in the active site funnel.

The paper by Vedani et al. (1989) demonstates that the structure of a metalloprotein can be simulated by molecular mechanics starting from the X-ray structure. The coordina-tion environment of the metal can be modeled, although how to do this correctly is still a subject of debate and continued work. The positions of the mechanistically important internal waters of CA I was determined using mechanics and the program SOLVGEN, and a new proton-relay network was discovered.

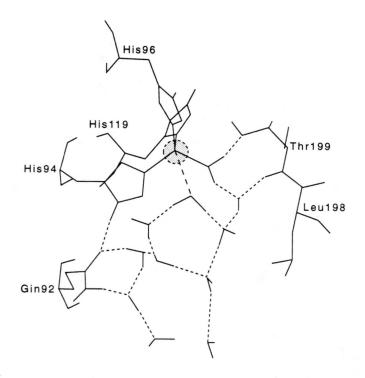

FIGURE 3.15.
Schematic representation of the proton relay network associated with the distal zinc-bound water molecule. The zinc is indicated by a sphere; hydrogen bonds are indicated by dotted lines. [Reproduced with permission from Vedani, A.; Huhta, D. W.; Jacober, S. P. (1989), *J. Am. Chem. Soc.* **111**, 4075. Copyright © (1989) American Chemical Society.]

3.4. SITE-DIRECTED MUTAGENESIS

3.4.1 Background

Insulin (**3.6**) is a polypeptide hormone that regulates carbohydrate metabolism.

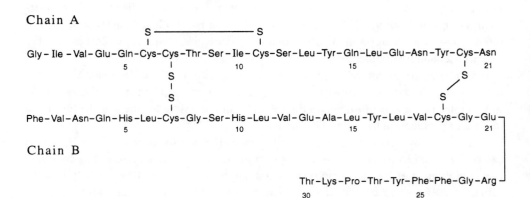

The hormone is produced by the pancreas and is secreted directly into the bloodstream. Inadequate insulin production, resulting in high glucose levels in the blood, characterizes the disease called Type I diabetes mellitus (Kolata, 1979). Type I diabetes usually develops before the age of 20. It is the leading cause of blindness in the United States and second most common cause worldwide. Persons suffering from Type I diabetes can be effectively treated by insulin injection (Burrow et al., 1982). This insulin therapy attempts to mimic the pattern of insulin release seen in nondiabetics: that is, constant secretion during 24 h, superimposed with a rapid increase after meals. It has been difficult to mimic the sharp increases in insulin peaks after eating with the current insulin preparations because the absorption of short-acting insulin from the subcutaneous tissue (the layer of loose connective tissue beneath the skin) is too slow. This causes hyperglycemia (high blood sugar) during meals and hypoglycemia (low blood sugar) between meals.

Although management of the acute symptoms of Type I, or insulin-dependent, diabetes can be achieved through conventional insulin therapies, the long-term secondary complications of diabetes are not well treated. Clinical work shows that strict metabolic control by continuous insulin infusion or multiple daily insulin injections may prevent the vascular (blood vessel) complications of diabetes such as blindness, impaired circulation, and high blood pressure (Hanssen et al., 1986).

Insulin exists in solution as monomers, dimers, tetramers, and hexamers in dynamic equilibrium (Blundell et al., 1972). The monomer is the biologically active form; it circulates in the bloodstream and binds to the cell-surface insulin receptor, which then initiates the responses of the cell to the hormone. In the insulin preparations used in current therapy, insulin molecules are assembled as zinc-containing hexamers, and this association limits the rate of absorption from subcutaneous tissue. If this equilibrium could be influenced toward the monomer, the time for bioabsorption might be decreased. This would constitute a very fast acting insulin, useful for mimicking the sharp increases in insulin secretion after eating.

3.4.2 Mutagenesis of Insulin

In this section we will describe a paper by Brange et al., of the Novo Research Institute, Denmark and the University of York, England (Brange, 1988) in which the 3-D structure of the insulin dimer was used as the basis to design a fast-acting monomeric insulin analog.

METHOD

The insulin analogs were modeled using computer graphics provided by the FRODO (Jones, 1985) and HYDRA programs. The HYDRA program is a commercial graphics display program distributed by POLYGEN corporation. No energy minimizations were carried out, although in some cases bonds were manually rotated to minimize steric clashes.

RESULTS

The Novo and York researchers set out to create a monomeric insulin that did not self-associate by analyzing the interactions between monomers in insulin dimers using modeling. The aim was to substitute amino acids that favored the monomer but did not

FIGURE 3.16.
Schematic representation of the insulin structure. Thin lines A chain; thick lines B chain. [Adapted with permission from Tager, H. S. (1990), in Cuatrecasas, P. and Jacobs, S., Eds., *Insulin,* Handbook of Experimental Pharmacology, Vol. 92, Springer-Verlag, Berlin, pp. 41–66.]

destabilize the 3-D structure of the monomer or interfere with receptor binding. By using DNA technology, mutant insulins were prepared in which the amino acids involved in the dimer interaction were substituted by other amino acids. The mutants were then evaluated for biological activity.

The manipulation of the structure of proteins by altering amino acid residues at specific locations with the use of genetic engineering is called site-directed mutagenesis (Ackers and Smith, 1985). This technique has been used enthusiastically in recent years to test theories of enzymatic catalysis and develop fundamental structure–function relationships for proteins. When the mutations are used to control a protein's therapeutic activity, as described here, site-directed mutagenesis is an example of the drug design approach called structure–activity relationships (SAR), discussed in more detail in Chapter 4.3.

The X-ray structures of monomer, dimer, and hexamer insulins have all been determined (Derewenda et al., 1990). A schematic of the monomer is illustrated in Figure 3.16. The monomer consists of two chains, the A chain of 21 amino acids, and the B chain of 30 amino acids. Disulfide bonds link the chain at CysA7 and CysB7 and CysA20 and CysB19; within the A chain there is a disulfide loop between CysA6 and CysA11. The A chain has two helices and the B chain has one helix. The surface of the molecule is covered by both polar and nonpolar residues.

The insulin dimer is organized with the two molecules disposed about a twofold axis. The two insulin molecules of the dimer are held together by van der Waals contacts and four hydrogen bonds arranged as a β-sheet structure between two antiparallel CO_2H terminal strands of the B chain. The hydrogen bonds are formed between residues of PheB24 of monomer 1 and TyrB26 of monomer 2, and the reverse, between TyrB26 of monomer 1 and PheB24 of monomer 2. The dimer is illustrated in Figure 3.17. The principal interactions between insulin monomers in the dimer involve residues GlyB8, SerB9, ValB12, GluB13, TyrB16, and GlyB23 through ProB28. Because some of these residues are thought to be involved in receptor binding, the targets chosen for mutation were those

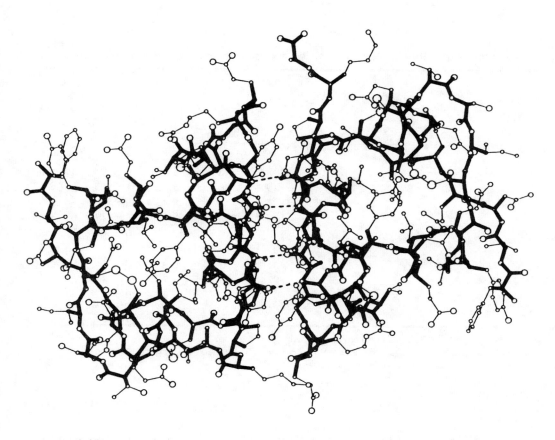

FIGURE 3.17.
The insulin dimer viewed down the twofold axis. Main chain atoms are drawn with thick bonds; side-chain atoms are drawn with thin bonds. The dashed lines show hydrogen bonds. [Reproduced with permission from Derewenda, U.; Derewenda, Z. S.; Dodson, G. G.; Hubbard, R. E. (1990), in Cuatrecasas, P.; Jacobs, S., Eds., *Insulin,* Handbook of Experimental Pharmacology, Vol. 92), Springer-Verlag, Berlin, pp. 23–40.]

amino acid residues located at the periphery of the putative receptor binding site: These targets were SerB9, HisB10, ValB12, TyrB26, ThrB27, and ProB28.

Figure 3.18 (see color plate) shows the six selected mutation sites on the insulin monomer. The researchers' strategy was to introduce charge repulsion into the monomer–monomer interface. In most cases, negative charges were introduced where they could oppose preexisting negatively charged carboxyl groups adjacent to the dimer-forming interface. The effect of steric hindrance in the monomers was also investigated by introducing an extra methyl group at ValB12 to give the mutant IleB12 insulin. The mutant insulins were prepared by site-directed mutagenesis, and the association, biological activity, and bioavailability of the analogs were determined. The results are collected in Table 3.6. The results in Table 3.6 show that the charge repulsion by side chains on the periphery of the dimer-forming surface reduced insulin association, but steric hindrance by an additional methyl group (ValB12 to IleB12) did not.

The most pronounced reduction in dimer formation was with the AspB9 mutant, where the neutral serine CH_2OH group at position B9 was replaced by the charged $CH_2CO_2^-$ group of aspartate. The reduced association in this mutant could be explained

TABLE 3.6. Identity, Association, Biological Activity, and Bioavailability of Insulin Analogs

Insulin	Association State Deduced from Osmometry		Biological Activity[a] (% of Insulin)		Subcutaneous Absorption[b] (% of human insulin ± s.e.m.)		
	$c = 0.2$ mM	$c = 1$ mM	FFC	MBGA	T_{25}	T_{50}	T_{75}
AspB9		1.1	26	79	69 ± 8	64 ± 6	75 ± 6
AspB10	2.1	2.2	207	98	53 ± 10	56 ± 8	59 ± 7
IleB12	2.3	3.3	29	86	69 ± 19	81 ± 19	88 ± 18
GluB12 + des-B30		1.0	0.04		64 ± 13	59 ± 8	58 ± 6
GluB26	1.4	2.0	125	104	52 ± 12	55 ± 11	63 ± 20
GluB27	2.9	4.0	108	110	79 ± 10	76 ± 6	77 ± 7
AspB28		1.3	101	104	49 ± 8	53 ± 7	66 ± 8
AspB9 + GluB27	1.1	1.1	31	93	40 ± 4	50 ± 5	60 ± 4
AspB27	1.3	1.5	87	61	50 ± 13	54 ± 12	67 ± 12
Zn-free human insulin	3.1	4.4			91 ± 15	90 ± 12	87 ± 12
Two Zn insulin (human)	≡6	≡6	≡100	≡100	≡100	≡100	≡100

[a] The FFC is an *in vitro,* and MBGA is an *in vivo* biological activity assay method.
[b] The parameters T_{25}, T_{50}, and T_{75} designate the times until 25, 50, and 75%, respectively, of the doses injected into pigs were absorbed relative to hexameric two Zn human insulin.

by examining a model of the mutant dimer (see Fig. 3.19) (see color plate). This model was generated by manually adjusting the side chains in order to maximize separation between groups. Figure 3.19 shows that the carboxyl groups of the two mutated B9 residues (Asp, R $=$ CH$_2$CO$_2^-$) and two B13 (Glu, R $=$ CH$_2$CH$_2$CO$_2^-$) residues were brought to within 4 Å of each other. Electrostatic repulsion between the AspB9 of monomer 1 and GluB13 of monomer 2, and vice versa, would be expected to inhibit dimer formation (Burley and Petsko, 1988). Similarly, dimer formation should also be inhibited by a repulsion between the charged carboxyl groups of a mutated GluB27 of monomer 1, and the C-terminal carboxyl group of AsnA21 of monomer 2, about 10 Å away.

Most of the amino acid substitutions, as listed in Table 3.6, did not significantly effect the *in vitro* (outside the organism) and *in vivo* (inside the organism) potency of the hormones. The rate of disappearance from the subcutaneous tissue after injection was monitored for the analogs, and absorption was faster than for human insulin. The initial rate of absorption (at 0–20 min) was approximately three times faster for the AspB9 + GluB27 mutant than for human insulin. The goal of developing a fast-acting insulin analog by site-directed mutagenesis was achieved.

The AspB9 + GluB27 human insulin analog has since undergone clinical trials in Europe (Vora et al., 1988). The Novo and York group has filed an investigational new drug (IND) application with the FDA to begin clinical testing of the insulin analog in the United States.

In summary, the work done by the researchers at the Novo Research Institute and the University of York shows that looking at a 3-D graphics model of a peptide, along with manual manipulations of the side chains and "eyeballing" the energy, can result in a fundamental understanding of peptide interactions. Here, the factors that might favor the insulin monomer over the insulin dimer were investigated, and the conclusions used to design mutant insulins. Several insulins were prepared by site-directed mutagenesis, and one appears to be a useful drug.

3.5. CONFORMATIONAL ANALYSIS

3.5.1 Background

A dogma of molecular biology is that the biological function of a bioactive peptide is determined by its conformation. The preferred, lowest energy, conformation of an isolated peptide may not necessarily be the same as the bioactive conformation (see Chapter 4 for further discussion); nonetheless, conformational studies of the peptide may provide clues about the nature of the bioactive form (Marshall et al., 1978; Hruby, 1984; Burt and Greer, 1988).

There is a continuum of polypeptide and peptide conformational stability. In aqueous solution, small peptides adopt many different conformations in dynamic equilibrium. Conformations of peptides are thus often defined only when bound to their receptors. Proteins, in contrast to peptides, usually adopt fairly well-defined 3-D structures in solution. Recently, it has been pointed out that small peptides as short as four or five residues can adopt nonrandom conformations in aqueous solution (Dyson et al., 1988). In fact, a short peptide (13 residues) with the same sequence as a segment of an intact protein, has been discovered to have the same ordered secondary structure in solution as that in the protein crystal (Kim and Baldwin, 1984). There is some evidence that large proteins consist of folding domains that can behave as independent structural entities, initiating the folding process and directing subsequent folding events. Fischer and Schmid (1990) suggest that the finite, but low stability of local structures (as modeled by short peptides with nonrandom conformations) may be the kinetic ''proofreading'' mechanism of protein folding. The determination of the low-energy conformations of small peptides is thus a part of the protein-folding problem.

The biggest handicap to computationally finding peptide conformations is the existence of many local minima on the conformational surface. There are several modeling strategies that attempt to overcome this problem as discussed in Chapters 1, 2, and Appendix E. One of the more common methods is the systematic search, or ''brute force'' method. It is based on the fact that the total energy of a peptide depends strongly on the rotable dihedral angles. The systematic search method uses a grid search of the conformational space in which only the torsional angles of the peptide are allowed to vary. Each rotable bond is varied sequentially by a fixed angle rotation. As the number of angles increases, the number of local energy minimum energy conformations grows exponentially. In the cases where a grid search is impossible because of the peptide size, other search methods must be employed. Monte Carlo searching involves exploring conformational space by producing new configurations by random displacements of atoms or torsional angles, see Appendix D for a discussion of Monte Carlo configuration selection. Another method for conformational searching is molecular dynamics (van Gunsteren and Berendsen, 1990). Two variants of molecular dynamics methods have been used for peptide conformational searching; these are simulated annealing and high-temperature quenched dynamics. In simulated annealing, dynamics simulations are carried out at high temperatures to aid in crossing energy barriers. The system is then gradually cooled and equilibrated with short dynamics runs so that the molecule is not trapped in local high-energy minima. In high-temperature quenched dynamics, the simulations are also carried out at high temperatures, but the high-energy structures are sampled and minimized. These conformational searching methods have been described in Chapter 1 and Appendix E.

3.5.2 Tuftsin

In this section we will discuss a paper by O'Connor, et al. (1992) at the University of Houston, who used high-temperature quenched molecular dynamics calculations to investigate the conformational properties of tuftsin and a cyclic analog. Tuftsin, named after Tufts, the university where it was discovered (Najjar and Nishioka, 1970), is the tetrapeptide Thr-Lys-Pro-Arg (**3.7**).

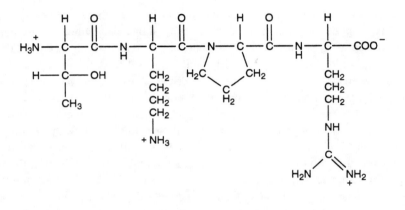

3.7

Tuftsin possesses a broad spectrum of activities related primarily to the stimulation of phagocytosis, the process of signaling certain cells to engulf and destroy bacteria. Tuftsin is active only in the free tetrapeptide state, and must be enzymatically liberated from the carrier molecule, the protein leukokinin, in order to exert its effects.

The peptide and its analogs are attractive candidates for immunotherapeutic drugs (Hahn, 1986; Fridkin and Najjar, 1989), and hence have attracted both theoretical (Fitzwater et al., 1978) and experimental interest (Verdini et al., 1991). Intensive structure–activity studies of tuftsin analogs have shown that the entire peptide sequence is vital for full bioactivity (Stabinsky et al., 1980; Siemion and Konopinska, 1981). Spectroscopic studies have been employed to determine the conformation of tuftsin in solution. Conflicting conclusions were reached. The ^{13}C NMR and circular dichroism (CD) work suggested that the tetrapeptide had a strong tendency for a type III β turn in neutral water solutions, with the proline in the $i + 2$ position (Siemion et al., 1983). In contrast, ^{1}H and ^{13}C NMR studies in water (Blumenstein et al., 1979) were interpreted as a random structure for tuftsin. These researchers found tuftsin was somewhat ordered in dimethyl sulfoxide (DMSO), but did not have a β turn. Recent 2-D NMR studies of tuftsin in DMSO gave results thought to be consistent with a folded structure containing a Lys-Pro three-residue inverse γ turn (D'Ursi et al., 1992).

METHOD

All calculations were performed with the CHARMM program using the united atom approximation for nonpolar hydrogen atoms (Brooks et al., 1983). The equations of motion were integrated using the Verlet algorithm (Verlet, 1967), with a time step of 1 fs and SHAKE (Ryckaert et al., 1977) to constrain all bond lengths. The influence of solvent was modeled by a dielectric continuum employing a dielectric constant of 45 for DMSO, and 80 for H_2O. The fully extended structure of tuftsin was minimized. The peptide was then heated to 1000 K over 10 ps (1 ps $= 10^{-12}$ s) of dynamics by incrementing the temperature 10 K every 0.1 ps. The peptide was equilibrated for an additional 10 ps at 1000 K, during which the temperature of the system was constrained to be within 13 K of 1000 K. The MD runs were then performed for a total time of 600 ps. The trajectories were saved every 1 ps. All 600 saved structures were energy minimized. Only the structures within 3–5 kcal/mol of the minimum energy structure were saved for analysis. Structures with an rms deviation of less than or equal to 0.6 Å between the backbone fragments were considered members of the same family.

RESULTS

The 42 thermally accessible structures obtained from the quenched dynamics of tuftsin in DMSO were divided into families according to their relative energies and conformational properties. Among the four families was one major family, F1, and three additional families all within 3 kcal/mol of the minimum. Representative dihedral angles for these families are given in Table 3.7. The F1 family had a β turn with one hydrogen bond

TABLE 3.7. Representative Dihedral Angles of Tuftsin (**3.7**) from Quenched Molecular Dynamics in DMSO

Residue	Dihedral	F1	F2	F3	F4
Thr1	ψ	129	146	-43	-56
	χ^1	-58	55	55	-58
	χ^2	-62	179	179	-62
Lys2	ϕ	-83	-85	-89	-81
	ψ	134	124	127	134
	χ^1	-66	-169	-173	42
	χ^2	-177	66	-179	69
Pro3	ϕ	-68	-65	-77	-70
	ψ	-22	94	-164	123
	χ^1	28	-28	35	-29
Arg4	ϕ	-82	-118	-71	-78
	χ^1	-61	-71	-66	-69
	χ^2	135	-66	-67	74
No. in family		38	1	2	1
Energy (kcal/mol)		-8.36	-6.75	-8.66	-7.89

F1

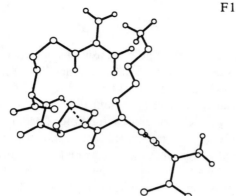
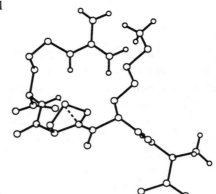

F2

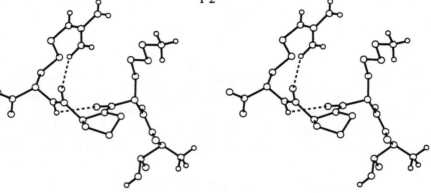

FIGURE 3.20.
Stereoplots of representative structures from families 1 and 2 of tuftsin (**3.7**) in DMSO generated by quenched dynamics. [Reproduced with permission from O'Connor, S. D.; Smith, P. E.; Al-Obeidi, F.; Pettitt, B. M. (1992), *J. Med. Chem.* **35**, 2870. Copyright © 1992 American Chemical Society.]

between Pro3 N and Arg4 NH. This structure, along with F2, is illustrated in Figure 3.20. Comparison of the dihedral angles in Table 3.7 with those listed in Table 3.1, the tuftsin turn does not fit into any of the common turn types. These miscellaneous turns, in which at least two of the four dihedral angles differ by more than 40° from the familiar bend types, are classified as Type IV β turns by Scheraga and co-workers (Lewis et al., 1973).

The 45 thermally accessible structures obtained from the quenched dynamics of tuftsin in an aqueous continuum revealed one major family (F1) and two additional ones. Representative dihedral angles for these families is given in Table 3.8. F1 had three hydrogen bonds, and the general conformation was also a Type IV β turn. This structure, along with F2, is illustrated in Figure 3.21. In general, the predicted backbone conformations of tuftsin in DMSO were similar to those in H_2O, except for the number of hydrogen bonds.

TABLE 3.8. Representative Dihedral Angles of Tuftsin (**3.7**) from Quenched Molecular Dynamics in H_2O

Residue	Dihedral	F1	F2	F3
Thr1	ψ	−48	129	147
	χ^1	55	−58	54
	χ^2	64	−62	64
Lys2	ϕ	−88	−91	−110
	ψ	126	122	113
	χ^1	−64	−174	−177
	χ^2	179	−176	80
Pro3	ϕ	−69	−75	−74
	ψ	−31	79	164
	χ^1	30	33	31
Arg4	ϕ	−105	−87	−86
	χ^1	−166	33	−59
	χ^2	−91	−169	−70
No. in family		40	4	1
Energy (kcal/mol)		−9.99	−8.92	−8.41

F1

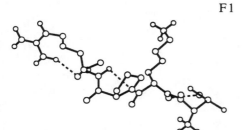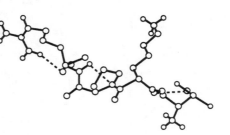

F2

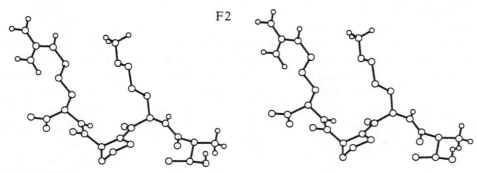

FIGURE 3.21.
Stereoplots of representative structures from families 1 and 2 of tuftsin (**3.7**) in water generated by quenched dynamics. The dotted lines represent hydrogen bonds. [Reproduced with permission from O'Connor, S. D.; Smith, P. E.; Al-Obeidi, F.; Pettitt, B. M. (1992), *J. Med. Chem.* **35**, 2870. Copyright © 1992 American Chemical Society.]

The major family in DMSO involved a Type IV β turn with the Lys and Arg side chains positioned on the same face of the molecule. The major water family also involved a Type IV β turn with the Thr, Lys, and Arg situated on one side of the molecule.

Statistically, it is known that proline rarely occupies the third position in β turns, as observed above in the conformation of F1. Nonetheless, though rare, the φ and ψ angles found for proline in these dynamics calculations are within the allowed region of conformational space for this residue. The dynamics prediction of a low-energy Type IV β turn backbone conformation is also not consistent with spectroscopic studies. The most recent spectroscopic study of tuftsin (D'Ursi et al., 1992) in DMSO employing 2-D NMR, was interpreted as the peptide having an inverse γ turn. This conformational assignment is (in contrast to the calculated Type IV β turn), at least consistent with statistical experimental proline preferences for this position in this type of turn. Interestingly, an inverse γ turn was observed in the dynamics studies for F2 in both solvents (see Tables 3.7 and 3.8).

Cyclo[Thr-Lys-Pro-Arg-Gly] (**3.8**), a cyclic pentapeptide analog of tuftsin, was designed and also subjected to a dynamics simulation in order to learn if the β turn found in the calculations of the linear peptide could be stabilized by cyclization.

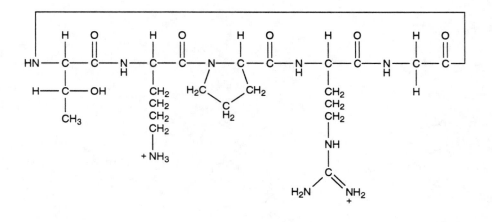

3.8

Representative dihedral angles for the cyclic peptide molecule, designated as ctuf (cyclic tuftsin) are collected in Table 3.9. The structures were all very similar and produced only one family as illustrated in Figure 3.22. A Type IV β turn at the Lys-Pro position was observed in all of the structures, analogous to the turn observed in F1 for linear tuftsin. Side-chain conformations of ctuf also had most of the characteristics of the linear tuftsin.

If a Type IV β turn is in fact important to the bioactivity of tuftsin, the conformationally constrained cyclic analog with the same turn might be expected to be even more active as a phagocytosis stimulator than its linear parent (see Section 4.3). This may be the case. In a note added in proof, Pettitt and co-workers (O'Connor et al. 1992) reported that experiments performed by Nishioka indicated the cyclic analog was many times more potent in a phagocytosis assay than tuftsin itself, in agreement with the theoretical predictions.

To summarize, the work by O'Connor et al. (1992) investigates the conformations of a short peptide tuftsin by high-temperature quenched MD. The solvent was modeled by a

TABLE 3.9. Representative Dihedral Angles of
ctuf **(3.8)** from Quenched Molecular Dynamics

Residue	Dihedral	ctuf F1
Thr1	ϕ	-90
	ψ	-57
	χ^1	-59
	χ^2	60
Lys2	ϕ	-156
	ψ	123
	χ^1	52
	χ^2	-176
Pro3	ϕ	-86
	ψ	-15
	χ^1	37
Arg4	ϕ	-94
	ψ	-59
	χ^1	-62
	χ^2	-66
Gly5	ϕ	-81
	ψ	-78

F1

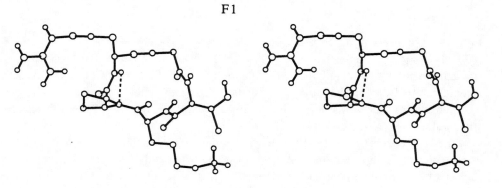

FIGURE 3.22.
Stereoplot of a representative structure of ctuf **(3.8)** in DMSO generated by quenched dynamics.
The dotted lines represent hydrogen bonds. [Reproduced with permission from O'Connor, S. D.;
Smith, P. E.; Al-Obeidi, F.; Pettitt, B. M. (1992), *J. Med. Chem.* **35**, 2870. Copyright © 1992
American Chemical Society.]

dielectric continuum. A set of low-energy structures with a Type IV β turn was found and used to design a cyclic form of the peptide whose conformational properties were also investigated. The dynamics results on the linear peptide do not appear to be consistent with current spectroscopic understanding of the backbone conformation in solution. This shows that, in general, the most populated low-energy family may not, in fact, be the experimental solution structure. The inability to reproduce the experimental structure may be due to force field or solvent modeling inadequacies, but, for this small peptide, it is probably not due to an incomplete sampling of conformational space. Alternatively, the experiments may be wrong. Interestingly, the structures of F1 do appear to be related to the bioactive conformation.

3.6. DYNAMICS OF PROTEINS

3.6.1 Background

Dynamics investigations of protein motion are normally carried out for a time interval of less than 100 ps, which is far less than the time scale of biological activity. Nonetheless, the fast vibrational motions of atoms that can be studied by dynamics partly determine the nature of the slower motions. A relatively short dynamics run can still tell us something about the slower, higher energy, collective distortions of groups of atoms. The first reported MD simulation of a protein was an 8.8-ps simulation of the 58 residue pancreatic trypsin inhibitor, in a vacuum. This work by McCammon et al. (1977) lead to a profound change in the understanding of the internal motions of proteins. The researchers found that the protein sampled a series of conformations near the X-ray structure. The time-averaged dynamics structure was not identical, but close to, the X-ray structure. Fluctuations in adjacent dihedral angles were correlated (dependent on each other or interrelated) in order to minimize disturbances of the backbone and side-chain atoms, particularly in secondary structural units. Concerted nonlocal collective motions, such as the oscillation of a flexible loop region, were also observed.

3.6.2 HIV-1 Protease

In this section, we will describe a 96-ps dynamics study of the enzyme HIV-1 protease (HPR) in water by Swaminathan et al. (1991) and Harte et al., (1990) of Wesleyan University and Bristol Meyers Squibb. The calculation required 100 h on a Cray Y-MP supercomputer. Analysis of the atomic fluctuations revealed interesting domain–domain communications. Recall that a protein domain is a compact, spatially distinct secondary structural unit.

The HPR enzyme is vital to the life cycle of the AIDS virus. It is the only significant macromolecular component of the AIDS virus to be structurally characterized, and is the subject of considerable pharmaceutical interest in the design of anti-AIDS drugs (see Section 4.2). A ribbon tracing of HPR is given in Figure 3.23. The enzyme has a quaternary structure, being a dimer composed of two 99-amino acid monomers (Lapatto et al., 1989; Navia et al., 1989). The protein is nearly C_2 symmetric. Each monomer contributes one Asp25-Thr26-Gly27 sequence to give an active site with a twofold axis of symmetry. Each HPR subunit also provides an extended β-hairpin loop, called a flap.

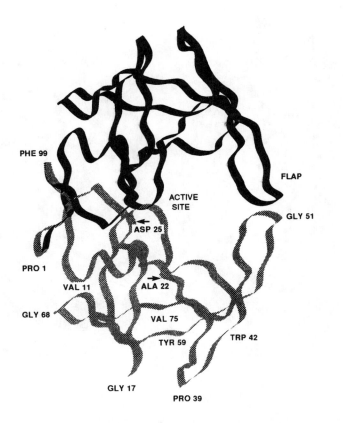

FIGURE 3.23.

The structure of HPR drawn as a kinked ribbon connecting the positions of α-carbon atoms. [Adapted with permission from Swaminathan, S.; Harte Jr., W. E.; Beveridge, D. L. (1991), *J. Am. Chem. Soc.* **113**, 2717. Copyright © 1991 American Chemical Society.]

X-ray studies of structures of HPR complexed with peptide or peptide analog inhibitors show that a substantial conformational change occurs in the enzyme upon binding. Most marked is a very large movement of the flaps, which close down on top of the inhibitor bound at the active site (Miller et al., 1989). These significant changes suggest structural dynamics should be considered in understanding HPR action.

METHOD

Molecular dynamics calculations were performed with the program WESDYN, using the GROMOS 86 force field, and the SPC model for water. The starting structure was the X-ray structure of HPR solved by Wlodawer et al. (1989). The protein was solvated with 6990 molecules of water in a hexagonal prism cell and treated in the simulation under periodic boundary conditions. The volume was chosen to maintain a density of 1 g/cm³. The solvent was relaxed by Monte Carlo simulation followed by minimization. The MD protocol involved heating to 300 K over 1.5 ps, an equilibration step of 2.5 ps, and a trajectory of 96 ps.

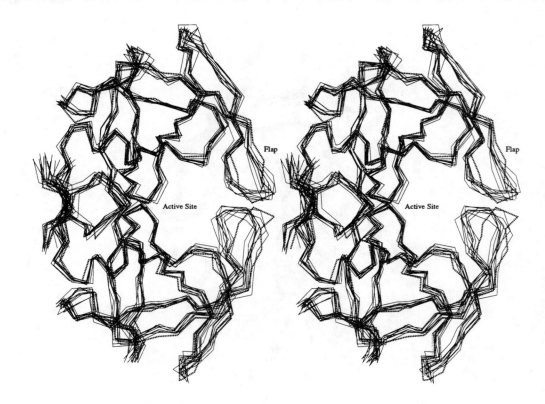

FIGURE 3.24.
Dynamical range of HPR dimer over the interval 40–90 ps in the MD simulation. Superposition of
C_α traces. The boldface broken line is the C_α trace of the X-ray crystal geometry. [Adapted with
permission from Swaminathan, S.; Harte Jr., W. E.; Beveridge, D. L. (1991), *J. Am. Chem. Soc.* **113**,
2717. Copyright © 1991 American Chemical Society.]

RESULTS

The rms deviation of the calculated MD structures from the crystal structure stabilized at about 1.3 Å, showing the calculated structure agreed well with the X-ray structure. The overall motion and dynamical structure of the protein is presented in Figure 3.24 as a superposition of nine snapshots obtained at equally spaced intervals from the MD trajectory. The dynamical range of the MD structures bracketed the crystal geometry. The main difference between the MD structures and the experimental X-ray structure was in the flap region. In the simulated structures, the flap formed intramolecular hydrogen bonds to the other flap; whereas in the crystal, the flap is known to be subject to crystal-packing contacts with another HPR molecule. The calculations predicted that the more closed-flap position should be favored in aqueous solution.

A topological diagram of the secondary structure of HPR is illustrated in Figure 3.25. The dynamical characteristics of the protein were analyzed to yield information about correlated motions. Correlated motions were found to occur among proximal (neighboring) residues composing well-defined domain regions of secondary structure. Such motions also occurred *between* regions, such as those involved in domain–domain communication.

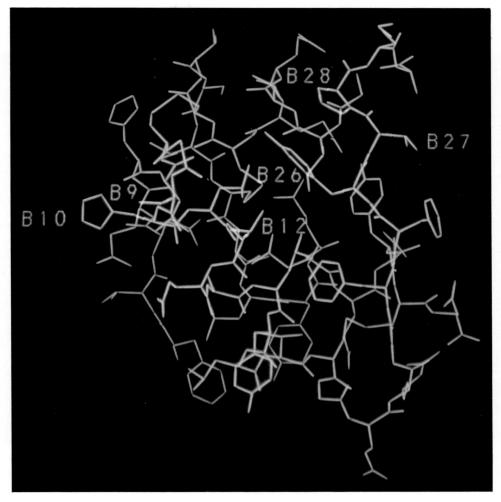

FIGURE 3.18.
Computer graphics of the human insulin monomer showing the side chains of the mutated residues. [Reprinted with permission from *Nature,* Brange, J.; Ribel, U.; Hansen, J. F.; Dodson, G.; Hansen, M. T.; Havelund, S.; Melberg, S. G.; Norris, F.; Norris, K.; Snel, L.; Sorensen, A. R.; Voigt, H. O. (1988), *Nature* (London) **333**, 679. Copyright © 1988 Macmillan Magazine Limited.]

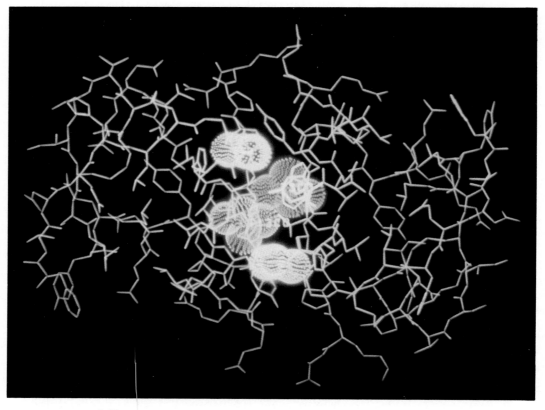

FIGURE 3.19.
Insulin dimer with GluB13 and mutated AspB9 superimposed on dots indicating the van der Waals surface of the carboxyl groups. [Reprinted with permission from *Nature,* Brange, J.; Ribel, U.; Hansen, J. F.; Dodson, G.; Hansen, M. T.; Havelund, S.; Melberg, S. G.; Norris, F.; Norris, K.; Snel, L.; Sorensen, A. R.; Voigt, H. O. (1988), *Nature* (London) **333**, 679. Copyright © 1988 Macmillan Magazine Limited.]

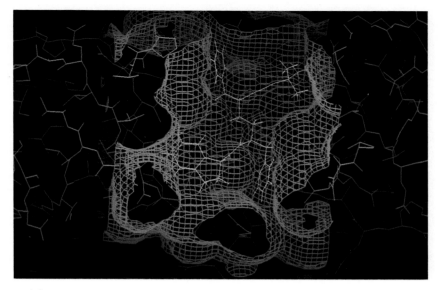

FIGURE 4.6.

One conformation of **4.4** modeled in the HPR active site steric surface map. [Reproduced with permission from Thompson, W. J.; Fitzgerald, P. M. D.; Holloway, M. K.; Emini, E. A.; Darke, P. L.; McKeever, B. M.; Schleif, W. A.; Quintero, J. C.; Zugay, J. A.; Tucker, T. J.; Schwering, J. E.; Homnick, C. F.; Nunberg, J.; Springer, J. P.; Huff, J. R. (1992), *J. Med. Chem.* **35**, 1685. Copyright © 1992 American Chemical Society.]

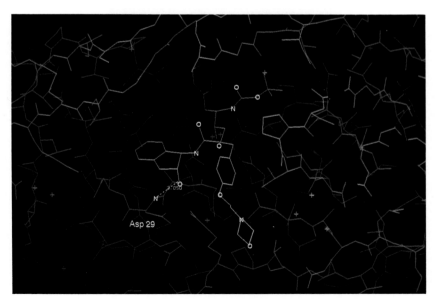

FIGURE 4.7.

Inhibitor **4.5** modeled in the native enzyme site. [Reproduced with permission from Thompson, W. J.; Fitzgerald, P. M. D.; Holloway, M. K.; Emini, E. A.; Darke, P. L.; McKeever, B. M.; Schleif, W. A.; Quintero, J. C.; Zugay, J. A.; Tucker, T. J.; Schwering, J. E.; Homnick, C. F.; Nunberg, J.; Springer, J. P.; Huff, J. R. (1992), *J. Med. Chem.* **35**, 1685. Copyright © 1992 American Chemical Society.]

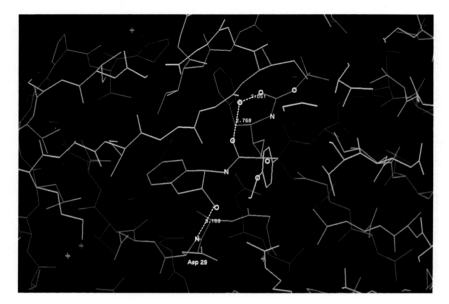

FIGURE 4.8.

X-ray structure of the **4.5**-HPR complex. [Reproduced with permission from Thompson, W. J.; Fitzgerald, P. M. D.; Holloway, M. K.; Emini, E. A.; Darke, P. L.; McKeever, B. M.; Schleif, W. A.; Quintero, J. C.; Zugay, J. A.; Tucker, T. J.; Schwering, J. E.; Homnick, C. F.; Nunberg, J.; Springer, J. P.; Huff, J. R. (1992), *J. Med. Chem.* **35**, 1685. Copyright © 1992 American Chemical Society.]

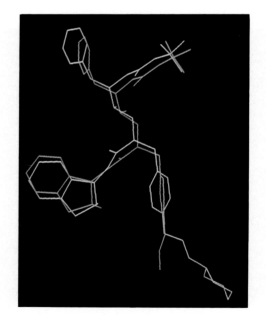

FIGURE 4.9.

Comparison of the active conformation of **4.5** by modeling (red) and X-ray diffraction (yellow). [Reproduced with permission from Thompson, W. J.; Fitzgerald, P. M. D.; Holloway, M. K.; Emini, E. A.; Darke, P. L.; McKeever, B. M.; Schleif, W. A.; Quintero, J. C.; Zugay, J. A.; Tucker, T. J.; Schwering, J. E.; Homnick, C. F.; Nunberg, J.; Springer, J. P.; Huff, J. R. (1992), *J. Med. Chem.* **35**, 1685. Copyright © 1992 American Chemical Society.]

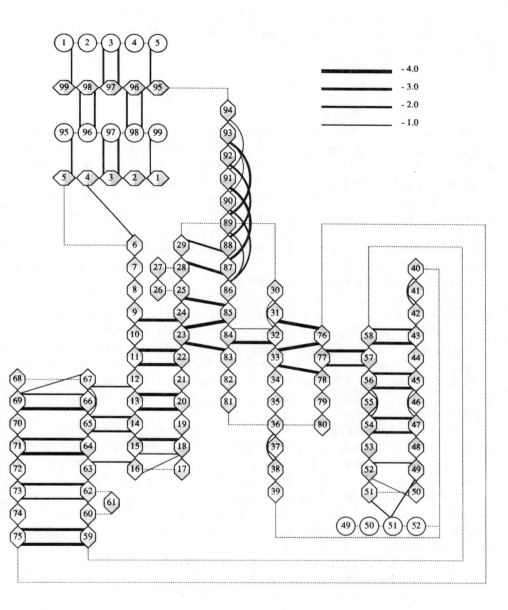

FIGURE 3.25.

Topological diagram of the HPR secondary structure. Solid lines indicate hydrogen bonds and their relative strengths; broken lines indicate the backbone peptide bonds with the residues in hexagons; residue numbers in circles represent the other chain in the dimer. The backbone hydrogen bonds are indicated by solid lines, with a thickness proportional to the calculated average bond strength. [Reproduced with permission from Harte Jr., W. E.; Swaminathan, S.; Mansuri, M. M.; Martin, J. C.; Rosenberg, I. E.; Beveridge, D. L. (1990), *Proc. Natl. Acad. Sci. USA* **87**, 8864.]

The extent of correlated motion can be indicated by the magnitude of the corresponding correlation coefficient. The cross-correlation coefficient for the displacement of any two atoms i and j is given by Eq. (3.24)

$$C_{ij} = \frac{\langle \Delta r_i \cdot \Delta r_j \rangle}{\sqrt{\langle \Delta r_i^2 \rangle \langle \Delta r_j^2 \rangle}} \tag{3.24}$$

where Δr_i is the displacement from the mean position of the ith atom. The elements C_{ij} can be collected in matrix form and displayed graphically as a dynamical cross-correlation map (DCCM). The C_{ij} were computed as averages over successive backbone N, C_α, and C atoms to give one entry per pair of amino acid residues. There is a time scale implicit in the C_{ij} as well. In Figure 3.26, the mean of two 40-ps block averages is represented. The intensity of shading is proportional to the magnitude of the coefficient and only correlations above a threshold value of 0.25 were included in Figure 3.26. Positive correlations are given in the upper triangle; negative correlations in the lower triangle. A positive correlation means the two residues move in the same direction; a negative correlation means the two residues move in opposite directions.

Regions of regular secondary structure are expected to move in concert. Thus domains of contiguous residues as in an α helix or β sheet should give rise to significant positive correlations. This can be seen in Figure 3.26. The three plumes emanating from the diagonal into the upper triangle are diagnostic of three distinct contiguous sequences of antiparallel β-sheet domains. Upon moving up the diagonal, the first region encountered is the β sheet formed by residues 9–24 (cf. the secondary structure diagram of Fig. 3.25). The second region is the β sheet composed of residues 42–58 (the flap). Residues 59–75 form the last plume.

Off-diagonal peaks in the DCCM indicate correlated motion in a domain composed of residues nonadjacent in the amino acid sequence. The major cross peaks between residues 22–25 and 81–85 and between residues 28–42 and 76–86 arise from the interaction between noncontiguous residues forming parallel β sheets (see Fig. 3.25).

Other cross peaks indicate domain–domain interactions. A very interesting cross-peak ensemble is formed by the flap (42–58), the two other plume domains of residues 9–24 and 59–75, and correlations between residues 9–21 and 59–75. A major cross peak of positive correlations is seen between residues 9–21 and 59–75; the first and third plume domains are correlated. The onsets of the flap region (42–55 and 55–58) correlate with certain areas of β sheet (59–65, 73–85) of the third plume domain. Negative domain–domain correlations are observed between the flap and both other plume domains.

These interdomain correlations suggest that flap motion occurs with compensatory changes in residues 59–75 in the manner of a lever. That is, the flap closes down as the β-sheet lever moves up. The first plume domain of residues 9–21 is correlated to both the flap and the lever, thus mediating the motion between the two domains. Swaminathan, et al. (1991) suggested that such domain–domain communications are likely to be involved in the action of the enzyme.

This paper by Swaminathan et al. (1991) shows that the large-scale motion of protein domains can be examined by MD. The dynamics of the protein dimer HPR was examined and the results displayed in a DCCM map. The time-averaged dynamics structure was close to the X-ray structure. Communication between domains was observed. The internal movement of HPR was not random; in fact, Swaminathan et al. likened the dynamics of this protein to the actions of a simple machine.

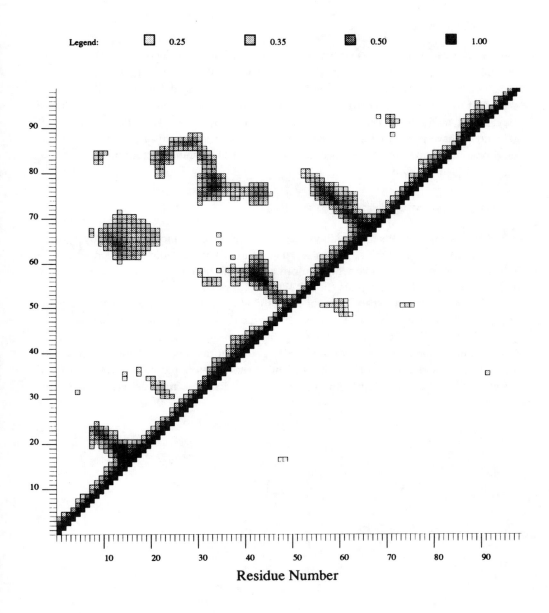

FIGURE 3.26.
Calculated DCCM for a monomeric unit of HPR. [Adapted with permission from Swaminathan, S.; Harte Jr., W. E.; Beveridge, D. L. (1991), *J. Am. Chem. Soc.* **113**, 2717. Copyright © 1991 American Chemical Society.]

3.7. PROTEIN STRUCTURE BY NMR AND MOLECULAR MODELING

3.7.1 Background

Our knowledge about the 3-D structure of proteins has been primarily derived from X-ray crystallography. Unfortunately, it is sometimes impossible to grow protein crystals. In recent years, a second method, ^1H NMR spectroscopy, has been developed for the determination of the complete 3-D structure of proteins (Bax, 1989). Unlike X-ray crystallography, NMR methods cannot be applied to large proteins ($>\sim 100$ residues), but NMR techniques have the advantage of giving a structure of the protein in its natural aqueous environment. Converting NMR spectral data into a 3-D protein structure involves several steps, outlined below.

SEQUENCE ASSIGNMENT

The first and perhaps most tedious step in determining a protein structure by NMR is to assign all of the individual proton resonances to specific protons in the protein. Normally, it is not possible to extract this information from the 1-D ^1H NMR of a protein due to the severe crowding of signals (Fig. 3.27).

The proton spectrum can, however, be fully assigned by using the 2-D NMR experiments COSY and NOESY. *CO*rrelation *S*pecroscop*Y* (COSY) yields information about pairs of protons that are coupled by through-bond spin–spin coupling, and nuclear Overhauser enhancement spectroscopy (NOESY) gives information about through-space dipole–dipole interactions. Both methods identify short proton–proton contacts. A COSY spectrum identifies the type of amino acid, and a NOESY spectrum defines the positions

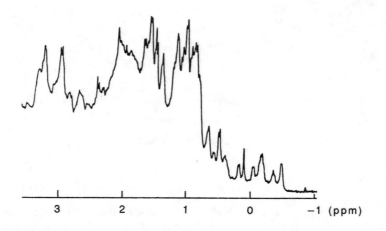

FIGURE 3.27.
A 270-MHz ^1H NMR spectrum (Campbell et al., 1973) of the methyl region of hen eggwhite lysozyme, a 129 residue protein. [Adapted with permission from Campbell, I. D.; Dobson, C. M.; Williams, R. J. P.; Xavier, A. V. (1973), *J. Magn. Reson.* **11**, 172.]

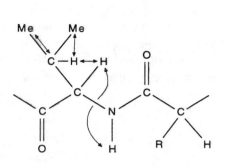

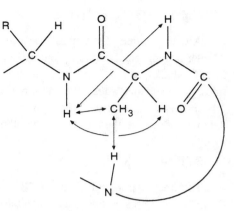

FIGURE 3.28.
Spin–spin interactions by COSY.

FIGURE 3.29.
Through-space interactions by NOESY.

of the amino acids within the protein or peptide. Important interactions used to assign the sequence of proteins by NMR are schematically illustrated by arrows between protons in Figure 3.28 and 3.29. The nuclear magnetization exchange between nucleii is referred to as the nuclear Overhauser effect (NOE). The rate of exchange is a function of interspacial distance and it is measured using NOESY.

For more information about the pulse sequences, mathematics and practice of 2-D NMR, and other techniques used for resonance assignment, we refer you to the Further Reading section at the end of the chapter.

NOE-DERIVED DISTANCES

Once the assignments are known, NOE cross-peak intensities can be translated into proton–proton distances [see Fig. 2.15(b) for an idealized NOESY contour plot]. Observed NOE cross-peak intensities are dependent not only on the proton–proton distances (the intensities of the cross peaks are proportional to the inverse sixth power of their inter-nuclear distance) but also on the local dynamics of the protein. Thus the intensities are interpreted in terms of upper limits or bounds on proton–proton distances rather than exact distances. Nonetheless, the mere presence of an NOE between two protons is taken to mean that the distance between them is less than 5 Å.

DIHEDRAL ANGLES

The interproton distance restraints derived from NOE measurements may be supplemented by backbone ϕ and χ_1 torsion angle restraints. These dihedral angle restraints are derived from a qualitative interpretation of the three-bond H_N-H_α and $H_\alpha-H_\beta$ couplings using the well-known Karplus relation [see Eq. (2.20)]. The relation between angle and coupling is unambiguous only for large J couplings (8–10 Hz) and for small J couplings (< 5.5 Hz).

HYDROGEN BONDS

Another source of structural constraints comes from hydrogen bonds. The NMR evidence for hydrogen bonds in proteins can be found by monitoring amide H−D exchange in D_2O; those protons involved in hydrogen bonds will exchange more slowly. For peptides, NH exchange rates are seldom measured. Rather the temperature dependence of the NH shift is used to give NMR evidence for hydrogen bonds (Kessler, 1982). The NMR data will not reveal the identity of the other atom involved in the hydrogen bond to amide.

STRUCTURE

The final step in converting NMR data into a 3-D protein structure is to construct 3-D model structures on the basis of available proton–proton distance information. This step is commonly carried out either by distance-geometry calculations, or distance restrained molecular dynamics (RMD) refinement. These calculations do not generate the same picture that a high-quality crystal structure provides; that is, the atomic coordinates are not known for all atoms in the protein. Rather, a set of structures, a statistical representative set of protein configurations, is given by the calculations. The similarity of the group of structures, given by the rms deviation, is a measure of the precision of the NMR structural determination. The number and size of constraint violations in the model structures is an indication of quality.

Distance geometry considers only the geometrical aspects of the model structure. Energy is not a factor. The goal is to obtain a random sample from the set of all conformations consistent with the distance constraints. The method is applied to the structure of tendamistat described in Section 3.7.2.

RMD considers energy. It involves carrying out MD calculations on a model structure with a force field that includes additional restraining energy terms of the form (Eq. 3.25):

$$V = \frac{1}{2}K(r - r_0)^2 \qquad r > r_0$$
$$V = 0 \qquad\qquad r \leq r_0 \qquad\qquad (3.25)$$

where r_0 is the experimentally determined proton–proton distance (from NOE data), r is the computed distance in the model structure, and K is a pseudobond force constant. The bond length restraint function of Eq. (3.25) has the effect of pulling protons within the distance r_0 in accordance with NOE observations but typically does not prevent the protons from coming closer. This approach requires a relatively large amount of computational time compared to distance geometry. RMD is applied to the structure of the *lac* repressor headpiece and discussed in Section 3.7.3.

3.7.2 Tendamistat

In this section we will describe a paper by Kline et al. (1988) at the ETH in Switzerland. These researchers determined the complete 3-D structure of the small protein tendamistat in aqueous solution by NMR and distance geometry. Simultaneously, researchers at the Max Planck Institute and Hoechst carried out an X-ray crystal structure analysis of the same molecule to 2-Å resolution (Pflugrath et al., 1986). Comparison of the two structures showed that the global molecular architecture for this protein was the same in the crystal-line solid and in solution (Billeter et al., 1989).

Tendamistat (**3.9**) consists of a polypeptide chain of 74 residues and 2 disulfide bonds (Vertesy et al., 1984).

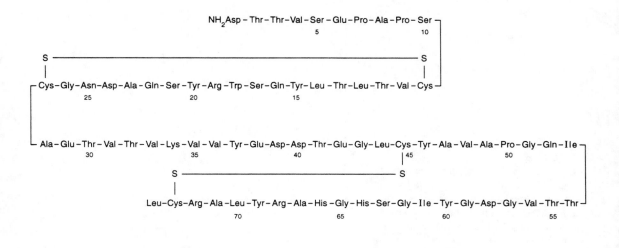

3.9

Tendamistat functions as an inhibitor of α-amylase, an enzyme that catalyzes the hydrolysis of sugar polymers. In recent years, a number of investigations have evaluated the possible efficiency of α-amylase inhibitors such as tendamistat in the treatment of diabetes and obesity. The inhibitors would work by diminishing the absorption of glucose by inhibiting the decomposition of starch by α-amylase.

METHODS AND RESULTS

The tendamistat structure determination was based on the complete ^{1}H NMR assignment using 2-D NMR (NOESY and COSY) described by Kline and Wüthrich (1986). A partial NOESY spectrum of tendamistat is illustrated in Figure 3.30.

The NOESY peak intensities were transformed to distance constraints by using the $1/r^6$ relationship and correcting for internal mobility of the protein. The proton–proton distance constaints were interpreted as upper bounds rather than absolute values. The upper limit of the constraints obtained from the relative NOESY cross-peak intensities are given in Table 3.10.

The cross-peak intensities of Table 3.10 are grouped in three categories, depending on the number of methyl groups involved. The values given in H_2O are for NOEs involving at least one amide proton, and the values in D_2O samples are for all other proton–proton interactions. The upper limit distance constraints are in parentheses; lower limit constraints for all proton pairs are given by the sum of the van der Waals radii of the two protons (~2 Å).

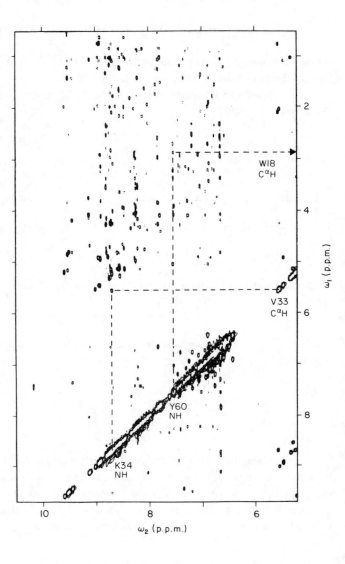

FIGURE 3.30.
Contour plot of the NOESY spectrum of tendamistat. The plot shows the cross peaks between amide and aromatic protons. As an illustration, two cross peaks are connected by broken lines with the corresponding diagonal peaks, which are labeled with the sequence-specific assignments. [Reproduced with permission from Kline, A. D.; Wüthrich, K. (1986), *J. Mol. Biol.* **192**, 869.]

TABLE 3.10. Relationship between NOESY Cross-Peak Intensity and Upper Limit Distance Constraints

Spins Related by NOE	Solvent	Intensity Range (Corresponding Upper Limit in Å)					
Proton–proton							
	^2H_2O	153–75 (2.3)	74–54 (2.5)	53–20 (3.3)	19–2 (4.5)		
	H_2O	168–120 (2.4)	119–72 (2.6)	71–40 (2.9)	39–24 (3.2)	23–8 (3.8)	8–2 (4.6)
Proton–methyl							
	^2H_2O	99–56 (3.3)	55–25 (3.8)	24–11 (4.5)	10–4 (6.0)		
	H_2O	134–76 (3.3)	75–34 (3.8)	33–15 (4.5)	14–5 (6.0)		
Methyl–methyl							
	^2H_2O	93–40 (4.0)	39–4 (5.0)				

The constraints are grouped into different types with respect to the relative positions of the interacting protons in the primary structure. They are intraresidual constraints (interactions between different protons of the same residue); sequential backbone constraints (interactions between backbone protons); medium-range backbone and long-range backbone constraints (interactions between the same proton types that are not nearest neighbors); and interresidual constraints with side-chain protons. The medium-range backbone and long-range backbone constraints are important to define the secondary structural elements, and interresidual constraints with side-chain protons are important to define the positions of the side chains in the structure. A total of 842 constraints were derived from the NOE data. The constraints for one residue, Asp58, are given in Table 3.11.

Supplemental constraints of the torsion angles about single bonds were obtained from measurement of J coupling constants between vicinal protons (protons separated by three bonds) using the well-known Karplus equation [see Eq. (2.20)]. Most of the ϕ angles were constrained.

TABLE 3.11. NOE Distance Constraints for Asp58

INTRARESIDUAL CONSTRAINTS				
Atom	Residue	Atom	Residue	Distance Constraint in Å
HN	Asp58	$HC_{\beta 2}$	Asp58	2.9
HN	Asp58	$HC_{\beta 3}$	Asp58	3.8
SEQUENTIAL BACKBONE CONSTRAINTS				
Atom	Residue	Atom	Residue	Distance Constraint in Å
HN	Asp58	HN	Gly57	4.6
HC_{α}	Asp58	HN	Gly59	2.6
MEDIUM-RANGE BACKBONE AND LONG-RANGE BACKBONE CONSTRAINTS				
Atom	Residue	Atom	Residue	Distance Constraint in Å
HC_{α}	Asp58	HC_{α}	Tyr20	4.0
HC_{α}	Asp58	HN	Ser21	4.6
HC_{α}	Asp58	HN	His64	6.5
INTERRESIDUAL CONSTRAINTS WITH SIDE CHAIN PROTONS				
Atom	Residue	Atom	Residue	Distance Constraint in Å
HN	Asp58	$Q_{\gamma 1}$[a]	Val35	6.0
HN	Asp58	$C_{\zeta 4}$	Tyr37	5.8[b]
$HC_{\beta 2}$	Asp58	$C_{\zeta 4}$	Tyr37	6.5[b]
$HC_{\beta 2}$	Asp58	HN	Gly59	3.8
$HC_{\beta 3}$	Asp58	HN	Gly59	3.2

[a] The pseudoatom Q represents more than one proton. Pseudoatoms are frequently needed to represent the prochiral protons of the methyl group of valine, for example, or the methylene group of tyrosine, where stereospecific assignments are not possible.

[b] Corrections were applied to the NOE distance constraint to allow for use of pseudoatoms.

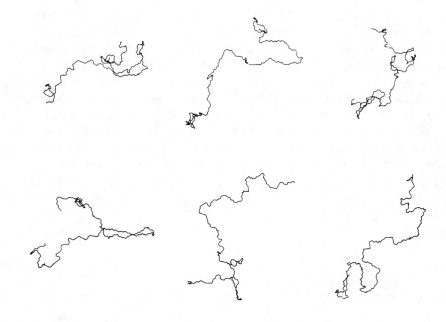

FIGURE 3.31.
Drawings of six random starting structures of the polypeptide backbone of tendamistat generated by DISMAN. [Adapted with permission from Kline, A. D.; Braun, W.; Wüthrich, K. (1988), *J. Mol. Biol.* **204**, 675.]

Further distance constraints were obtained by the assignment of hydrogen and disulfide bonds. Hydrogen bonds were assigned based on the observation that a particular regular secondary structure can be associated with a certain pattern of NOEs (Wüthrich et al., 1984). An additional, independent, criterion for the presence of hydrogen bonds was the observation of slowed amide exchange in the NMR spectrum.

The molecular structures of tendamistat were calculated with the distance geometry approach using the program DISMAN (Braun and Gō, 1985). Because each constraint described an allowed distance range rather than a precise value for the distance, there was likely more than one structure that would satisfy the distance constraints. Therefore many starting peptide conformations or trial structures were studied by distance geometry. One hundred separate calculations were started, using initial dihedral values generated randomly with the use of the local constraints defined by dihedral angle constraints, and the sequential constraints between backbone protons. Six of the initial structures are shown in Figure 3.31.

The DISMAN program then adjusted the conformation of the trial structure by changing the dihedral angles in order to generate a structure that best satisfied the distance constraints obtained by NMR and van der Waals distances. The progressions for the six starting structures given in Figure 3.31 are illustrated in Figure 3.32.

Of the 100 starting structures, only nine final solutions survived. Many structures at the intermediate level failed to satisfy the distance constraints and so were thrown out.

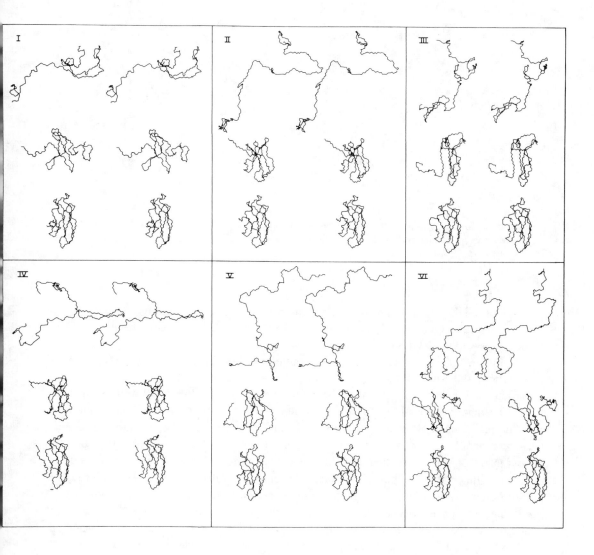

FIGURE 3.32.
Stereodrawings showing the progression of six starting tendamistat backbone structures to the completely folded final backbone structures. [Reproduced with permission from Kline, A. D.; Braun, W.; Wüthrich, K. (1988), *J. Mol. Biol.* **204**, 675.]

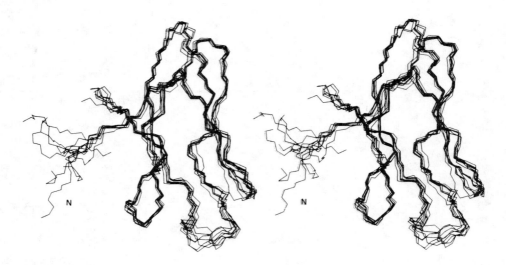

FIGURE 3.33.
Stereoview of the nine superimposed DISMAN structures. The N identifies the amino-terminal end of the polypeptide chain. [Reproduced with permission from Kline, A. D.; Braun, W.; Wüthrich, K. (1988), *J. Mol. Biol.* **204**, 675.]

None of the final solutions satisfied all of the constraints. Certain regions of the tendamistat structure had consistently high violations of the constraints, perhaps as a consequence of artifacts in the NOESY spectra.

Figure 3.33 shows the complete polypeptide backbone of all nine DISMAN structures superimposed with a best-fit orientation. The average of the rms deviations for the nine structures was 0.85 Å for the backbone atoms, when only residues 5–73 were considered. The average rms deviations for all heavy atoms was 1.53 Å. Although none of the final solutions satisfied all of the constraints, residual violations of NOE distance constraints were generally less than 0.5 Å, and violations of van der Waals constraints were usually less than 0.3 Å. Low-residual violations indicate that the calculations have converged.

As illustrated in Figure 3.33, distance geometry results are presented as a group of conformers, similar but not identical. Note the region of residues 1–4, which were virtually unconstrained by NMR data, and for that reason were poorly defined. The interior of the polypeptide was well defined, and was described as an antiparallel β barrel made up of two triple stranded β sheets. Superpositions of distinct peptide segments are illustrated in Figure 3.34, giving a detailed view of the side chain and backbone conformations.

The set of distance geometry structures (Kline et al., 1988) were compared (Billeter et al., 1989) to the structure obtained by X-ray crystallography (Pflugrath et al., 1986). The average of the nine DISMAN structures superimposed with the X-ray structure is illustrated in Figure 3.35.

In summary, Kline et al. (1988) shows the structure of a polypeptide in aqueous solution can be determined by NMR and distance geometry. Here the small protein tendamistat was examined. The global molecular architecture for this polypeptide was the same in the crystal and in aqueous solution, with localized differences seen primarily near the protein surface, including the active site (thought to be Trp18-Arg19-Tyr20). If we consider only the residues 5–73, the rms deviation between the average NMR structure and the X-ray structure was 1.05 Å for the backbone atoms.

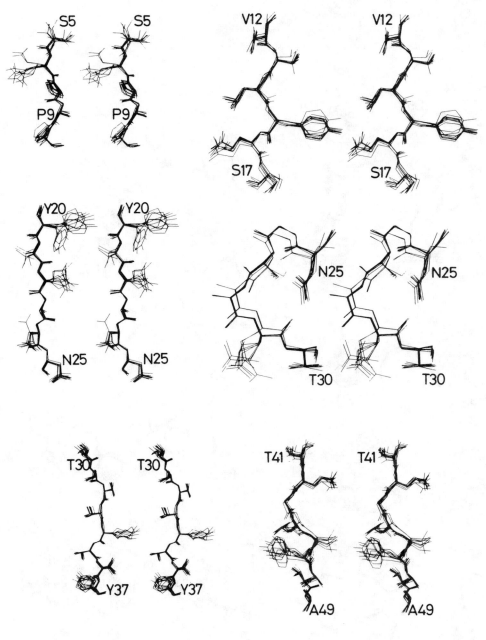

FIGURE 3.34.
The tendamistat polypeptide chain 5–73. [Reproduced with permission from Kline, A. D.; Braun, W.; Wüthrich, K. (1988), *J. Mol. Biol.* **204**, 675.]

(continued)

FIGURE 3.34. (*continued*)
The tendamistat polypeptide chain 5–73.

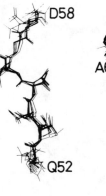
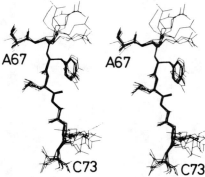
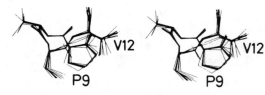
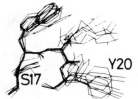

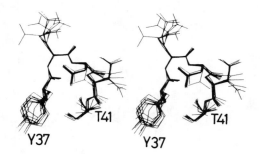
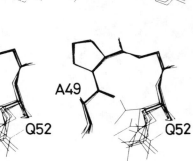
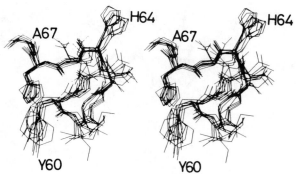

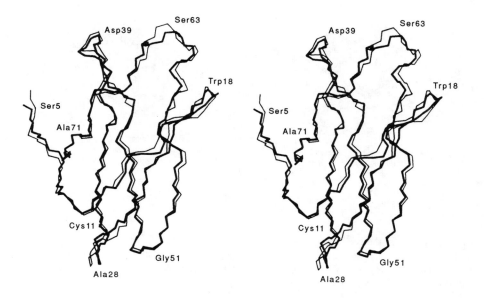

FIGURE 3.35.
Stereoviews of the backbone of tendamistat, residues 5–73. The heavy line corresponds to the average coordinate positions of the backbone of the nine DISMAN structures, and the thin line corresponds to the X-ray structure. [Reproduced with permission from Billeter, M.; Kline, A. D.; Braun, W.; Huber, R.; Wüthrich, K. (1989), *J. Mol. Biol.* **206**, 677.]

3.7.3 *Lac* Repressor Headpiece

In this section we will describe a paper by Kaptein et al. (1988) of the University of Utrecht, and of the University of Gronigen, in the Netherlands. These researchers determined a low-resolution solution structure of a small protein called the *lac* repressor headpiece by NMR and restrained MD (RMD).

Repressors are proteins that regulate gene expression by binding to specific sequences of double-stranded DNA. The *lac* repressor binds to *Escherichia coli* DNA and turns off the lactose metabolism genes. The *lac* repressor is composed of four identical subunits, with a total molecular weight of 148,000. The monomeric subunits of the protein consist of two domains, the "headpiece" and "core." The headpiece is the part of the repressor that actually binds to DNA, and consists of 51 amino acid residues in the N-terminal region. The core is made up of 288 C-terminal amino acid residues. The X-ray structure of the headpiece is unknown, and only a very low (3.5 Å) resolution X-ray structure of the *lac* repressor has been reported (Pace et al., 1990).

METHODS

The complete proton spin system was assigned using NOE, *J*-coupling, and amide H−D exchange data (Zuiderweg et al., 1983a, 1985a, 1985b). The tertiary protein structure was determined using 169 NOE-derived distance constraints and 17 hydrogen-bond constraints (Zuiderweg et al., 1983b; Kaptein et al., 1985; de Vleig et al., 1986). The

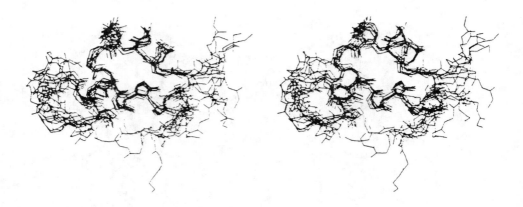

FIGURE 3.36.
Stereodiagrams of the conformations of the *lac* repressor headpiece obtained after restrained molecular dynamics refinement. Only the backbone atoms are shown. [Reproduced with permission from Kaptein, R.; Boelens, R.; Scheek, R. M.; van Gunsteren W. F. (1988), *Biochemistry* **27**, 5389. Copyright © (1988) American Chemical Society.]

initial set of structures was obtained by distance geometry, followed by a distance bounds driven dynamics calculation (Kaptein et al., 1988). The distance bounds driven dynamics algorithm has the effect of "shaking up" the distance geometry conformations, improving the sampling of conformational space; it does not correspond to any physical process. A RMD refinement was applied to the distance bounds driven dynamics structures. All atoms were treated explicitly except for the hydrogens bound to carbon atoms. Bond lengths were kept rigid using SHAKE. SHAKE is an MD method in which internal coordinates such as bond distances are held fixed to save computer time. The potential energy function consisted of bond-angle bending, dihedral torsions, electrostatic, and van der Waals interactions. Solvent was not included. A bond distance constraint term and dihedral bond angle constraint term were added to the potential energy function. The value of the bond distance constraint force constant was varied between 250 and 4000 kJ/mol-nm², increasing as the refinement proceeded. The value of the dihedral angle constraint force constant was 6.8 kJ/mol.

RESULTS

Figure 3.36 shows the polypeptide backbones of the family of 10 structures after RMD. The NMR-derived structure showed three helical regions: residue 6–13 (helix I), residue 17–25 (helix II), and residue 34–45 (helix III). The average of the rms deviations for the 10 structures was 1.7 Å for the C_α atoms, when only residues 4–47 were included. The rms deviations for the C_α atoms of the helixes was 0.8 Å, showing that the helical core of the protein was especially well determined. The average rms deviations for all C_α atoms was 3.0 Å, a consequence of the large spread in conformations in the terminal regions. The structure is illustrated schematically in Figure 3.37.

Kaptein et al. (1988) used NMR and RMD to predict the structure of a polypeptide for which there was no high-resolution X-ray structure. The low-resolution NMR-derived

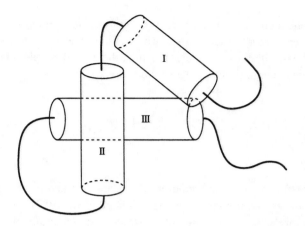

FIGURE 3.37.
Schematic of the *lac* repressor headpiece.

structure of the *lac* repressor headpiece in aqueous solution showed a helix–turn–helix type arrangement for helixes I and II. This motif is characteristic of many DNA-binding proteins (Brennan and Matthews, 1989).

3.7.4 Discussion

Both distance geometry and distance RMD approaches give an ensemble of structures consistent with the NOE constraints. The distance geometry method, as applied to tendamistat, was used to demonstrate that an NMR-derived structure can be as precise as a high-resolution X-ray structure. The NMR data showed that the interior of the peptide was well defined, whereas the chain ends were poorly defined. A large spread in the terminal regions was also observed in the distance RMD derived NMR structure of the *lac* repressor headpiece. The precision of *lac* repressor headpiece structure was much less than that of the tendamistat structure. The difference in precision of the structures described above has nothing to do with the modeling methods, and everything to do with the number of experimental NMR constraints per residue (an average of 11 constraints per residue in tendamistat vs. a little more than 3 constraints per residue in the headpiece).

A weakness of both the distance geometry and distance RMD modeling methods is that the final protein structures are not energy minima, and are therefore unnatural. The final structures are obtained by forcing these structures to obey the experimental NOE distance constraints at the expense of the protein's own intrinsic geometry. This may induce a great deal of strain energy into the protein. Several approaches to solve this problem have been described (Billeter et al., 1990; Olafson et al., 1991).

Another possible problem with NMR-derived structures is that the protein is assumed to be static. Violation of this assumption can result in poor structures, or misrepresentation of the system altogether (Torda et al., 1989). It is important to note that, although the ensemble of structures generated from the NMR modeling work may look like it tells you about the motion of the protein, it does not. The ensemble is simply a collection of solutions to a theoretical equation, not a trajectory. For example, although the NMR-derived structure of the ends of a protein is less well defined than the interior, and one might expect that the ends might also be more dynamic, there is no simple relationship between the ensemble spread and intramolecular motions.

3.8. PROTEIN HOMOLOGY MODELING

3.8.1 Background

The most successful method for predicting the tertiary structure of a protein based on amino acid sequence is homology modeling (Blundell et al., 1987). This technique uses a known 3-D structure as a scaffold to build the structure of a homologous, or evolutionarily related, protein. Homology modeling works because the amino acid sequence that codes for protein architecture is degenerate (Bowie et al., 1990; Overington et al., 1990): that is, many different sequences can code for proteins with essentially the same structure and function. Substituting one amino acid for another will not have a large effect on protein structure if the new amino acid is similar enough to the old one so that critical interactions are preserved, or if some interactions are not very important to protein structure. Typically, the solvent exposed or flexible site amino acids can be substituted with little consequence. Other amino acids cannot be replaced without destabilizing the structure. These include the buried residues forming the hydrophobic core, and residues with unusual stereochemical properties such as Gly and Pro. Key residues are ''conserved,'' that is, evolutionarily related proteins will have the same residues at this position.

If only a few of the residues in a protein sequence contribute to the stability and structure of the protein, it is important to be able to identify these residues. There is experimental evidence that just a few conserved residues in the hypervariable region in immunoglobulins determine the main chain conformation (Padlan and Davies, 1975; Kabat et al., 1977; de la Paz, et al., 1986; Chothia and Lesk, 1987). In section 3.8.2. we will describe a paper that exploits this knowledge to predict the structure of new immunoglobulins.

3.8.2 Hypervariable Loop Region of Immunoglobulins

Immunoglobulins, or antibodies, are the first line of defense against bacterial and viral infection used by vertebrates. These soluble proteins circulate in the blood, where they recognize and bind foreign macromolecules. Each foreign macromolecule, or antigen, can evoke its own specific antibody, which recognizes and combines with only the antigen that evoked it. An antibody is a Y shaped protein with a molecular weight of about 160,000. It is made of four polypeptide chains; designated as light (a short chain) and heavy (a long chain). The chains are attached to each other by S−S cross links; there are also intrachain cross links. The heavy chains contain a covalently bound oligosaccharide component. Each heavy and light chain contains a region of invariant amino acid sequence, characteristic of each vertebrate species, called the constant or C region. Each chain also has a variable or V region in which the amino acid sequence appears to be different for each specific antibody (Fig. 3.38). Antibodies have two antigen-binding sites, each located at the ends

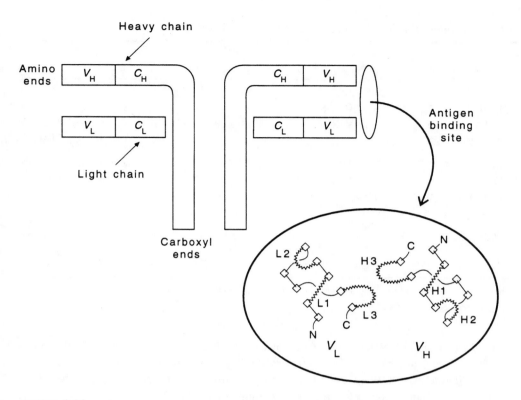

FIGURE 3.38.

Schematic structure of an antibody molecule. Inset: Antigen-binding sites of the variable region. The wavy lines represent loops, and the squares represent strands of a conserved β sheet. [Inset reprinted with permission from *Nature,* Chothia, C.; Lesk, A. M.; Tramontano, A.; Levitt, M.; Smith-Gill, S. J.; Air, G.; Sheriff, S.; Padlan, E. A.; Davies, D.; Tulip, W. R.; Colman, P. M.; Spinelli, S.; Alzari, P. M.; Poljak, R. J. (1989), *Nature (London)* **342**, 877. Copyright © 1989 Macmillan Magazines Limited.]

of the variable regions of the heavy and light chain. The antigen-combining site is composed of six loops (see Fig. 3.38 inset), called hypervariable loops, where there is an especially high frequency of amino acid replacement. The loops are dissimilar in their folding; some are longer and shorter, and some are extended, and some are helical. The end points of these loops have a structural role and are fixed in a β-sheet framework that varies little from one antibody to another. It is the loops that give the antibody an almost unlimited capacity to recognize millions of different antigens.

METHODS AND RESULTS

By examining the data base of immunoglobulin X-ray structures and the amino acid sequences of immunoglobulins of unknown structure, Chothia and Lesk (1987) discovered that for at least five of the hypervariable regions there was only a small repertoire of canonical (or idealized) main-chain conformations. The loop conformation could often be

TABLE 3.12. Sequences and Conformation of L1 Hypervariable Regions[a]

Canonical Structure	Protein	26	27	28	29	30	31	a	b	c	d	e	f	32	2	25	33	71
						*									*	*	*	*
1	J539	S	S	S	V	S								S	I	A	L	Y
	HyHEL-5	S	S	S	V	N								Y	I	A	M	Y
	NQ10	S	S	S	V	R								Y	I	A	M	Y
2	REI	S	Q	D	I	I	K							V	I	A	L	Y
	D1.3	S	G	N	I	H	N							Y	I	A	L	Y
	HyHEL-10	**S**	**Q**	**S**	**I**	**G**	**N**							**N**	**I**	**A**	**L**	**F**
	NC41	S	Q	D	V	S	T							A	I	A	L	Y
3	McPC603	S	E	S	L	L	N	S	G	N	E	K	N	F	I	S	L	F
4	4-4-20	S	Q	S	L	V	H	S		N	G	N	T	Y	V	S	L	F

[a] The sequences and conformation assignment of HyHEL-10 is written in boldface.

predicted from the sequence by the presence of specific residues responsible for the canonical structures and by loop size comparisons (Chothia and Lesk, 1987).

Chothia et al. (1989) tested the "canonical structure" model by using it to predict the hypervariable loop structures of four immunoglobulins before their structures were experimentally determined by X-ray crystallography.

We will illustrate the procedure for the prediction of the structures of five of the hypervariable regions of the immunoglobulin HyHEL-10 and compare them to the subsequently determined X-ray structure.

Table 3.12 presents the sequences and canonical main-chain structure assignments for L1 hypervariable and framework regions from known X-ray structures, including structures as determined by Chothia et al. (1989). The residues in the hypervariable and β-chain regions that are mainly responsible for the main-chain conformation, either through packing, hydrogen bonding, or the ability to assume unusual ϕ, ψ, or ω dihedral conformations are indicated by an asterisk in this and the following tables. A comparison of the sequences of the HyHEL-10 L1 region with the L1 regions in the X-ray data base showed that purely based on length arguments, the new sequence was assigned canonical structure 2.

The sequences of L2 hypervariable and framework regions from the known X-ray structures are given in Table 3.13. There is only one canonical main-chain structure. Based on length, and the presence of I (isoluecine) and G (glycine) at residues 48 and 64 of the β-chain framework, the L2 region of the antibody HyHEL-10 was assigned canonical Structure 1.

The sequences of the L3 hypervariable and framework regions from the known X-ray structures are given in Table 3.14. Based on loop size and the presence of proline in position 95, the L3 region of the antibody HyHEL-10 was assigned canonical structure 1. Although the position of the proline and size determines the type of canonical structure, framework residue 90 is also vital. Glutamine, or Q must be in this position, presumably

150

TABLE 3.13. Sequences and Conformations of L2 Hypervariable Regions[a]

Canonical Structure	Protein	50	51	52	48	64
					*	*
1	REI	E	A	S	I	G
	McPC603	G	A	S	I	G
	J539	E	I	S	I	G
	D1.3	Y	T	T	I	G
	HyHEL-5	D	T	S	I	G
	HyHEL-10	**Y**	**A**	**S**	**I**	**G**
	NC41	W	A	S	I	G
	NQ10	D	T	S	I	G
	4-4-20	K	V	S	I	G

[a] The sequences and conformation assignment of HyHEL-10 is written in boldface.

TABLE 3.14. Sequences and Conformations of L3 Hypervariable Regions[a]

Canonical Structure	Protein	91	92	93	94	95	96	90
						*		*
1	REI	Y	Q	S	L	P	Y	Q
	McPC603	D	H	S	Y	P	L	N
	D1.3	F	W	S	T	P	R	H
	HyHEL-10	**S**	**N**	**S**	**W**	**P**	**Y**	**Q**
	NC41	H	Y	S	P	P	W	Q
	4-4-20	S	T	H	V	P	W	Q
	NQ10	W	S	S	N	P	L	Q
					*			*
2	J539	W	T	Y	P	L	I	Q
						*		*
3	HyHEL-5	W	G	R	N	P		Q

[a] The sequences and conformation assignment of HyHEL-10 is written in boldface.

because of side-chain hydrogen-bond formation. The residue can be replaced by N (asparagine) or H (histidine), two other residues with side chains that can hydrogen bond.

The sequences of the H1 hypervariable and framework regions from the known X-ray structures are given in Table 3.15. There are two possible canonical structures, 1 and 1'. The structures do not differ in length, but the sequences of 1' only partially satisfy the structural requirements of structure 1; thus 1' has a distorted view of the conformation of structure 1. Canonical structure 1 demands a phenylalanine (F) or structurally related residue such as tyrosine (Y) at position 27, because of side-chain packing. In HyHEL-10, this residue is an aspartic acid (D), which is more similar to the serine at the same position

TABLE 3.15. Sequences and Conformations of H1 Hypervariable Regions[a]

Canonical Structure	Protein	26	27	28	29	30	31	32	34	94
		*	*		*				*	*
1	McPC603	G	F	T	F	S	D	F	M	R
	KOL	G	F	I	F	S	S	Y	M	R
	J539	G	F	D	F	S	K	Y	M	R
	D1.3	G	F	S	L	T	G	Y	V	R
	HyHEL-5	G	Y	T	F	S	D	Y	I	R
	NC41	G	Y	T	F	T	N	Y	M	R
	NQ10	G	F	T	F	S	S	F	M	R
	4-4-20	G	F	T	F	S	D	Y	M	G
1′	NEW	G	S	T	F	S	N	D	Y	R
	HyHEL-10	**G**	**D**	**S**	**I**	**T**	**D**	**D**	**W**	**N**

[a] The sequences and conformation assignment of H1 region residues of HyHEL-10 is written in boldface.

of the immunoglobulin NEW. Similar arguments can be made for the canonical structure 1 requirement for a methionine (M) at position 34 not being fulfilled by the tryptophan (W) of HyHEL-10. The HyHEL-10 was assigned canonical structure 1′.

The sequences of the H2 hypervariable and framework regions from the known X-ray structures are given in Table 3.16. The symbol (*) indicates that residues at positions 55 or 54 in canonical structures 2, 3, and 4 have residues with positive values for ϕ and ψ. The residues Gly (G), Asn (N), or Asp (D) are usually found at these sites. Based on

TABLE 3.16. Sequences and Conformations of H2 Hypervariable Regions[a]

Canonical Structure	Protein	52a	b	c	53	54	55	71
							*	
1	NEW				Y	H	G	
	D1.3				G	D	G	
	HyHEL-10				**Y**	**S**	**G**	
		*					(*)	*
2	HyHEL-5	P			G	S	G	A
	NC41	T			N	T	G	L
						(*)		*
3	KOL	D			D	G	S	R
	J539	P			D	S	G	R
	NQ10	S			G	S	S	R
						(*)	*	*
4	McPC603	N	K	G	N	K	Y	R
	4-4-20	N	K	P	Y	N	Y	R

[a] The sequences and conformation assignment of H2 region residues of HyHEL-10 is written in boldface.

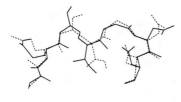
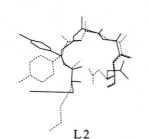
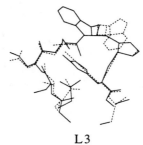

L1   L2   L3

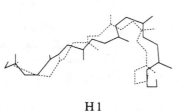
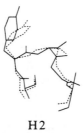

H1   H2

FIGURE 3.39.

The predicted (broken line) and observed (continuous line) conformations of the hypervariable regions of HyHEL-10. Predicted and observed side-chain conformations are shown. [Reprinted with permission from *Nature,* Chothia, C.; Lesk, A. M.; Tramontano, A.; Levitt, M.; Smith-Gill, S. J.; Air, G.; Sheriff, S.; Padlan, E. A.; Davies, D.; Tulip, W. R.; Colman, P. M.; Spinelli, S.; Alzari, P. M.; Poljak, R. J. (1989), *Nature (London)* **342**, 877. Copyright © 1989 Macmillan Magazines Limited.]

loop size and the presence of glycine at position 55, the H2 region of HyHEL-10 was assigned canonical structure 1.

No prediction of conformation of the H3 region was possible based on known canonical structures.

Once the loops of HyHEL-10 were assigned to canonical structure types, an immunoglobulin loop of known structure and of the same canonical type was used as a starting point or parent to develop a model for the structure of the same region in HyHEL-10. If the HyHEL-10 side chains were different than the parent, the side chains of the parent loop were replaced with the side chains of HyHEL-10, and then the model was subjected to limited energy minimization. The results of this model were compared to the X-ray structure of the hypervariable loops of HyHEL-10.

Figure 3.39 shows the predicted and observed (by X-ray crystallography) structures of each of the hypervariable regions for HyHEL-10, superimposed by a least-squares fit of their main chain atoms. Side-chain conformations are not shown for H1 and L3 because they obscure the main chain in this view. The rms differences in atomic positions of main-chain atoms after optimal superposition are given in Table 3.17. The main-chain conformations of the predicted and observed structures of each of the hypervariable regions are very similar.

TABLE 3.17. Differences in Predicted and Observed Hypervariable Regions in Angstroms

Hypervariable Region	Root Mean Square Difference
L1:26–32	1.1
L2:49–53	0.8
L3:90–97	0.3
H1:26–32	1.3
H2:52–56	1.0

The relative positions of the hypervariable regions in the predicted and observed structures are shown in Figure 3.40. This figure was produced by a least-squares superposition of the framework residues. The range of differences in atomic positions of C_α atoms after the superposition of framework residues are given in Table 3.18.

The research by Chothia et al. (1987, 1989) demonstrated that the conformations of immunoglobulin hypervariable regions were given by a small repertoire of main-chain conformations. [A few new canonical structures were added as the result of the work of Chothia (1989).] The canonical structures for the hypervariable regions of the light and heavy chains are illustrated in Figures 3.41 and 3.42.

If only a few main-chain conformations describe the hypervariable regions, how can immunoglobulins be specific to so many antigens? The sequences given in Tables 3.12–3.16 show that there are still sequence variations in the loops. Although the main-chain atoms may be described by a few canonical structures, the side chains are not all the same and can modulate the surface that the canonical structures present to the antigens.

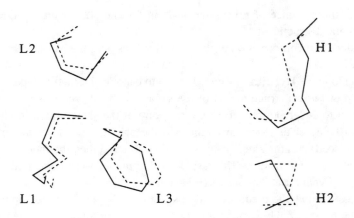

FIGURE 3.40.
The predicted (broken line) and observed (continuous line) C_α structures of the hypervariable regions of HyHEL-10. [Reprinted with permission from Chothia, C.; Lesk, A. M.; Tramontano, A.; Levitt, M.; Smith-Gill, S. J.; Air, G.; Sheriff, S.; Padlan, E. A.; Davies, D.; Tulip, W. R.; Colman, P. M.; Spinelli, S.; Alzari, P. M.; Poljak, R. J. (1989), *Nature* (*London*) **342**, 877. Copyright © 1989 Macmillan Magazines Limited.]

TABLE 3.18. Differences in C_α Atom Position Relative to Framework in Angstroms

Hypervariable Region	C_α Difference Range
L1	0.7–1.6
L2	0.6–1.3
L3	0.8–1.5
H1	1.3–3.5
H2	0.6–2.9

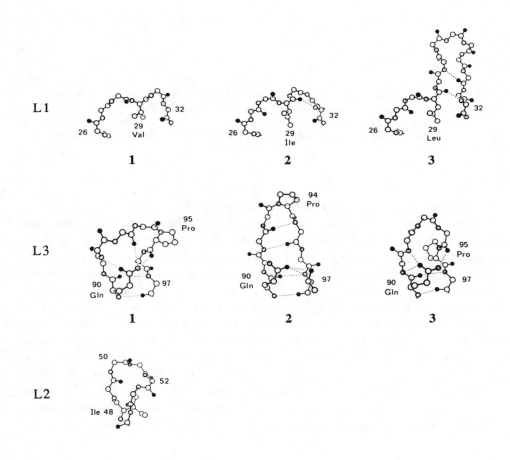

FIGURE 3.41.

Canonical structures for the hypervariable regions of the V_L domain of immunoglobulins. The main-chain conformation and some of the side chains that determine this conformation are shown.
[Reprinted with permission from Chothia, C.; Lesk, A. M.; Tramontano, A.; Levitt, M.; Smith-Gill, S. J.; Air, G.; Sheriff, S.; Padlan, E. A.; Davies, D.; Tulip, W. R.; Colman, P. M.; Spinelli, S.; Alzari, P. M.; Poljak, R. J. (1989), *Nature (London)* **342**, 877. Copyright © 1989 Macmillan Magazines Limited.]

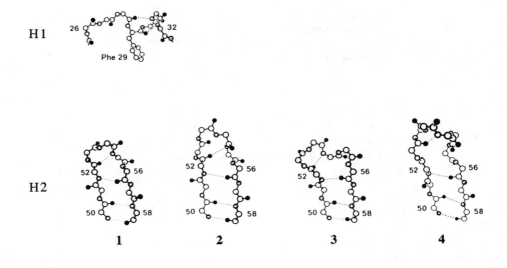

FIGURE 3.42.
Canonical structures for the hypervariable regions of the V_H domain of immunoglobulins. The main-chain conformation and some of the side chains that determine this conformation are shown. [Reprinted with permission from Chothia, C.; Lesk, A. M.; Tramontano, A.; Levitt, M.; Smith-Gill, S. J.; Air, G.; Sheriff, S.; Padlan, E. A.; Davies, D.; Tulip, W. R.; Colman, P. M.; Spinelli, S.; Alzari, P. M.; Poljak, R. J. (1989), *Nature (London)* **342**, 877. Copyright © 1989 Macmillan Magazines Limited.]

To summarize, Chothia et al. (1989) shows that the structures of parts of proteins can be accurately predicted by homology modeling. Here the canonical structure model was used to predict the conformations of the hypervariable regions of immunoglobulins. The relationship between homology modeling and site-directed mutagenesis points out the limitations of predicting the structures of mutant proteins by modeling. As described in Sections 3.4 and 3.8, both homology modeling and site-directed mutagenesis use the X-ray structure of a protein as a starting point. In homology modeling, old protein parts are then replaced by new protein parts. The old and new sequences are similar enough so that critical interactions are maintained. In site-directed mutagenesis, single residues, rather than sequences, are replaced. The old and new residues may not be related at all. Molecular mechanics minimization of mutated proteins works only if the structure or function of the old and new sequences are closely related, or there are only few amino acids replaced, and the replacement does not significantly alter the conformation of the rest of the protein. Minimization makes only limited rearrangements in the positions of atoms.

Homework

3.1. Use a pencil and paper to draw the primary structure of the tripeptide His-Pro-Cys.

3.2. Do a conformation search on dialanine. Vary the angles ϕ and ψ, keeping the amide bond planar. Use a grid of 10°, and make a table of ϕ, ψ, and energy. Generate a Ramachandran plot for dialanine by plotting ϕ versus ψ for only the low-energy conformations. Compare your allowed conformations to the areas within the solid lines in the Ramachandran plot of Figure 3.3.

3.3. Create an α helix from eight residues of leucine (Leu). Locate the hydrogen bonds between the peptide oxygen of residue i and the NH of residue $i + 4$.

3.4. Form an antiparallel β-pleated sheet by docking two extended five-residue segments of tyrosine (Tyr).

3.5. Form a Type III β turn for a glycine tetrapeptide using the dihedral angles given in Table 3.1. Form an inverse γ turn for a glycine tripeptide. Find the relevant hydrogen bonds. Compare your turns to the structures in **3.4** and **3.5**.

3.6. Calculate the energy of a hydrogen bond using Eq. (3.21) ($D_{IJ} = 4.25$ kcal/mol and $\rho_{IJ} = 2.75$ Å). Assume the distance between the donor and the acceptor is 2.9 Å, and the donor–hydrogen-acceptor angle is 150°.

3.7. Calculate the energy of the van der Waals interaction between two hydrogen atoms using Eq. (3.7) ($D_{IJ} = 0.0152$ kcal/mol and $\rho_{IJ} = 3.195$ Å). Assume the distance between the two hydrogen atoms is 2.0 Å. What is the energy if the distance is 3.0 Å?

3.8. Extract the crystal structure of crambin from the Brookhaven Protein Data Bank (File 1CRN), and find the α helixes formed from roughly residues 7–20 and 22–31. Using a force field you have available, minimize the structure of the polypeptide. Compare your minimized crambin structure to the crystal structure. Superimpose the two structures and use your program to determine rms differences.

3.9. Read in the X-ray structure of carbonic anhydrase II from the Brookhaven Protein Data Bank (File 4CA2) and find the active site Zn^{2+}, the active site histidines, and the internal waters. Find the external waters.

3.10. Extract the crystal structure of the insulin dimer from the Brookhaven Protein Data Bank (File 4INS). Identify the A and B chains of each monomer. Locate the four important hydrogen-bonding interactions in the dimer. Find SerB9 and carry out the computer site-directed mutagenesis of both insulin monomers to AspB9. Minimize only the mutated residues. Find the shortest distance between the O atoms of AspB9 and GluB13. With the use of this distance, use Coulomb's law [Eq. (3.16)] to calculate the electrostatic repulsion energy of two partial charges of -0.2. Assume ε is 80 (H_2O). Compare this number to the attractive energy of four average hydrogen bonds. Repeat the electrostatic calculation using one-half, and then twice, your measured distance. Repeat the first calculation using an ε of 1 (vacuum). Repeat with the distance dependent electrostatic potential of Eq. (3.22), assuming ε^* is 1. Why should you use a distance dependent dielectric parameter when doing an *in vacuo* calculation?

3.11. Sketch in, minimize, and carry out a "brute force" or grid conformational search on Thr-Lys-Pro-Arg (tuftsin, **3.7**). To make your task managable, reduce the degrees of freedom by setting the side chain dihedrals χ to those of Table 3.7 for conformational family F1. Vary the ψ and ϕ of each residue from 0° to 360° by 30° increments. Take the lowest energy structure from the search and minimize it. Compare your best structure to the structures of Figures 3.20 and 3.21, and the dihedrals listed in Tables 3.7 and 3.8.

3.12. Carry out a simple dynamics run on your lowest energy, minimized tuftsin, at 150 K for 2 ps. Has equilibrium been reached after 2 ps? How can you tell? Make a movie of the dynamics trajectory. Repeat the calculation at 600 K. Compare the trajectories at 150 and 600 K. Do the trajectories show a relationship between temperature and motion? Is this physically reasonable? What kind of atomic motion do you see in the 150 K tuftsin trajectory? If the experimental time scale of lysine side-chain rotation is on the order of 30 ps, is it possible to see this motion during a 2-ps dynamics calculation? The room temperature interconversion between different backbone conformations of a peptide occurs experimentally at time scales of around 10^{-7} s. Convert 10^{-7} s to ps. If the backbone conformational dynamics is so slow compared to normal dynamics calculation times, what would you need to do to use dynamics to carry out a peptide conformational search?

3.13. Extract HPR from the Brookhaven Protein Data Bank (File 8HVP). This HPR is bound to a peptide-like inhibitor. Please ignore the inhibitor for the time being. Find the external waters of crystallization. Find the internal water. Find the β-sheet flap domain (residues 42–58). Find the β-sheet lever domain (residues 59–75). Find residues 9–21, which mediate the motion between the flap and lever.

3.14. Extract the X-ray structure of tendamistat from the Brookhaven Protein Data Bank (File 1HOE) and find the β-pleated sheets. Why is the polypeptide called a β barrel? Find residue Thr32 in the β strand formed by residues 30–37. Would you expect to observe long-range interstrand NOEs between the backbone hydrogen atoms of Thr32 and other backbone hydrogen atoms? Measure the closest interstrand backbone proton–backbone proton distances. (You will need to add hydrogen atoms to the backbone, because they were not determined in the X-ray structure.) Assign the spins to residues as in Table 3.11, and using Table 3.10, qualitatively predict whether the NOEs in ^2H_2O might be very strong, strong, medium, or weak.

For example,

Atom	Residue	Atom	Residue	Distance	NOE
HN	Thr32	HN	Ala47	2.3	Very strong

3.15. Below, in Figure 3.43, is part of a schematic NOESY spectrum of a hypothetical peptide with a residue sequence of *abcdefghijkl* (**3.10**). The resonances have been assigned to

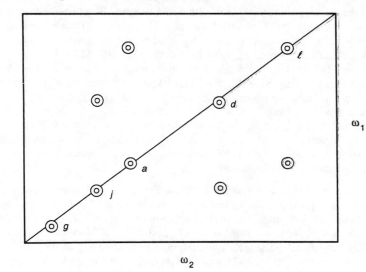

FIGURE 3.43.
Schematic NOESY spectrum of hypothetical peptide.

3.10

specific residues. Based on what the NOE tells you about the proximity of some of the residues, draw a possible folded structure for the peptide.

3.16. Determine the φ torsion angles for an α helix and the extended-chain structure of the β-pleated sheet generated in Problems 3.3 and 3.4. Given the following Karplus relation

$$^3J_{HH} = 7.76 \cos^2\phi - 1.0 \cos\phi + 1.40$$

would you expect the three bond H_N–H_α coupling constants (between amide and C_α protons) to be useful for assigning secondary structure? What are your predicted H_N–H_α coupling constants for the α helix and the extended chain?

3.17. Rules for amino acid replacement are important in homology modeling. Sometimes an amino acid can be interchanged in structurally conserved regions with another amino acid because the two amino acids are similar enough. For example, T can often be replaced by S; and V can often be replaced by L or I. What makes the members of each of these sets similar?

T, S
V, L, I
W, Y, F
N, D
Q, E

3.18. As a first step in simulating the structure of a protein using homology modeling, you must decide, based on amino acid sequences, where the proteins are homologous. This process is normally done by computer because only small sections of the protein, often connected by loops of varying size and structure, are homologous. Use a pencil and paper to examine these short sequences of three enzymes, and define the structurally conserved amino acids.

Hint: Put each sequence on a strip of paper, and slide the papers by each other until you find amino acid alignment. Use the replacement rules above. Note that Pro is *not* a key residue for this problem.

  Enzyme 1: TPAQTLNLDFDTGSSDLWVFSSETTA
  Enzyme 2: GTTLNLNFDTGSADLWVFSTELPASQQSGH
  Enzyme 3: KSTSIDGIADTGTTLLYLPATVVSAY

References

Ackers, G. K.; Smith, F. R. (1985), Effects of site-specific amino acid modification on protein interactions and biological function, *Annu. Rev. Biochem.* **54**, 597.

Allen, F. H.; Kennard, O.; Watson, D. G.; Brammer, L.; Orpen, A. G.; Taylor, R. (1987), Tables of bond lengths determined by X-ray and neutron diffraction. Part 1. Bond lengths in organic compounds, *J. Chem. Soc. Perkin Trans. 2*, S1.

Antonaroli, S.; Bianco, A.; Brufani, M.; Cellai, L.; Lo Baido, G.; Potier, E.; Bonomi, L.; Perfetti, S.; Fiaschi, A. I.; Segre, G. (1992), Acetazolamide-like carbonic anhydrase inhibitors with topical ocular hypotensive activity, *J. Med. Chem.* **35**, 2697.

Bax, A. (1989), Two dimensional NMR and protein structure, *Annu. Rev. Biochem.* **58**, 223.

Billeter, M.; Kline, A. D.; Braun, W.; Huber, R.; Wüthrich, K. (1989), Comparison of the high-resolution structures of the α-amylase inhibitor tendamistat determined by nuclear magnetic resonance in solution and by X-ray diffraction in single crystals, *J. Mol. Biol.* **206**, 677.

Billeter, M.; Schaumann, T.; Braun, W.; Wüthrich, K.(1990), Restrained energy refinement with two different algorithms and force fields of the structure of the α-amylase inhibitor tendamistat determined by NMR in solution, *Biopolymers* **29**, 695.

Blumenstein, M.; Layne, P. P.; Najjar, V. A. (1979), Nuclear magnetic resonance studies on the structure of the tetrapeptide tuftsin, L-Threonyl-L-lysyl-L-prolyl-L-arginine, and its pentapeptide analogue L-Threonyl-L-lysyl-L-prolyl-L-prolyl-L-arginine, *Biochemistry* **18**, 5247.

Blundell, T. L.; Dodson, G.; Hodgkin, D.; Mercola, D. (1972), Insulin: The structure in the crystal and its reflection in chemistry and biology, *Adv. Protein Chem.* **26**, 279.

Blundell, T. L.; Sibanda, B. L.; Sternberg, M. J. E.; Thornton, J. M. (1987), Knowledge-based prediction of protein structures and the design of novel molecules, *Nature (London)* **326**, 347.

Botre, F.; Gros, G., and Storey, B. T., Eds., (1991) *Carbonic Anhydrase. From Biochemistry and Genetics to Physiology and Clinical Medicine*, VCH, Weinheim, Germany.

Bowie, J. U.; Reidhaar-Olson, J. F.; Lim, W. A.; Sauer, R. T. (1990), Deciphering the message in protein sequences: Tolerance to amino acid substitutions, *Science* **247**, 1306.

Brange, J.; Ribel, U.; Hansen, J. F.; Dodson, G.; Hansen, M. T.; Havelund, S.; Melberg, S. G.; Norris, F.; Norris, K.; Snel, L.; Sorensen, A. R.; Voigt, H. O. (1988), Monomeric insulins obtained by protein engineering and their medical implications, *Nature (London)* **333**, 679.

Braun, W.; Gō, N. (1985), Calculation of protein conformations by proton–proton distance constraints. A new efficient algorithm, *J. Mol. Biol.* **186**, 611.

Brennan, R. G.; Matthews, B. W. (1989), The helix–turn–helix DNA binding motif, *J. Biol. Chem.* **264**, 1903.

Brooks, B. R.; Bruccoleri, R. E.; Olafson, B. D.; States, D. J.; Swaminathan, S.; Karplus, M. (1983), CHARMM: A program for macromolecular energy, minimization, and dynamics calculations, *J. Comp. Chem.* **4**, 187.

Burley, S. K.; Petsko, G. A. (1988), Weakly polar interactions in proteins, *Adv. Protein Chem.* **39**, 125.

Burrow, G. N.; Hazlett, B. E.; Phillips, M. J. (1982), A case of diabetes mellitus, *New Eng. J. Med.* **306**, 340.

Burt, S. K.; Greer, J. (1988), Search strategies for determining bioactive conformers of peptides and small molecules, *Annu. Rep. Med. Chem.* **23**, 285.

Campbell, I. D.; Dobson, C. M.; Williams, R. J. P.; Xavier, A. V. (1973), Resolution enhancement of protein PMR spectra using the difference between a broadened and normal spectrum, *J. Magn. Reson.* **11**, 172.

Chothia, C.; Lesk, A. M. (1987), Canonical structures for the hypervariable regions of immunoglobulins, *J. Mol. Biol.* **196**, 901.

Chothia, C.; Lesk, A. M.; Levitt, M.; Amit, A. G.; Mariuzza, R. A.; Phillips, S. E. V.; Poljak, R. J. (1986), The predicted structure of immunoglobulin D1.3 and its comparison with the crystal structure, *Science* **233**, 755.

Chothia, C.; Lesk, A. M.; Tramontano, A.; Levitt, M.; Smith-Gill, S. J.; Air, G.; Sheriff, S.; Padlan, E. A.; Davies, D.; Tulip, W. R.; Colman, P. M.; Spinelli, S.; Alzari, P. M.; Poljak, R. J. (1989), Conformations of immunoglobulin hypervariable regions, *Nature (London)* **342**, 877.

Clark, M.; Cramer III, R. D.; Van Opdenbosch, N. (1989), Validation of the general purpose Tripos 5.2 force field, *J. Comp. Chem.* **10**, 982.

Derewenda, U.; Derewenda, Z. S.; Dodson, G. G.; Hubbard, R. E. (1990), Insulin structure. In Cuatrecasas, P.; Jacobs, S., Eds., *Insulin*, Handbook of Experimental Pharmacology, Vol. 92, Springer-Verlag, Berlin, pp. 23–40.

D'Ursi, A.; Pegna, M.; Amodeo, P.; Molinari, H.; Verdini, A.; Zetta, L.; Temussi, P. A. (1992), Solution conformation of tuftsin, *Biochemistry* **31**, 9581.

Davis, M. E.; McCammon, J. A. (1990), Electrostatics in biomolecular structure and dynamics, *Chem. Rev.* **90**, 509.

de Vlieg, J.; Boelens, R.; Scheek, R. M.; Kaptein, R.; van Gunsteren, W. F. (1986), Restrained molecular dynamics procedure for protein tertiary structure determination from NMR data: a *lac* repressor headpiece structure based on information on *J*-coupling and from presence and absence of NOE's, *Israel J. Chem.* **27**, 181.

de la Paz, P.; Sutton, B. J.; Darsley, M. J.; Rees, A. R. (1986), Modelling of the combining sites of three anti-lysozyme monoclonal antibodies and of the complex between one of the antibodies and its epitope, *EMBO J.* **5**, 415.

Dyson, H. J.; Rance, M.; Houghten, R. A.; Lerner, R. A.; Wright, P. E. (1988), Folding of immunogenic peptide fragments of proteins in water solution. I. Sequence requirements for the formation of a reverse turn, *J. Mol. Biol.* **201**, 161.

Fischer, G. Schmid, F. X. (1990), The mechanism of protein folding. Implications of *in vitro* refolding models for *de novo* protein folding and translocation in the cell, *Biochemistry* **29**, 2205.

Fitzwater, S.; Hodes, Z. I.; Scheraga, H. A. (1978), Conformational energy study of tuftsin, *Macromolecules* **11**, 805.

Fridkin, M.; Najjar, V. A. (1989), Tuftsin: its chemistry, biology, and clinical potential, *Criti. Rev. Biochem. Mol. Biol.* **24**, 1.

Gelin, B. R.; Karplus, M. (1979), Side-Chain torsional potentials: effects of dipeptide, protein, and solvent environment, *Biochemistry* **18**, 1256.

Hahn, G. S. (1986), Immunoglobulin-derived drugs, *Nature (London)* **324**, 283.

Hanssen, K. F.; Dahl-Jorgensen, K.; Lauritzen, T.; Feldt-Rasmussen, B.; Brinchmann-Hansen, O.; Deckert, T. (1986), Diabetic control and microvascular complications: the near-normoglycaemic experience, *Diabetologia* **29**, 677.

Harte Jr., W. E.; Swaminathan, S.; Mansuri, M. M.; Martin, J. C.; Rosenberg, I. E.; Beveridge, D. L. (1990), Domain communication in the dynamical structure of human immunodeficiency virus 1 protease, *Proc. Natl. Acad. Sci. USA* **87**, 8864.

Hartman, G. D.; Halczenko, W.; Prugh, J. D.; Smith, R. L.; Sugrue, M. F.; Mallorga, P.; Michelson, S. R.; Randall, W. C.; Schwam, H.; Sondey, J. M. (1992), Thieno[2,3-*b*]furan-2-sulfonamides as topical carbonic anhydrase inhibitors, *J. Med. Chem.* **35**, 3027.

Hendrickson, W. A.; Teeter, M. M. (1981), Structure of the hydrophobic protein crambin determined directly from the anomalous scattering of sulphur, *Nature (London)* **290**, 107.

Hendrickson, W. A.; Teeter, M. M. (1986), unpublished work cited in (Whitlow and Teeter, 1986).

Hoops, S. C.; Anderson, K. W.; Merz Jr., K. M. (1991), Force field design for metalloproteins, *J. Am. Chem. Soc.* **113**, 8262.

Hruby, V. J. (1984), Design of pepetide superagonists and antagonists. Conformational and dynamic considerations. In Vida, J. A.; Gordon, M., Eds., *Conformationally Directed Drug Design.*, (ACS Symposium Series 251), ACS, Washington, DC, pp. 9–28.

Jacober, S. P. (1988), *SOLVGEN: An Approach to Protein Hydration*, M. S. Thesis, Department of Computer Sciences, University of Kansas, Lawrence.

Jones, T. A. (1985), Interactive computer graphics: FRODO, *Methods Enzymol.* **115**, 157.

Jorgensen, W. L.; Chandrasekhar, J.; Madura, J. D.; Impey, R. W.; Klein, M.L. (1983), Comparison of simple potential functions for simulating liquid water, *J. Chem. Phys.* **79**, 926.

Jorgensen, W. L.; Tirado-Rives, J. (1988), The OPLS potential functions for proteins. Energy minimizations for crystals of cyclic peptides and crambin, *J. Am. Chem. Soc.* **110**, 1657.

Kabat, E. A.; Wu, T. T.; Bilofsky, H. (1977), Unusual distributions of amino acids in complementarity-determining (hypervariable) segments of heavy and light chains of immunoglobulins and their possible roles in specificity of antibody-combining sites, *J. Biol. Chem.* **252**, 6609.

Kannan, K. K.; Ramanadham, M.; Jones, T. A. (1984), Structure, refinement, and function of carbonic anhydrase isozymes: refinement of human carbonic anhydrase I. *Ann. NY Acad. Sci.* **429**, 49.

Kaptein, R.; Boelens, R.; Scheek, R. M.; van Gunsteren W. F. (1988), Protein structures from NMR, *Biochemistry* **27**, 5389.

Kaptein, R.; Zuiderweg, E. R. P.; Scheek, R. M.; Boelens, R.; van Gunsteren W. F. (1985), A protein structure from nuclear magnetic resonance data *lac* repressor headpiece, *J. Mol. Biol.* **182**, 179.

Kessler, H. (1982), Conformation and biological activity of cyclic peptides, *Angew. Chem. Int. Ed. Engl.* **21**, 512.

Kim, P. S.; Baldwin, R. L. (1984), A helix stop signal in the isolated S-peptide of ribonuclease A, *Nature (London)* **307**, 329.

Kline, A. D.; Braun, W.; Wüthrich, K. (1988), Determination of the complete three-dimensional structure of the α-amylase inhibitor tendamistat in aqueous solution by nuclear magnetic resonance and distance geometry, *J. Mol. Biol.* **204**, 675.

Kline, A. D.; Wüthrich, K. (1986), Complete sequence-specific ¹H nuclear magnetic resonance assignments for the α-amylase polypeptide inhibitor tendamistat from *Streptomyces tendae*, *J. Mol. Biol.* **192**, 869.

Knowles, J. R.; Albery, W. J. (1977), Perfection in enzyme catalysis: the energetics of triosephosphate isomerase, *Acc. Chem. Res.* **10**, 105.

Kolata, G. B. (1979), Blood sugar and the complications of diabetes, *Science* **203**, 1098.

Lapatto, R.; Blundell, T.; Hemmings, A.; Overington, J.; Wilderspin, A.; Wood, S.; Merson, J. R.; Whittle, P. J.; Danley, D. E.; Geoghegan, K. F.; Hawrylik, S. J.; Lee, S. E.; Scheld, K. G.; Hobart, P. M. (1989), X-ray analysis of HIV-1 proteinase at 2.7 Å resolution confirms structural homology among retroviral enzymes, *Nature (London)* **342**, 299.

Lau, K. F.; Dill, K. A. (1989), A lattice statistical mechanics model of the conformational and sequence spaces of proteins, *Macromolecules* **22**, 3986.

Lennard-Jones, J. E. (1924), On the determination of molecular fields-II. From the equation of state of a gas, *Proc. R. Soc. London, Ser. Ser. A* **106**, 463.

Lewis, P. N.; Momany, F. A.; Scheraga, H. A. (1973), Chain reversals in proteins, *Biochim. Biophys. Acta* **303**, 211.

Lii, J.-H.; Allinger, N. L. (1991), The MM3 force field for amides, polypeptides and proteins. *J. Comp. Chem.* **12**, 186.

Lii, J.-H.; Gallion, S.; Bender, C.; Wikstrom, H.; Allinger, N. L.; Flurchick, K. M.; Teeter, M. M. (1989), Molecular mechanics (MM2) calculations on peptides and on the protein crambin using the Cyber 205, *J. Comp. Chem.* **10**, 503.

Lim, C.; Bashford, D.; Karplus, M. (1991), Absolute pK_a calculations with continuum dielectric methods, *J. Phys. Chem.* **95**, 5610.

Lindahl, M.; Vidgren, J.; Eriksson, E.; Habash, J.; Harrop, S.; Helliwell, J.; Liljas, A.; Lindeskog, M.; Walker, N. (1991), Crystallographic studies of carbonic anhydrase inhibition. In Botre, F.; Gros, G.; Storey, B. T., Eds., *Carbonic Anhydrase. From Biochemistry and Genetics to Physiology and Clinical Medicine*, VCH, Weinheim, Germany pp. 111–118.

Marshall, G. R.; Gorin, F. A.; Moore, M. L. (1978), Peptide conformation and biological activity, *Annu. Rep. Med. Chem.* **13**, 227.

Mayo, S. L.; Olafson, B. D.; Goddard III, W. A. (1990), DREIDING: A generic force field for molecular simulations, *J. Phys. Chem.* **94**, 8897.

McCammon, J. A.; Gelin, B. R.; Karplus, M. (1977), Dynamics of folded proteins, *Nature (London)* **267**, 585.

Miller, M.; Schneider, J.; Sathyanarayana, B. K.; Toth, M. V.; Marshall, G. R.; Clawson, L.; Selk, L.; Kent, S. B. H.; Wlodawer, A. (1989), Structure of complex of synthetic HIV-1 protease with a substrate-based inhibitor at 2.3 Å resolution, *Science* **246**, 1149.

Momany, F. A.; Rone, R. (1992), Validation of the general purpose QUANTA® 3.2/CHARMm® force field, *J. Comp. Chem.* **13**, 888.

Najjar, V. A.; Nishioka, K. (1970), 'Tuftsin': a natural phagocytosis stimulating peptide, *Nature (London)* **228**, 672.

Navia, M. A.; Fitzgerald, P. M. D.; McKeever, B. M.; Leu, C.-T.; Heimbach, J. C.; Herber, W. K.; Sigal, I. S.; Darke, P. L.; Springer, J. P. (1989), Three-dimensional structure of aspartyl protease from human immunodeficiency virus HIV-1, *Nature (London)* **337**, 615.

Nilsson, L.; Karplus, M. (1986), Empirical energy functions for energy minimization and dynamics of nucleic acids, *J. Comp. Chem.* **7**, 591.

O'Connor, S. D.; Smith, P. E.; Al-Obeidi, F.; Pettitt, B. M. (1992), Quenched molecular dynamics simulations of tuftsin and proposed cyclic analogues, *J. Med. Chem.* **35**, 2870.

Olafson, B. D.; Marusin, J.; Ary, M. L.(1991), Techniques to determine macromolecular structure, *Sci. Comp. Automation* 47.

Overington, J.; Johnson, M. S.; Sali, A.; Blundell, T. L. (1990), Tertiary structural constraints on protein evolutionary diversity: templates, key residues and structure prediction, *Proc. R. Soc. London B* **241**, 132.

Pace, H. C.; Lu, P.; Lewis, M. (1990), *lac* repressor: crystallization of intact tetramer and its complexes with inducer and operator DNA, *Proc. Natl. Acad. Sci. USA* **87**, 1870.

Padlan, E. A.; Davies, D. R. (1975), Variability of three-dimensional structure in immunoglobulins, *Proc. Natl. Acad. Sci. USA* **72**, 819.

Pflugrath, J. W.; Wiegand, G.; Huber, R.; Vertesy, L. (1986), Crystal structure determination, refinement and the molecular model of the α-amylase inhibitor Hoe-467A, *J. Mol. Biol.* **189**, 383.

Ramachandran, G. N.; Ramakrishnan, C.; Sasisekharan, V. (1963), Stereochemistry of polypeptide chain configurations, *J. Mol. Biol.* **7**, 95.

Reidhaar-Olson, J. F.; Sauer, R. T. (1988), Combinatorial cassette mutagenesis as a probe of the informational content of protein sequences, *Science* **241**, 53.

Richardson, J. S.; Richardson, D. C. (1988), Amino acid preferences for specific locations at the ends of α helices, *Science* **240**, 1648.

Rose, G.D.; Gierasch, L. M.; Smith, J. A. (1985), Turns in peptides and proteins, *Adv. Protein Chem.* **37**, 1.

Rupley, J. A.; Careri, G. (1991), Protein hydration and function, *Adv. Protein Chem.,* **41**, 37.

Ryckaert, J. P.; Ciccotti, G.; Berendsen, H. J. C. (1977), Numerical integration of the Cartesian equations of motion of a system with constraints: Molecular dynamics of *n*-alkanes, *J. Comp. Phys.* **23**, 327.

Saenger, W. (1987), Structure and dynamics of water surrounding biomolecules, *Annu. Rev. Biophys. Biophys. Chem.* **16**, 93.

Siemion, I. Z.; Konopinska, D. (1981), Tuftsin analogs and their biological activity, *Mol. Cell. Biochem.* **41**, 99.

Siemion, I. Z.; Lisowski, M.; Sobczyk, K. (1983), Conformational investigations in the tuftsin group, *Ann. NY Acad. Sci.* **419**, 56.

Singh, U. C.; Kollman, P. (1984), An approach to computing electrostatic charges for molecules *J. Comp. Chem.* **5**, 129.

Singh, U. C.; Weiner, P. K.; Caldwell, J. W.; Kollman, P. A. (1986), AMBER Program, University of California, San Franscisco.

Smith, J. C.; Karplus, M. (1992), Empirical force field study of geometries and conformational transitions of some organic molecules, *J. Am. Chem. Soc.* **114**, 801.

Stabinsky, Y.; Gottlieb, P.; Fridkin, M. (1980), The phagocytosis stimulating peptide tuftsin: further look into structure–function relationships, *Mol. Cell. Biochem.* **30**, 165.

Swaminathan, S.; Harte Jr., W. E.; Beveridge, D. L. (1991), Investigation of domain structure in proteins via molecular dynamics simulation: Application to HIV-1 protease dimer, *J. Am. Chem. Soc.* **113**, 2717.

Tager, H. S. (1990), Mutant human insulins and insulin structure-function relationships. In Cuatrecasas, P. and Jacobs, S., Eds., *Insulin*, Handbook of Experimental Pharmacology, Vol. 92, Springer-Verlag, Berlin, pp. 41–66.

Teeter, M. M. (1984), Water structure of a hydrophobic protein at atomic resolution: pentagon rings of water molecules in crystals of crambin, *Proc. Natl. Acad. Sci. USA* **81**, 6014.

Teeter, M. M. (1991), Water-protein interactions: theory and experiment, *Annu. Rev. Biophys. Biophys. Chem.* **20**, 577.

Torda, A. E.; Scheek, R. M.; van Gunsteren, W. F. (1989), Time-dependent distance restraints in molecular dynamics simulations, *Chem. Phys. Lett.* **157**, 289.

Tu, C.; Silverman, D. N.; Forsman, C.; Jonsson, B.-H.; Lindskog, S. (1989), Role of histidine 64 in the catalytic mechanism of human carbonic anhydrase II studied with a site-specific mutant, *Biochemistry* **28**, 7913.

van Gunsteren, W. F.; Berendsen, H. J. C. (1987), ''Groningen molecular simulation (GROMOS) library manual,'' Biomos, Nijenborgh 16, Groningen, The Netherlands.

van Gunsteren, W. F.; Berendsen, H. J. C. (1990), Computer simulation of molecular dynamics: methodology, applications, and perspectives in chemistry, *Angew. Chem. Int. Ed. Engl.* **29**, 992.

Vedani, A. (1988), YETI: an interactive molecular mechanics program for small-molecule protein complexes, *J. Comp. Chem.* **9**, 269.

Vedani, A.; Huhta, D. W.; Jacober, S. P. (1989), Metal coordination, H-bond network formation, and protein-solvent interactions in native and complexed human carbonic anhydrase I: a molecular mechanics study, *J. Am. Chem. Soc.* **111**, 4075.

Verdini, A. S.; Silvestri, S.; Becherucci, C.; Longobardi, M. G.; Parente, L.; Peppoloni, S.; Perretti, M.; Pileri, P.; Pinori, M.; Viscomi, G. C.; Nencioni, L. (1991), Immunostimulation by a partially modified *retro-inverso*-tuftsin analogue containing $Thr^1\psi[NHCO](R,S)Lys^2$ modification, *J. Med. Chem.* **34**, 3372.

Verlet, L. (1967), Computer "experiments" on classical fluids. I. Thermodynamical properties of Lennard-Jones molecules, *Phys. Rev.* **159**, 98.

Vertesy, L.; Oeding, V.; Bender, R.; Zepf, K.; Nesemann, G. (1984), Tendamistat (HOE 467), a tight-binding α-amylase inhibitor from *Streptomyces tendae* 4158, *Eur. J. Biochem.* **141**, 505.

Vora, J. P.; Owens, D. R.; Dolben, J.; Atiea, J. A.; Dean, J. D.; Kang, S.; Burch, A.; Brange, J. (1988), Recombinant DNA derived monomeric insulin analogue: comparison with soluble human insulin in normal subjects, *Br. Med. J.* **297**, 1236.

Whitlow, M.; Teeter, M. M. (1986), An empirical examination of potential energy minimization using the well-determined structure of the protein crambin, *J. Am. Chem. Soc.* **108**, 7163.

Weiner, P. K.; Kollman, P. A. (1981), AMBER: Assisted model building with energy refinement. A general program for modeling molecules and their interactions, *J. Comp. Chem.* **2**, 287.

Weiner, S. J.; Kollman, P. A.; Case, D. A.; Singh, U. C.; Ghio, C.; Alagona, G.; Profeta, S.; Weiner, P. (1984), A new force field for molecular mechanical simulation of nucleic acids and proteins, *J. Am. Chem. Soc.* **106**, 765.

Weiner, S. J.; Kollman, P. A.; Nguyen, D. T.; Case, D. A. (1986), An all atom force field for simulations of proteins and nucleic acids, *J. Comp. Chem.* **7**, 230.

Wlodawer, A.; Miller, M.; Jaskolski, M.; Sathyanarayana, B. K.; Baldwin, E.; Weber, I. T.; Selk, L. M.; Clawson, L.; Schneider, J.; Kent S. B. H. (1989), Conserved folding in retroviral proteases: crystal structure of a synthetic HIV-1 protease, *Science* **245**, 616.

Wüthrich, K.; Billeter, M.; Braun, W. (1984), Polypeptide secondary structure determination by nuclear magnetic resonance observation of short proton–proton distances, *J. Mol. Biol.* **180**, 715.

Zuiderweg, E. R. P.; Boelens, R.; Kaptein, R. (1985a), Stereospecific assignments of ¹H-NMR methyl lines and conformation of valyl residues in the *lac* repressor headpiece, *Biopolymers* **24**, 601.

Zuiderweg, E. R. P.; Kaptein, R.; Wüthrich, K. (1983a), Sequence-specific resonance assignments in the ¹H nuclear-magnetic-resonance spectrum of the *lac* repressor DNA-binding domain 1–51 from *Escherichia coli* by two-dimensional spectroscopy, *Eur. J. Biochem.* **137**, 279.

Zuiderweg, E. R. P.; Kaptein, R.; and Wüthrich, K. (1983b), Secondary structure of the *lac* repressor DNA-binding domain by two-dimensional ¹H nuclear magnetic resonance in solution, *Proc. Natl. Acad. Sci. USA* **80**, 5837.

Zuiderweg, E. R. P.; Scheek, R. M.; Kaptein, R. (1985b), Two-dimensional ¹H-nmr studies on the *lac* repressor DNA binding domain: further resonance assignments and identification of nuclear Overhauser enhancements, *Biopolymers* **24**, 2257.

Further Reading

Peptides

Blundell, T.; Wood, S. (1982), The conformation, flexibility, and dynamics of polypeptide hormones, *Annu. Rev. Biochem.* **51**, 123.

Dyson, H. J.; Wright, P. E. (1991), Defining solution conformations of small linear peptides, *Annu. Rev. Biophys. Biophys. Chem.* **20**, 519.

König, W. (1993), *Peptide and Protein Hormones*, VCH, Weinheim, Germany.

Smith, P. E.; Al-Obeidi, F.; Pettitt, B. M. (1991), Aspects of the design of conformationally constrained peptides, *Methods Enzymol.* **202**, 411.

Proteins

Chadwick, D. J.; Widdows, K., Eds. (1991), *Protein Conformation*, Wiley, Chichester, UK.

Hamaguchi, K. (1992), *The Protein Molecule*, Japan Scientific Societies Press, Japan.

Rawn, J. D. (1989), *Proteins, Energy, and Metabolism*, Neil Patterson Publishers, Burlington, NC.

Enzymes

Fersht, A. (1984), *Enzyme Structure and Mechanism,* Second Edition, Freeman, New York.

Glusker, J. P. (1991), Structural aspects of metal liganding to functional groups in proteins, *Adv. Protein Chem.* **42**, 1.

Protein Dynamics

Daggett, V.; Levitt, M. (1993), Realistic simulations of native-protein dynamics in solution and beyond, *Annu. Rev. Biophys. Biomol. Struct.* **22**, 353.

Karplus, M. (1986), Internal dynamics of proteins, *Methods Enzymol.* **131**, 283.

McCammon, J. A.; Harvey, S. C. (1987), *Dynamics of Proteins and Nucleic Acids* Cambridge University Press, Cambridge, UK.

van Gunsteren, W. F.; Luque, F. J.; Timms, D.; Torda, A. E. (1994), Molecular mechanics in biology: from strucutre to function, taking account of solvation, *Annu. Rev. Biophys. Biomol. Struct.* **23**, 847.

Protein Structure and Structure Prediction

Chothia, C. (1984), Principles that determine the structure of proteins, *Annu. Rev. Biochem.* **53**, 537.

Creighton, T. E. (1990), Protein folding, *Biochem. J.* **270**, 1.

DeGrado, W. F. (1988), Design of peptides and proteins, *Adv. Protein Chem.* **39**, 51.

Fasman, G. D., Ed. (1989), *Prediction of Protein Structure and the Principles of Protein Conformation*, Plenum, New York.

Kim, P. S.; Baldwin, R. L. (1990), Intermediates in the folding reactions of small proteins, *Annu. Rev. Biochem.* **59**, 631.

Lesk, A. M. (1991), *Protein Architecture: A Practical Approach*, IRL Press, Oxford, UK.

Rose, G. D.; Wolfenden, R. (1993), Hydrogen bonding, hydrophobicity, packing, and protein folding, *Annu. Rev. Biophys. Biomol. Struct.* **23**, 847.

Schulz, G. E. (1988), A critical evaluation of methods for prediction of protein secondary structures, *Annu. Rev. Biophys. Biophys. Chem.* **17**, 1.

Skolnick, J.; Kolinski, A. (1989), Computer simulations of globular protein folding and tertiary structure. *Annu. Rev. Phys. Chem.* **40**, 207.

Wright, P. E.; Dyson, H. J.; Lerner, R. A. (1988), Conformation of peptide fragments of proteins in aqueous solution: implications for initiation of protein folding, *Biochemistry* **27**, 7167.

NMR and NMR of Biomolecules

Bertini, I.; Molinari, H.; Niccolai, N., Eds. (1991), *NMR and Biomolecular Structure*, VCH, Weinheim, Germany.

Clore, G. M.; Gronenborn, A. M., Eds. (1993), *NMR of Proteins*, CRC Press, Boca Raton, FL.

Neuhaus, D.; Williamson M. P. (1989), *The Nuclear Overhauser Effect in Structural and Conformational Analysis.* VCH, New York.

Williams, K. R.; King, R. W. (1990), The Fourier transform in chemistry—NMR. Part 4. Two dimensional methods, *J. Chem. Educ.* **67**, A125.

Williamson, M. P.; Waltho, J. P. (1992), Peptide structure from NMR, *Chem. Soc. Rev.* **21**, 227.

Wüthrich, K. (1986), *NMR of Proteins and Nucleic Acids*, Wiley-Interscience, New York.

Supplemental Case Studies

Greer, J. (1990), Comparative modeling methods: Application to the family of the mammalian serine proteases, *PROTEINS: Structure, Function, Genetics* **7**, 317.

Harte Jr., W. E.; Beveridge, D. L. (1993), Mechanism for the destabilization of the dimer interface in a mutant HIV-1 protease: A molecular dynamics study, *J. Am. Chem. Soc.* **115**, 1231.

Levitt, M.; Sharon, R. (1988), Accurate simulation of protein dynamics in solution, *Proc. Natl. Acad. Sci. USA* **85**, 7557.

McCammon, J. A.; Gelin, B. R.; Karplus, M. (1977), Dynamics of folded proteins, *Nature (London)* **267**, 585.

Moore, J. M.; Case, D. A.; Chazin, W. J.; Gippert, G. P.; Havel, T. F.; Powls, R.; Wright, P. E. (1988), Three-dimensional solution structure of plastocyanin from the green alga *Scenedesmus obliquus*, *Science* **240**, 314.

Mosberg, H. I.; Sobczyk-Kojiro, K.; Subramanian, P.; Crippen, G. M.; Ramalingam, K.; Woodard, R. W. (1990), Combined use of stereospecific deuteration, NMR, distance geometry, and energy minimization for the conformational analysis of the highly opioid receptor selective peptide [D-Pen^2,D-Pen^5]enkephalin, *J. Am. Chem. Soc.* **112**, 822.

CHAPTER 4

Drug Design

4.1. INTRODUCTION

Conventional drug discovery is a long and expensive process: The average new drug takes 12 years and $230 million to develop. Possible therapeutic agents are first identified by randomly selecting natural products and synthetic chemicals, and screening them for biological activity. When a lead, or a structural class with potential in a therapeutic area, is found, it is modified in an attempt to increase its activity and reduce its toxicity. From a small set of the most active and least toxic compounds one of them might be tested in clinical trials, perhaps reaching the market as a drug. For every pharmaceutical product that reaches the market, nearly 20,000 compounds have been synthesized and evaluated. Thus, methods that could make drug discovery more "rational," and thereby faster, are of enormous commercial interest.

Rational approaches to drug design have profoundly affected lead discovery. Identifying the cause of the disease is the first step. Most diseases, or their symptoms, arise from infection by a foreign organism, or from abnormal cell growth, or from an imbalance of chemicals in the cells. Diseases arising from invasion by microorganisms might be treated by inhibition of a catalytic receptor essential to the microorganism, or by interfering with the microorganism's DNA. Blocking or altering DNA biosynthesis or function also can offer a way to treat abnormal cell growth. A metabolite imbalance could be corrected by drug binding to specific receptors.

The drug receptor, or the site of drug action, is a concept crucial to contemporary drug design. A drug works because its unique geometric pattern of atoms can form attractive intermolecular interactions with a structurally and chemically complimentary receptor site. The lock and key model of receptor–drug interaction, in which the receptor is a rigid negative of the drug, has evolved into the "induced fit" model. The drug still fits into the receptor, but it fits because of structural changes induced in the flexible receptor site by the drug itself.

Receptor families include enzymes (proteins with catalytic properties), hormone receptors (proteins that cause transduction of a chemical messenger), neurotransmitter receptors (synapse proteins involved in impulse triggering), and DNA. Molecules that stimulate a receptor in a way similar to the natural effectors or ligands are called agonists; those that block the effect by binding to the receptor are called antagonists. In rational drug design, antagonists and the natural receptor agonists are often lead compounds.

167

After understanding the pathophysiology of disease processes, the next step in contemporary drug design is to modify the lead compound based on a structural understanding of the relevant drug receptor. Ways to do this by molecular modeling are the subjects of the case studies in this chapter. Here we will only describe the design of protein-binding drugs, deferring discussion of DNA-binding drugs to Chapter 5.

In Section 4.2, the structure of the drug receptor was obtained directly by X-ray crystallography and used to design human immunodeficiency virus (HIV) protease inhibitors as anti-acquired immunodeficiency syndrome (AIDS) drugs. A receptor active site structure can also be determined indirectly by carrying out a comparative analysis of ligands and deducing a receptor active site model. The pharmacaphore model developed for the central nervous system (CNS)-active drugs of Section 4.5 is one example of this method. Here the properties of biologically active ligands were used to deduce the properties of the receptor. Some parts of the molecule were found to be critical for activity, other parts were not. The receptor site was determined by deduction, using 3-D mimicry. In Section 4.3, a conformational structure–activity relationship (SAR) for somatostatin analogs is described. The structure of the peptide was varied, the parts not needed for bioactivity were removed, and some critical parts were replaced with chemically and physically similar groups. Synthesis, bioassays, and modeling were carried out iteratively to design an analog of the polypeptide hormone. A related but more mathematical concept is the quantitative structure–activity relationship (QSAR) given in Section 4.4. The QSAR assumes that the biological activity of a series of compounds is related to one or more physicochemical properties and can be outlined mathematically. This technique was used in the design of a species-specific dihydrofolate reductase inhibitor.

Before we proceed on to the case studies, it is appropriate to mention the limitations of molecular modeling in drug design. Molecular modeling will allow a researcher to visualize and analyze complex molecular systems, to be sure, but the computer will not burp out the best drug. Indeed, there are many examples of designed drugs in search of a good disease. Even if a medicinal modeler were to design a molecule that binds well to a receptor, the researcher must also be concerned about improving absorption, reducing loss through the kidneys, controlling sequestering by nonspecific binding, reducing loss through metabolism, as well as reducing toxicity and undesirable side effects by the original molecule or its metabolites. A great deal of human intervention, iteration, intuition, synthesis, biological evaluation, and creativity is still necessary. Molecular modeling is simply another powerful tool to be used by researchers in the cooperative efforts of chemists, biologists, pharmacokineticists, toxicologists, and drug delivery experts to understand and manipulate nature.

If you compare the complexity of the modeling methods described in the protein chapter to those used in drug design, you will discover the modeling in drug design is somewhat more primitive. In drug design modeling you will not often find computationally expensive explicit waters or dynamics. Instead, you will find qualitative computer graphics, simple force fields, and empirical QSAR. This occurs because the protein modeler and the drug designer have different goals. The protein modelers are often found in academic laboratories, and generally want to define the limits of new theoretical techniques. In contrast, the drug design modelers are frequently found in industrial laboratories, and they are there to create useful drugs that make money; drug design modelers want to understand which drug–receptor interactions are important in the fastest, easiest, way possible.

4.2. X-RAY CRYSTAL STRUCTURE ASSISTED DESIGN OF HIV-1 PROTEASE INHIBITORS

Human immunodeficiency virus is the agent that causes AIDS. Fighting AIDS has proven difficult as the HIV virus develops mutations that make it resistant to zidovudine (AZT), the principal drug used clinically to treat this incurable disease. The limited benefits of AZT has prompted the extensive investigation of other chemotherapeutic agents. Several new strategies for potential therapeutic intervention have been suggested by study of the molecular events critical to the viral life cycle. Among them is the blockade of HIV-1 protease (HPR). The HPR is a virally derived enzyme responsible for the hydrolysis of viral proteins before their assembly as new viruses. Because inactivation of HPR renders the virus noninfectious, inhibition of HPR has been identified as a promising approach in the design of anti-AIDS drugs (Huff, 1990).

Crystallographic work has shown HPR is a nearly C_2 symmetric dimer (Fig. 4.1), composed of two 99 amino acid monomers (Lapatto et al., 1989; Navia et al., 1989). The

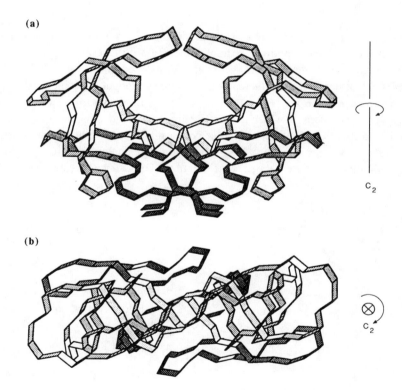

FIGURE 4.1.
Two views of the structure of HPR drawn as a kinked ribbon connecting the positions of α-carbon atoms. (a) The twofold axis relating one monomer to the other is vertical and in the plane of the page. The flaps are at the top of the figure. (b) The twofold axis is perpendicular to the plane of the page. [Adapted with permission from Huff, J. R. (1990), *J. Med. Chem.* **34**, 2305. Copyright © American Chemical Society.]

enzyme is structurally related to other well-characterized cellular aspartic acid proteases such as pepsin and renin. Each monomer of the viral protease contributes one Asp25-Thr26-Gly27 sequence to give an active site with a twofold axis of symmetry. Each HPR subunit also provides an extended β-hairpin loop, called a flap, which enhances the binding of inhibitors.

A valuable strategy for designing enzyme inhibitors is the transition state analog or mimetic approach. The concept, first recognized by Pauling (1948), suggests that a transition state analog should bind tightly to an enzyme active site because it resembles the structure of the transition state for the enzymatic reaction. The hydrolysis of amides catalyzed by proteases proceeds through an intermediate tetrahedral diol species (**4.1**) resulting from the nucleophilic attack at the carbonyl carbon atom of the cleavage site [Eq. (4.1)]:

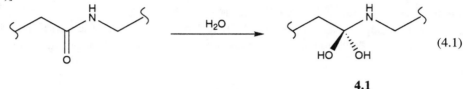

$$(4.1)$$

4.1

Several transition state analogs in which the amide is replaced by a nonhydrolyzable mimetic of the tetrahedral hydrolysis transition state are illustrated in Figure 4.2. These include the hydroxyethylene, hydroxyethylamine, reduced amide, and statine-containing "isosteres." Isosteres are substituents or groups that are about the same size and produce a similar effect in some key parameter; in this case, binding to protease enzymes. The hydroxyethylene isostere, in which the scissile dipeptide bond is replaced by $CH(OH)CH_2$, has been used successfully in the design of potent inhibitors of other aspartic proteases.

In structural studies of peptide-based transition state analog inhibitors bound to HPR, the inhibitor binds in the extended conformation. The exact C_2 symmetry of the HPR

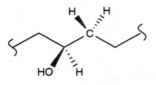

Hydroxyethylene

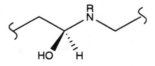

Hydroxyethylamine

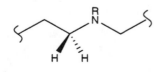

Reduced amide

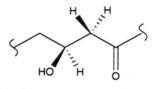

Statine

FIGURE 4.2.
Scissile bond modifications.

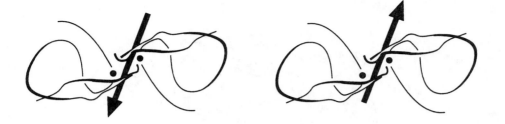

FIGURE 4.3.
Schematic of the topography of HPR complexed with an asymmetric peptide inhibitor. The polarity of the inhibitor (N → C) is represented by the thick arrow. The twofold axis is perpendicular to the page. [Adapted with permission from Jaskolski, M.; Tomasselli, A. G.; Sawyer, T. K.; Staples, D. G.; Heinrikson, R. L.; Schneider, J.; Kent, S. B. H.; Wlodawer, A. (1991), *Biochemistry* **30**, 1600. Copyright © American Chemical Society.]

active site must be broken during the binding of the intrinsically asymmetric peptide inhibitors, see Figure 4.3. Because of the symmetric nature of the active site, there are two possible orientations of an asymmetric inhibitor.

The inhibitor is held in the active site cleft by extensive van der Waals contacts, and by hydrogen bonds to the active site loops and to residues in the flaps. Hydrogen-bond patterns common to hydroxyethylene isostere inhibitors are illustrated in Figure 4.4 for a general peptide-based inhibitor. By convention (Schechter and Berger, 1967), the sub-

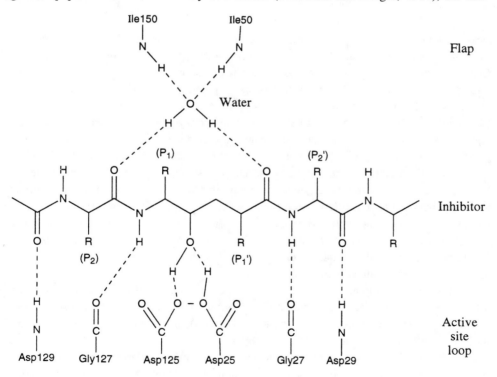

FIGURE 4.4.
Schematic representation of the hydrogen-bonding interactions between hydroxyethylene containing inhibitors and HPR.

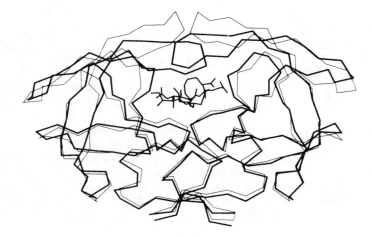

FIGURE 4.5.
Superimposed C_α backbones of the native (thin lines) and inhibited (thick lines) protease dimers. The inhibitor is a peptide with a reduced amide isostere (medium lines). [Reproduced with permission from Miller, M.; Schneider, J.; Sathyanarayana, B. K.; Toth, M. V.; Marshall, G. R.; Clawson, L.; Selk, L.; Kent, S. B. H.; Wlodawer, A. (1989a) Structure of complex of synthetic HIV-1 protease with a substrate-based inhibitor at 2.3 Å resolution, *Science* **246**, 1149. Copyright © 1989 by the AAAS.]

strate residues are called P (for peptide), and the binding subsites S. Cleavage occurs at the scissile bond, in between P_1 and P_1', where P and P′ refer to residues on the NH_2 and CO_2H terminal sides, respectively. Binding to the protease is stereoselective in that strong inhibition requires that the configuration of the tetrahedral $-CH(OH)-$ group of the inhibitor be S. The inhibitor hydroxyl (OH) is positioned between the two active site aspartatic acid residues (Asp25 and Asp125). In the native enzyme there is a water bound at an equivalent position. A second set of hydrogen bonds are formed between the inhibitor amide nitrogen at P_1 and P_2' with the carbonyl oxygen atoms of active site Gly127 and Gly27. In a third pair of interactions, the amide of residue Asp29 (or Asp129) donates a hydrogen bond to the carbonyl oxygen of the inhibitor in position P_2' (or P_3). Very similar hydrogen-bonding networks have been found in the structures of hydroxyethylene isostere complexes with other apartate proteases (Wlodawer and Erickson, 1993). A tightly bound water molecule is positioned between the inhibitor and the leading strands of the two flaps. It is tetrahedrally coordinated to the carbonyls of P_2 and P_1' of the inhibitor and the amide nitrogen atoms of residues Ile50 and Ile150.

X-ray crystallographic work on both native and complexed HPR has shown the binding of the inhibitor introduces substantial conformational changes in the enzyme. Most marked is a very large movement of the flaps (Fig. 4.5), which close down on top of the inhibitor bound at the active site (Miller, 1989a).

There have been two general approaches taken by pharmaceutical companies in the design of HPR inhibitors. The first involves modifying peptide-based inhibitors in conjunction with the transition state analog concept, and the second capitalizes on the unique C_2 symmetry of the HPR active site to design new symmetric drugs. In this section we will study the two strategies, one employed by researchers at Merck and the other by scientists at Abbott Laboratories. Both sets of papers describe the design of HIV-1 protease inhibitors by using molecular modeling in concert with knowledge of enzyme active site structure. We will also discuss the application of structural data base searching to this problem.

The HPR inhibitors ritonavir, designed by Abbott, and indinavir, designed by Merck, were both approved by the FDA in the Spring of 1996. The development of these new AIDS drugs can be traced back to the early research described in the section below.

4.2.1 Peptide-Based Inhibitors

Due to the wealth of information known about inhibitors of human renin (Kleinert et al., 1991), an aspartic protease like HPR, HPR inhibitor discovery approaches are often based on using renin inhibitors as lead compounds. Many potent renin inhibitors have been designed by using the transition state analog or mimetic concept.

As a first step toward identifying promising inhibitors of HPR, researchers at Merck screened a number of known renin inhibitors against HPR (Vacca et al., 1991). The renin inhibitor L-364,505 (**4.2**) contains a hydroxyethylene isostere (BOC is a *tert*-butoxycarbonyl protecting group; Phe is phenylalanine; Leu is leucine; and Ph is phenyl.).

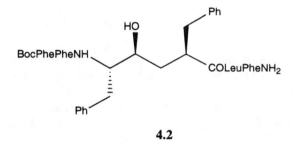

4.2

The compound **4.2** was a potent inhibitor of HPR, but stopped the spread of viral infection in cell culture only at high concentrations.

Modification of **4.2** gave **4.3**.

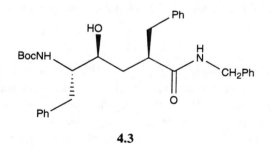

4.3

This smaller peptide-based compound was a weak HPR inhibitor, but it served as a starting point for further modifications (Lyle et al., 1991).

Replacement of the terminal benzylamide group of **4.3** with a series of amides derived from aminobenzocycloalkanes gave new compounds that were evaluated as HPR inhibitors. Compound L-685,434 (**4.4**), containing the novel 1(*S*)-amino-2(*R*)-hydroxyindan P_2' ligand, was a potent HPR inhibitor and showed substantial anti-HIV activity *in vitro* (outside a live organism).

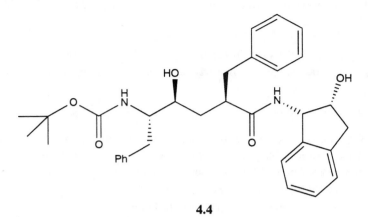

4.4

Lead **4.4** was further modified, and this work was described by Thompson et al. (1992) of Merck, and is discussed below. By using a combination of X-ray, synthetic, and modeling approaches, these researchers developed a new anti-HIV drug with better cell penetration.

METHODS

The inhibitor structures were built using the Merck molecular modeling program advanced modeling facility (AMF) and energy minimized using the Merck molecular force field OPTIMOL. The OPTIMOL force field is a variant of the MM2 force field, differing from MM2 in that partial charges are used on atoms rather than dipoles, and unshared pairs of electrons are absent on certain nitrogen and oxygen atoms. All titratable residues were charged in the calculations with the exception of Tyr59 and one of the pair of catalytic aspartic acids (Asp25).

Graphics visualization and molecular surface calculations were performed using Quanta. Comparison of the modeled and X-ray structures of L-689,502 (**4.5**) was accomplished by aligning the native and inhibited forms of the enzyme using the atoms of the conserved Asp-Thr-Gly motif, not by aligning the conformations of L-689,502.

RESULTS

Although inhibitor **4.4** blocked the spread of HIV-1 in cell culture, the compound suffered from aqueous insolubility. The Merck researchers reasoned that enhancement of cell penetration by **4.4**, and thereby the antiviral potency, could be achieved by adding polar groups to the compound. A search for molecular sites to append the hydrophilic (water-loving) groups was carried out by using molecular mechanics.

When this research was begun, the Merck researchers did not have the coordinates of an X-ray structure of HPR bound to an inhibitor (as in Fig. 4.5). Therefore a model of the complexed HPR enzyme was derived from the X-ray structure of the native (uncomplexed) HPR (Navia et al., 1989). A model of inhibitor **4.4** bound to the HPR active site was derived from the X-ray structures of renin inhibitors bound in the active sites of the aspartic proteases endothiapepsin (Foundling et al., 1987) and rhizopuspepsin (Suguna et al., 1987). With the use of computer graphics, the C and N termini of the hydroxymethy-lene group of the inhibitor were docked to the HPR active site by visual inspection.

This model of **4.4** was then energy minimized in the static (unminimized) native HPR enzyme active site by molecular mechanics (Fig. 4.6) (see color plate). The minimization resulted in two low-energy conformations differing only in the position of the P_1' phenyl group. This was a consequence of modeling the inhibitor complex with the open structure of the static native enzyme active site; as seen in the X-ray structures given in Figure 4.5, in a real HPR-inhibitor complex, the flaps would move and tightly embrace the inhibitor. Clearly, modeling the inhibitor complex by the native enzyme cannot correctly describe the interactions of the inhibitor with the flap. Nonetheless, in this model, for both confor-mations, the P_1' phenyl groups were near the exterior surface of the enzyme active site, pointing toward solvent, suggesting that a polar substituent could be tethered to this ring without interfering with the binding to HPR.

Analogs of **4.4** in which polar hydrophilic groups were tethered to the para positions of the P_1 or P_1' substituents were prepared. The antiviral potency of the inhibitors was examined *in vitro* and found to be improved over the lead compound. Thus the tethering of polar groups to the lead compound did not interfere with inhibitor activity, consistent with the modeling predictions. The optimum enhancement of anti-HIV activity was ob-served for L-689,502 (**4.5**), with a 4-morpholinylethoxy substituent on P_1'.

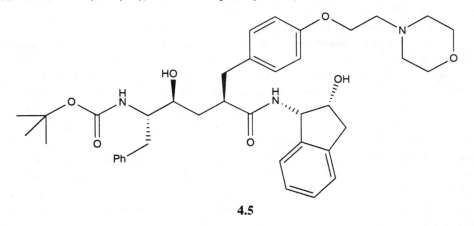

4.5

Interestingly, **4.5** was a less potent enzyme inhibitor (as determined using purified enzyme) compared to the parent compound, consistent with increased cell penetration as the mechanism leading to higher antiviral potency of **4.5** in cell culture. The inhibitor was highly specific against HIV-1 protease; the compound showed no inhibition of the pro-teases human renin, pepsin, or papain. In contrast to AZT and other inhibitors of the reverse transcription process, the protease inhibitor was effective against chronically in-fected cells. Figure 4.7 (see color plate) shows an energy minimized model of **4.5** in the native enzyme site. Mechanics simulation predicted a hydrogen bond between the hy-droxyl of the P_2' indanol and the NH of Asp29. In this model the 2-hydroxyindan group acted as a P_2' carbonyl surrogate; the hydrogen-bonding interaction of the 2-hydroxy group of the indanol with the amide of residue Asp29 (or Asp129) is analogous to the hydrogen bond between the carbonyl oxygen of HPR inhibitors in position P_2' or P_3 illus-trated in Figure 4.4.

An X-ray crystallographic analysis of the enzyme–inhibitor complex of **4.5** bound to HPR was also carried out. The crystal structure of the active site region is illustrated in

Figure 4.8 (see color plate). There were two orientations of the unsymmetrical inhibitor with the nearly symmetric active site: one orientation right to left, the other orientation left to right. There was also disorder in the morpholino group of the P_1' substituent. The hydroxymethylene hydroxyl group of **4.5** was within hydrogen-bonding distance of all the side-chain oxygen atoms of the catalytic Asp25 and Asp125 residues. Two solvent medi- ated interactions were found. A tightly bound water bridging between the BOC and P_1' carbonyl of the inhibitor and the amide nitrogen atoms of Ile50 and Ile150 was observed. A water also bridged the hydroxyl of the P_2' indanol ligand of the inhibitor, the carbonyl oxygen of Gly27, and the side-chain oxygen atoms of Asp29. The latter interaction has been observed in several other X-ray structures of related inhibitor–HPR complexes (Swain et al., 1991). Hydrogen bonding between the inhibitor amide nitrogen at P_1 and P_2' with the carbonyl oxygen atoms of active site Gly127 and Gly27 were also observed. The hydrogen-bonding interaction between the hydroxyl of the P_2' indanol and the NH of Asp29 predicted by molecular modeling was observed in the X-ray structure.

A comparison of the active conformation determined by modeling and by X-ray dif- fraction is shown in Figure 4.9 (see color plate). The modeled structure of **4.5** was in good agreement with the X-ray structure. The only major difference between the two structures was the orientation of the *p*-4-morpholinylethoxy substituent on the P_1' phenyl. This dif- ference seems to be mostly due to a change in the position of the side chain of Arg8 between the native and inhibited enzyme structures.

In summary, this work modeled the binding of a peptide-based drug (**4.4**) to the HPR enzyme in order to determine good sites of attachment for polar groups by molecular mechanics. The X-ray structure of the native (uncomplexed) HPR was used in the simula- tion. The modeling results lead to the synthesis and *in vitro* antiviral screening of several derivatives. The structure of one of these derivatives (**4.5**), a potent antiviral compound, was modeled in the active site, again using the X-ray coordinates of the native HPR en- zyme. The simulated structure of the inhibitor was found to be in good agreement with the X-ray structure of the same inhibitor as bound to the HPR active site. There were some differences between predicted and experimental structures, probably due to the large approximation in using the native HPR active site to model the structure of the inhibited active site. The Merck work points out that it is necessary to accurately model the confor- mational changes in the protein structure resulting from ligand binding, particularly when such structural accommodations are significant, as in HPR.

4.2.2 C_2 Symmetric Inhibitors

The development of peptide-based inhibitors into effective drugs has been slowed by the inherently poor pharmacalogic properties of peptides, such as low absorption after oral ingestion, rapid proteolytic breakdown in the bloodstream and gastrointenstinal tract, and rapid excretion through liver and kidneys. Design of synthetic HPR inhibitors with some- what less peptide character was described in three papers by researchers at Abbott labora- tories (Erickson et al., 1990; Kempf et al., 1990; Hosur et al., 1994). Here knowledge of the C_2 symmetry of the HPR active site was exploited to create C_2 symmetric inhibitors. This symmetry-based design strategy was expected to lead to inhibitors with high speci- ficity for the viral HPR, because the active sites of mammalian aspartic proteases are only approximately symmetric. Selective and potent HPR inhibitors with anti-HIV activity did indeed result, as described below.

METHODS AND RESULTS

When the Abbott researchers (Erickson et al., 1990) began this work, no HPR crystal structure was known, though the C_2 symmetry characteristics of the enzyme had been proposed by Pearl and Taylor (1987) based on sequence homology and the structures of other aspartic proteases. Erickson and co-workers (1990) modeled HPR by using the X-ray crystal coordinates of the Rous sarcoma virus (RSV) protease (Miller et al., 1989b). Like HPR, the RSV protease is a viral aspartic protease and is a highly twofold symmetric dimer. The structure of the reduced peptide inhibitor D-His-Pro-Phe-His-Phe-Ψ[CH$_2$−NH]-Phe-Val-Tyr was derived from the crystal structure of the rhizopuspepsin–inhibitor complex (Suguna et al., 1987). (The same X-ray structure was used in the Merck work to develop a model of **4.4** bound in the HIV protease active site.) The reduced peptide inhibitor was then docked into the active site of RSV protease by superimposing the active sites of rhizopuspepsin and the viral protease. One-half of the inhibitor was deleted; the P_1' portion of the reduced peptide was truncated after the reduced methylene carbon of the P_1 Phe. Finally, a C_2 symmetry operation was applied to the remaining structure of the inhibitor. Depending on whether the orientation of the symmetry axis was through the carbonyl carbon, or through the middle of the sissile bond, two distinct pseudosymmetric (**4.6**) or symmetric (**4.7**) core units were generated (see Fig. 4.10).

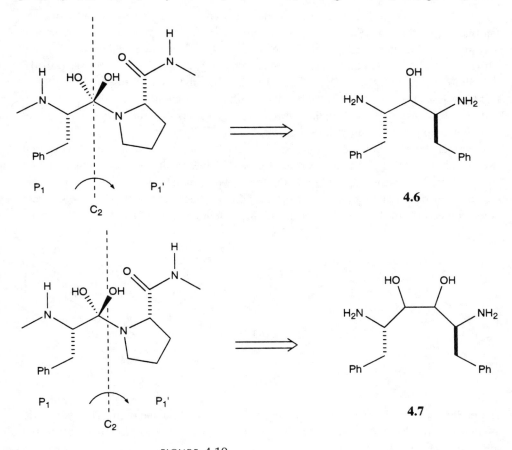

FIGURE 4.10.
Design of C_2 symmetric HPR inhibitors.

Pseudosymmetric molecule **4.6** was a weak HPR inhibitor and did not exhibit significant anti-HIV activity *in vitro*. In order to increase the activity of the pseudosymmetric inhibitors, the molecule was modified by the symmetric addition of NH_2-blocked amino acids. These efforts culminated in the synthesis of A-74704 (**4.8**), which had a carbobenzyloxy–valine attached to both ends of the core.

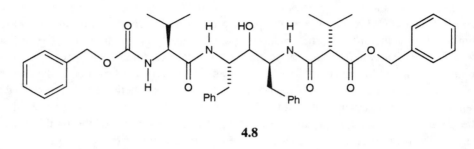

4.8

The HPR inhibitory activity of **4.8** was evaluated, and the molecule was found to be both potent and highly specific, with no significant inhibition of other cellular aspartic proteases. The compound inhibited HIV infection *in vitro*, and was markedly resistant to proteolytic digestion.

An X-ray structure analysis of the HPR–**4.8** complex was carried out, showing the inhibitor was bound to HPR in a very symmetric fashion. The hydrogen-bonding pattern was similar to that illustrated in the scheme of Figure 4.4: Hydrogen bonding was observed between the inhibitor and the catalytic Asp carboxylates, the Asp29 NH and Asp129 NH, the Gly27 CO and Gly127 CO. Additional hydrogen bonds were formed by Gly48 CO and Gly148 CO. A water was tetrahedrally coordinated to the amide NH atoms of Ile50 and Ile150 of the flaps and to the carbonyl oxygen atoms of the P_2 valyl groups of A-74704, analogous to the water in Figure 4.4.

Further experimental work was also carried out using the symmetric **4.7** core unit (Kempf et al., 1990). *Tert*-butoxycarbonyl protected intermediates, which contained no amino acids, were respectable inhibitors in their own right, and were more effective than the corresponding inhibitors derived from lead **4.6**. Replacement of BOC in the **4.7** core units with carbobenzyloxy–valine yielded highly potent inhibitors (**4.9**).

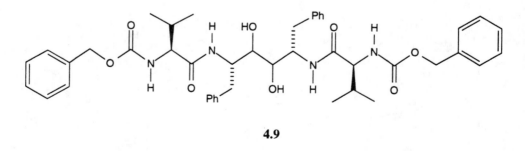

4.9

The inhibitors were highly specific, showing no inhibition of human renin. The symmetric inhibitors were also evaluated for anti-HIV activity, and found to effectively block the spread of HIV *in vitro*.

Further modification of **4.9** gave **4.10** (Hosur et al., 1994).

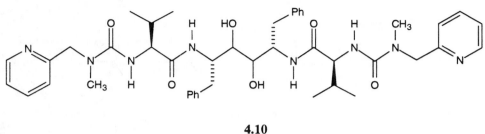

4.10

The (R, S) diol diastereomer of **4.10** (A-77003) had favorable antiviral properties and solubility, and was selected for clinical trials.

To summarize, the first of the Abbott papers (Erickson et al., 1990) used simple molecular graphics to dock a reduced peptide inhibitor into the active site of HPR, as modeled by the active site of RSV protease. An artificial symmetric inhibitor was generated using the resulting coordinates, and was observed to have nearly equivalent interactions with the enzyme subsites in the simulation. The inhibitor served as a starting point for further synthetic and crystallographic work. (Kempf et al., 1990) and (Hosur et al., 1994). Here, just as in the development of insulin analogs (Section 3.4), rather simple modeling efforts have been used to supplement the more important structural knowledge to design new drugs.

4.2.3 Non-Peptide Inhibitors by Structural Data Base Searching

Three-dimensional structural data base searching attempts to generate biologically active compounds by searching large chemical data bases for structures that have a particular 3-D geometry. The 3-D requirements for the active compound might be obtained from a pharmacaphore model, an X-ray structure of the receptor, or an X-ray structure of the inhibitor–receptor complex. This drug discovery method often assumes that bioactivity is primarily a function of size and shape rather than specific chemical interactions. Here we describe the application of structural data base searching to non-peptide HPR inhibitor development by researchers at the School of Pharmacy at the University of California San Francisco.

METHODS AND RESULTS

DesJarlais et al., (1990) created a negative image of the HIV-1 protease active site cavity using the X-ray structure of native HPR. This image was compared for steric complimentarity with a subset (10,000) of the structures found in the Cambridge Structural Database. (The Cambridge Structural Database is the repository of X-ray crystallographic data for small organic and organometallic molecules.) The molecules that best fit the protease active site cavity were further evaluated by interactive graphics using chemical criteria such as potential to form hydrogen bonds with the active site.

One interesting candidate was the molecule haloperidol (**4.11**).

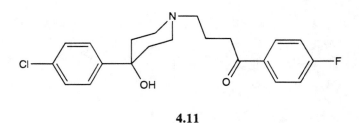

4.11

4.11 was evaluated for inhibition activity against HPR, and found to be active and highly selective. Unfortunately, the compound is also a known antipsychotic agent and is toxic at high concentrations. Although **4.11** is not useful as a treatment for AIDS, it holds promise as a nonpeptide lead compound.

Replacement of the ketone of haloperidol with a thioketal ring gave **4.12**.

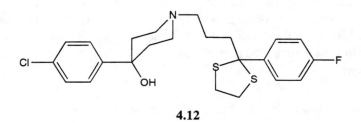

4.12

The X-ray structure of **4.12** bound to HPR was recently reported by Stroud and co-workers (Rutenber, 1993). The orientation of the inhibitor was quite different than that predicted from the structural data base search: **4.12** binds 4.8 Å away and is rotated by 79°. Furthermore, the bound conformation of the modified haloperidol was not the same as the structure of haloperidol from the data base.

Just as in the previous two HPR-drug case studies, which used the structure of uncomplexed, native HPR (or RSV protease) to model the inhibited active site, native HPR was used to generate the active site cavity of inhibited HPR for the data base search. This is incorrect, though expedient, given that an X-ray structure of the inhibited HPR enzyme was unavailable.

Data base searching promises to be a popular drug discovery method, but there are serious problems inherent to the technique, and they are amply illustrated in this case study. The data base searches are based on shape, rather than chemical complementarity, which seems an unreasonable simplification. In addition, the structures in conventional data bases are rigid, having a single conformation. Real molecules are conformationally flexible, and it is known that the biologically active form is not necessarily the lowest energy form. Thus a 3-D data base search will likely miss many possible bioactive molecules, and suggest nonbioactive molecules.

4.3 CONFORMATIONAL ANALYSIS IN HORMONE DRUG DESIGN

4.3.1 Background

Hormones are chemical messengers that regulate communication between cells. Binding between hormone and receptor leads to a conformational change in the receptor, which in turn causes transduction of a signal, causing formation of another chemical messenger, leading to a biological response. Characterizing the structural basis of the hormone–receptor interaction is of prime importance to rational hormone analog drug design. The early lock and key model, which hypothesized that the shape of a hormone was complimentary to the shape of its receptor, has been superseded by the ''induced fit'' model (Koshland, 1958). In this model a flexible hormone binds to its dynamic receptor, inducing complimentarity, relaxing the requirement for exact preexisiting fit between hormone and receptor (Williams, 1977). The induced-fit type mechanism has led to the proposal that incorporation of conformational restrictions in a hormone might result in an increase in affinity because there would be little conformational entropy loss upon binding to the receptor (Rizo and Gierasch, 1992). Such constrained molecules might also be more receptor selective as these analogs may interact favorably with one type of receptor but not others.

   In the absence of direct information about the bioactive form(s) of a flexible hormone, or structure of the receptor(s), a classical approach to hormone drug design is to synthesize and biologically test a great many analogs. The resulting structure–activity information is then used to design even more analogs, which are then screened. These synthetic and biological studies are carried out in an iterative fashion, in order to ultimately draw conclusions about the bioactive form, culminating in the design of a useful drug. The objective of rational drug design is to reduce the number of derivatives that must be synthesized and assayed to ultimately yield a drug candidate.

   By combining molecular modeling and systematic structure–activity studies of the peptide hormone somatostatin (SMS) and its analogs, several research groups successively eliminated those amino acids that were not required for biological activity of SMS and increased the rigidity of the molecule. This work, described in Section 4.3.2, resulted in the development of the clinically useful drug Sandostatin.

4.3.2 Somatostatin Analogs

Somatostatin, first isolated in 1973 by researchers at the Salk Institute (Brazeau, 1973), is a 38-membered cyclic tetradecapeptide containing a disulfide bridge (**4.13**).

H$_2$N–Ala 1 – Gly 2 – Cys 3 – Lys 4 – Asn 5 – Phe 6 – Phe 7 – Trp 8 – Lys 9 – Thr 10 – Phe 11 – Thr 12 – Ser 13 – Cys 14 – COOH

4.13

This hormone has a broad profile of gastrointestinal and endocrine effects (Veber and Saperstein, 1979): In the stomach, SMS inhibits the secretion of hydrochloric acid, gastrin, and pepsin. In the pituitary gland, SMS inhibits the secretion of growth hormone, and in the pancreas, SMS suppresses the secretion of insulin and glucagon. Because of its discovery in nerve cells, SMS may also play a role as a neurotransmitter. In view of its physiologic functions, native SMS might be expected to have useful therapeutic properties. However, SMS has a very short duration of action after intravenous injection (3 min) due to enzyme degradation, and it interacts simultaneously with several organ systems, making its clinical use impractical. The design of SMS analogs with improved specificity and stability is the subject of this section.

METHODS AND RESULTS

Soon after the discovery of SMS, the Salk research group synthesized and evaluated the biological activity of the first SMS analogs (Rivier et al., 1975; Brown et al., 1977). Experimental conformational studies of the analogs were carried out by Holladay et al. (1977) and (Holladay and Puett, 1976), leading to a proposal for the bioactive conformation. Researchers at Merck continued the structure–activity work on SMS analogs. Veber et al. (1978) used these results to propose that the biologically active form of SMS had residues Trp8-Lys9 at the corners of a type II' β turn.

A β turn is a segment of four amino acids (**4.14**; residues i to $i + 3$) that occurs when a peptide chain reverses direction (Wilmot and Thornton, 1988). A carbonyl of the ith residue is aligned to form an intrachain hydrogen bond with the amide hydrogen at residue $i + 3$. Beta turns have been classified as Types I, II, and III according to the torsion angles of the corner residues at positions $i + 1$ and $i + 2$ (see Table 3.1). Such turns are ideal sites for receptor recognition because they present amino acid side chains in a highly accessible arrangement around a compact folding of the peptide backbone.

Heterochiral pairs (DL or LD) of amino acids are known to stabilize β-turn structures (Rose et al., 1985); thus a β-turn conformation in the bioactive form of SMS was supported by finding increased potency of SMS analogs containing the unnatural amino acid D-tryptophan at position 8 instead of L-tryptophan. The proposed turn structure was also

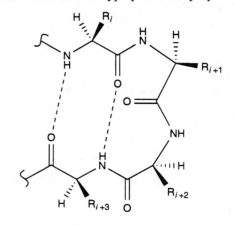

4.14

consistent with the loss of bioactivity on replacement of Thr10 by proline (in which case
the intramolecular hydrogen bond between residue i and $i + 3$ would be lost).

Interestingly, NMR studies of natural SMS in both H_2O and DMSO suggested the
cyclic peptide was highly flexible in solution (Deleuze and Hull, 1982; Buffington et al.,
1983). No evidence was found for a β turn. The biologically active conformation of SMS
is thus not the preferred conformation in solution. A word of caution comes from the
solution structure: Conformational studies of flexible peptides, whether by theoretical or
experimental methods, are of limited value in determining the conformation at the receptor. (See Section 3.4 for further discussion of the difficulties in modeling flexible peptides). Modeling of flexible peptides must be guided by extensive structure–activity data
in order to be used productively in rational drug design.

By using the conformational constraints of the proposed β turn and results from other
structure–activity studies, Veber and co-workers (1978) of Merck constructed a computer-
projected 3-D model of an analog of somatostatin (**4.15**).

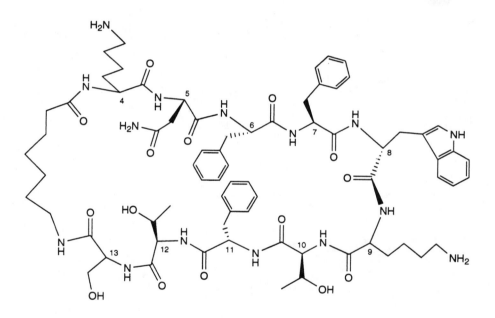

4.15

For future ease in preparing new analogs, the sulfur atoms of the disulfide bridge of SMS
were replaced by methylenes in the analysis. The backbone coordinates for residues 6–11
of SMS were obtained from a molecular mechanics minimization of the decapeptide
LHRH (Momany, 1976).

Several interesting features were observed in the conformational model of the dicarba
analog (**4.15**) of SMS. The proximity of the β-carbon atoms of Asn5 and Thr12 suggested
that the distance between these methylene groups was correct for a disulfide bridge. In
addition, the β-carbon atoms of Ph6 and Phe11 also appeared to be close; a hydrophobic

bonding of the two aromatic rings seemed possible.

By using the results of this model, the Merck researchers synthesized two new bicy-clic analogs of SMS (**4.16** and **4.17**).

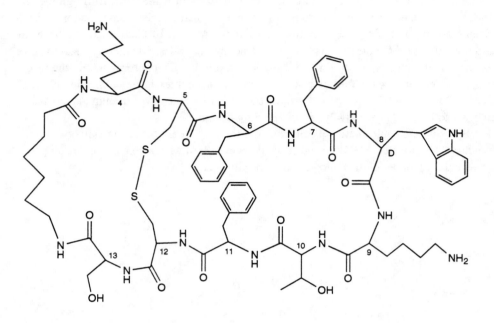

4.16

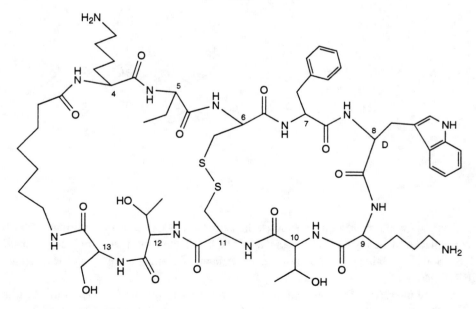

4.17

In Compound **4.16**, a disulfide link between the residues 5 and 12 of **4.15** was created by replacing the Asn5 and Thr12 by Cys. The Trp8 was replaced by a D-Trp 8 to stabilize a β turn. Compound **4.17** contained a disulfide link between the residues 6 and 11, and L-Trp was replaced by a D-Trp residue. The Asn 5 of **4.15** was also replaced by α-aminobutyric acid, a modification that previously has been found to have little effect on the biological potency. The biological activity of both analogs **4.16** and **4.17** was evaluated. High biological activity for inhibition of glucagon, insulin, and growth hormone was observed for both compounds.

Veber et al. (1979) reported on the further modification of analog **4.16**. Based on the observation that residues 4 and 13 of SMS could be deleted without loss of activity, the amino acids to the left of the disulfide bridge were replaced by an aminoheptanoic acid (Aha). In addition, the disulfide bridge was replaced by methylene groups to make another bicyclic analog. The SMS model (**4.15**) suggested the close proximity of residues 6 and 11 (see analog **4.17**), so the Phe residues were replaced by cystines to give a disulfide bridge. Analog **4.18** is illustrated below.

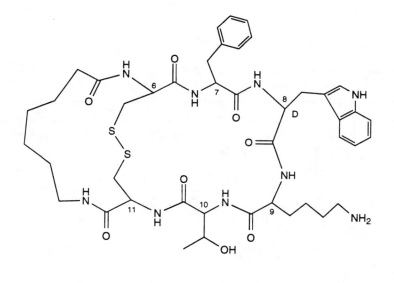

4.18

Improved biological activity for inhibition of glucagon, insulin, and growth hormone was observed for this bicyclic SMS analog. Analog **4.18** was also found to be more resistant to enzymatic cleavage than SMS.

These experiments all supported the hypothesis that amino acids 7–10 of SMS contained the essential receptor-binding elements; the remainder of the amino acids were there only to constrain those four amino acids into the correct β-turn conformation.

Bauer et al. (1982) reported on their search for a clinically useful growth hormone inhibitor. As a starting point the Sandoz researchers carried out structure–activity studies on the SMS analog **4.19**, a molecule similar to the Merck analog **4.16**:

Cys – Phe 7 – D –Trp 8 –Lys 9 –Thr 10 – Cys

4.19

Although **4.19** contained the essential four residue fragment elucidated by Veber and co-workers (1978), the cyclic analog had less than 1000th of the activity of SMS itself for *in vitro* growth hormone inhibition.

By using both physical and computer-assisted model building, Bauer et al. (1982) designed the N-terminal D-phenylalanyl analog **4.20**:

D – Phe – Cys – Phe 7 – D –Trp 8 –Lys 9 –Thr 10 – Cys

4.20

The model indicated that the aromatic side chain of the D-phenylalanyl residue might occupy some of the conformational space available to Phe6 in the parent SMS, and also protect the disulfide linkage from enzymatic attack. The analog had about 4% of the activity of SMS *in vitro* and was about equally active *in vivo* (in the living organism).

Addition of the C-terminal amino acid Thr(ol), the alcohol derivative of threonine, to **4.20** gave Compound SMS 201–995 (**4.21**).

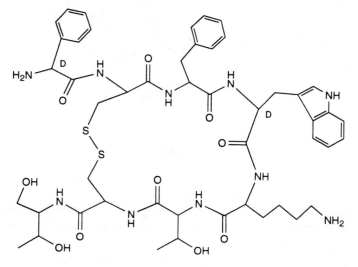

4.21

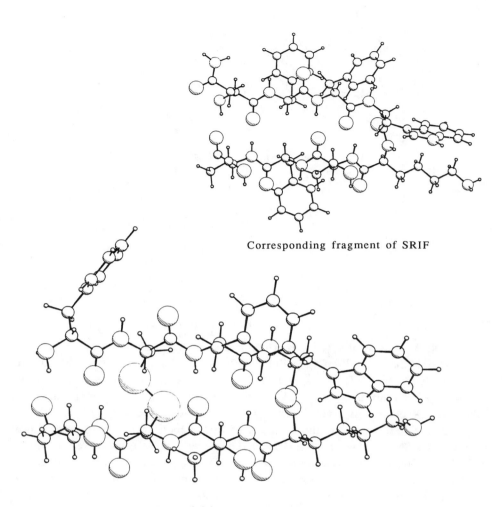

Corresponding fragment of SRIF

4.21 SMS 201-995

FIGURE 4.11.
Comparison of a possible low-energy conformation of **4.21** with the corresponding fragment of the parent SMS (inset). [Reproduced from *Life Sciences,* **31**, Bauer, W.; Briner, U.; Doepfner, W.; Haller, R.; Huguenin, R.; Marbach, P.; Petcher, T. J.; Pless, J., A very potent and selective octopeptide analogue of somatostatin until prolonged action, 1133. Copyright © 1982 with kind permission from Elsevier Science Ltd., The Boulevard, Langford Lane, Kidlington OX5 1GB, UK.]

The Thr(ol) was thought to take the place of the corresponding Thr12 in the native hormone (see Fig. 4.11). *In vivo* studies of **4.21** by the Sandoz researchers (Bauer et al, 1982) demonstrated that the SMS analog was up to 70 times more active than SMS itself for the inhibition of growth hormone release, and inhibited gastric secretion about as effectively as SMS. The peptide was more selective in the inhibition of secretion of growth hormone than that of insulin. Most importantly, the SMS analog was highly resistant to enzyme degradation and thus was much longer acting than the native peptide.

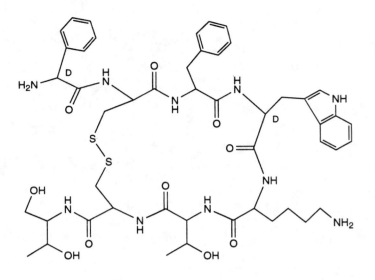

4.21

The NMR analysis of **4.21** in water (Wyants et al., 1985) showed that the molecule was more rigid than SMS itself, but still rather flexible, and did not have β-turn structure. In DMSO, however, NMR studies of **4.21** were in favor of a predominant conformation with a Type II′ β turn involving residues Phe7–Thr10 (Wyants et al., 1988), supporting Veber and co-workers' (1978) original proposal on the bioactive conformation of SMS.

Since the 1982 report on the synthesis of **4.21**, hundreds of *in vitro* and clinical studies of this SMS analog have been carried out (Moreau and DeFeudis, 1987; Lamberts et al., 1991). The drug has been approved by the FDA for clinical use in the treatment for the control of symptoms such as severe diarrhea in patients with certain types of intestinal tumors (Lamberts, 1987). Although it does not cure the tumor, it controls the hypersecretion of hormones by them and improves the patient's quality of life. The drug is also beneficial in the medical management of acromegaly (excessive growth of nose, ears, chin, hands, and feet due to hypersecretion of growth hormone by pituitary), certain pancreatic tumors, and AIDS related diarrhea (Gorden, 1989). Somatostatin analog **4.21** is marketed by Sandoz under the brand name Sandostatin, with the generic name of octreotide.

To summarize, this section described systematic structure–activity studies of somatostatin analogs, guided in part by the results of computer modeling work. The amino acids that were not required for biological activity of somatostatin were successively eliminated, and the rigidity of the molecule was increased by cyclization. The work resulted in a commercial peptide drug.

4.4. QUANTITATIVE STRUCTURE–ACTIVITY RELATIONSHIP

4.4.1 Background

Quantitative structure–activity relationship is one of the oldest forms of rational, empirical drug design (Martin, 1981). At its heart is the assumption that the biological activity of a chemical compound is determined by its physical properties (Osman et al., 1979). In general mathematical terms, Eq. (4.2), QSAR asserts,

$$\log\left(\frac{1}{C}\right) = b_0 + \sum_i b_i f(D_i) \qquad (4.2)$$

where C is the concentration of a compound needed to elicit a biological response, b_0 is a constant, b_i are coefficients, and D_i are the molecular descriptors or physical properties. Because there are normally more equations than coefficients, the values of the coefficients are determined by least-squares multiple regression analysis to find the best values for b_i.

The most popular QSAR method was introduced in the early 1960s by Hansch et al., (1962) and Hansch and Fujita (1964). The Hansch analysis is based on the assumption that the effect of substituents on the strength of interactions between the drug and its receptor is an additive combination of the effects of substituents on simpler model inter-molecular interactions. One widely used equation in QSAR assumes that the physical properties responsible for biological activity can be separated into hydrophobic (π), electronic (σ), and steric volume effects (MR), as shown in Eq. (4.3):

$$\log\left(\frac{1}{C}\right) = a\pi + b\pi^2 + \rho\sigma + m(\text{MR}) + c \qquad (4.3)$$

The π, σ, and MR parameters are a function of the substituent, and independent of receptor; they are determined from spectroscopy, and by quantifying the effect of substituents on the rates or equilibria of simple model organic reactions. The values of a great many substituent parameters have been established and collected in data bases. The a, b, ρ and m coefficients are a function of the specific receptor, and are determined in the QSAR work.

An attractive feature of QSAR is that it requires no knowledge of the structure of the active site of a receptor, or even the structure of the inhibitors, although the terms derived in the QSAR equations often bear a relationship to the static structure of the receptor active site (Hansch and Klein, 1986). The QSAR method is inexpensive and easy to use, and it can quantify differences in binding free energy of less than 0.5 kcal/mol, a feat not easily accomplished by more sophisticated theoretical techniques.

4.4.2 Dihydrofolate Reductase Inhibitors

In this section, we will describe QSAR studies of inhibitors of the enzyme dihydrofolate reductase (DHFR), which predicts new benzylpyrimidine antibacterials and explains the selectivity of the popular antibiotic trimethoprim toward the bacterial enzyme in an intuitively satisfying way.

Dihydrofolate reductase is an enzyme crucial for the biosynthesis of DNA. The enzyme catalyzes the reduction of dihydrofolic acid to tetrahydrofolic acid using nicotinamide adenine dinucleotide phosphate (NADPH) as an electron donor (Fig. 4.12). Inhibition of DHFR results in cessation of DNA synthesis and in cell death.

There has been enormous effort by medicinal chemists to find new DHFR inhibitors, or antifolates, as chemotherapeutic agents, stimulated by the 1949 discovery of the clini-

METHOD

The apparent inhibition constants K_i for reduction of dihydrofolic acid were experimentally measured for a series of 2,4-diamino-5-(substituted-benzyl)pyrimidines (**4.27**) for *L. casei*, *E. coli*, and chicken liver DHFR in the presence of NADPH.

As indexes of the hydrophobicity of substituents, the hydrophobic constants π, were derived from the relationship in Eq. (4.4),

$$\pi = \log P_X - \log P_H \tag{4.4}$$

where P_H is the partition coefficient of the unsubstituted reference compound (benzene) and P_X is the partition coefficient of the substituted compound, in an octanol–water system. The partition coefficient is determined experimentally by placing a compound in a separatory funnel with 1-octanol and water, and determining the concentration of the compound dissolved in each layer after mixing. For a nonionizable compound, Eq. (4.5) is used to derive P:

$$P = (\text{compound})_{\text{octanol}}/(\text{compound})_{\text{water}} \tag{4.5}$$

The values of π for this work were taken from the Pomona College Medicinal Chemistry Databank. A positive value of π indicates the substituted compound has higher hydrophobicity than the benzene reference, and a negative value indicates the compound is less hydrophobic than benzene.

The electronic properties of the substituents were evaluated in terms of Hammett's σ, given in Eq. (4.6),

$$\sigma = \log K_X - \log K_H \tag{4.6}$$

where K_H is the ionization constant of the unsubstituted reference compound (benzoic acid) in water and K_x is that of a substituted benzoic acid, obtained from the Pomona College Medicinal Chemistry Databank. A positive sign with σ indicates electron withdrawl by the substituent, a negative value denotes electron donatation by substituent.

The molar refractivity, which describes the volume and electronic polarizability of a substituent, was given by Eq. (4.7),

$$MR = \left[\frac{(n^2 - 1)}{(n^2 + 2)}\right]\frac{MW}{d} \tag{4.7}$$

where n is the refractive index for the sodium D line, MW is the molecular weight, and d is the density of the compound. Since n does not vary much for most organic compounds, MR is primarily a measure of volume. Molar refractivity does not contain significant information on the geometry of the substituent, and so it cannot graphically pinpoint steric problems. A negative sign with MR in the QSAR indicates a negative steric effect by the substituent, and a positive value is thought to indicate contact between a polar receptor and substituent. Molar refractivity also measures polarizability and may reflect dipole–dipole interactions at the receptor site.

The best QSAR correlations resulted when the meta substituents were analyzed based on the assumption that the DHFR enzyme has a hydrophobic meta 3-space and a hydrophilic meta 5-space, and that phenyl ring rotation could occur, allowing hydrophilic meta substituents to align with hydrophilic space, and hydrophobic meta substituents to align with hydrophobic space, maximizing the binding potential.

RESULTS

The QSAR for the inhibitory effects of benzylpyrimidines, **4.27**, in *E. coli* DHFR was based on 68 compounds *(nc)*, and is given in Eq. (4.8).

$$\log\left(\frac{1}{K_i}\right) = 0.95MR_5' + 0.89MR_3' + 0.80MR_4 - 0.21MR_4^2 +$$

$$1.58\pi_3' - 1.77 \log(\beta_3 \times 10^{\pi_3'} + 1) + 6.65 \quad (4.8)$$

$$nc = 68 \quad r = 0.890 \quad s = 0.290 \quad MR_4^0 = 1.85 \quad \pi_3^0 = 0.73 \quad \beta_3 = 1.50$$

The apparent inhibition constant is K_i, the correlation coefficient is r, and the standard deviation is s. The correlation coefficient is a measure of how much of the variation in the $\log\left(\frac{1}{K_i}\right)$ is explained by the QSAR equation, or put in another way, it is a measure of how correct the QSAR is in predicting $\log\left(\frac{1}{K_i}\right)$ for a compound. The standard deviation is a measure of precision, or spread, of the data.

The parameter MR (scaled by 0.1) is the experimental molar refractivity of the substituents in the corresponding positions in the benzyl ring of **4.27**, and is primarily a measure of volume or bulk. These terms were clearly the most important for this bacterial enzyme. Positive values of the MR coefficient, as observed in Eq. (4.8), implied that the larger the substituents the stronger the binding. This increase in binding was thought to result from contact of inhibitor with a polar receptor surface. The primes on MR_5 and MR_3 indicated that these terms were truncated; the maximum value of MR' for any mono-substituent at position 3 or 5 was 0.79, regardless of the actual value of MR. Because MR is primarily a measure of substituent volume, truncation implied parts of groups (having MR' > 0.79) do not effectively contact the enzyme. Most of the activity enhancement of the meta substituents came from a positive steric effect for the first one or two atoms attached to the benzene ring. The activity of the inhibitors depended parabolically on the size of the 4 substituent. An optimum bulk for para substituents was found to be $MR_4^0 = 1.85$.

The contribution of hydrophobicity, or π, to inhibitory potency at position 3 was accounted for in Eq. (4.8) by using the bilinear model of Kubinyi (1977) [Eq. (4.9)].

$$\log\left(\frac{1}{K_i}\right) = a\pi_3' - b \log(\beta_3 \times 10^{\pi_3'} + 1) \quad (4.9)$$

The parameters a (= 1.58), b (= 1.77), and β_3 (= 1.50) were evaluated by an iterative procedure. This model showed that potency first rose linearly with a slope of 1.58 to π_3^0 = 0.73 (the optimum value of π_3), and then decreased linearly with a slope of -0.19; (1.58–1.77). After reaching the optimum value of 0.73, or the breaking point, the more hydrophobic substituents had essentially no effect. The flat slope after the breaking point indicated that larger substituents projected beyond the enzyme in the aqueous phase. There was a slight collinearity between π_3' and MR_3'; thus in 3-space, steric and hydrophobic effects are not independent of each other.

In conclusion, the bulk of the initial atoms of the substituent adjacent to the phenyl ring, as given by the truncated $MR_{3,5}$, was crucial in the inhibition of the *E. coli* DHFR. Hydrophobic effects as given by π, except in the 3 position, were only of modest importance.

The corresponding QSAR for *L. casei* DHFR is given in Eq. (4.10):

$$\log\left(\frac{1}{K_i}\right) = 1.24\,MR_4' + 0.52MR_3' + 0.42MR_5 - 0.13MR_5^2 + 0.46\pi_4$$
$$+ 0.31\pi_3' - 0.92\log(\beta_4 \times 10^{\pi_4} + 1) - 0.71\log(\beta_3 \times 10^{\pi_3'} + 1) + 5.45 \quad (4.10)$$

$$nc = 65 \qquad r = 0.894 \qquad s = 0.245 \qquad \beta_4 = 0.032 \qquad \beta_3 = 0.037$$
$$\pi_4^0 = 0.49 \qquad \pi_3^0 = 1.33 \qquad MR_5^0 = 1.66$$

The QSAR for *L. casei* DHFR [Eq. (4.10)] was more complex than for *E. coli* DHFR [Eq. (4.8)], but the results were qualitatively similar for the two bacterial enzymes. As with *E. coli*, the positive steric effect of substituents was critical in the inhibition of the *L. casei* DHFR, and hydrophobic effects, except in the 3 position, were only of modest importance. The optimum hydrophobicity of the 3 substituents for *E. coli* was smaller than that of *L. casei*—0.73 versus 1.33, suggesting a smaller binding region for *E. coli* DHFR.

The QSAR for chicken liver DHFR is given in Eq. (4.11):

$$\log\left(\frac{1}{K_i}\right) = 0.39\pi_3 + 0.44\pi_4 - 0.75MR_5 + 0.44\sigma \times$$
$$1.04\log(\beta_3 \times 10^{\pi_3} + 1) + 0.37\pi_5 - 0.32\log(\beta_4 \times 10^{\pi_4} + 1) + 4.70 \quad (4.11)$$

$$nc = 65 \qquad r = 0.906 \qquad s = 0.207 \qquad \pi_3^0 = 2.45 \qquad \pi_4^0 \approx 3.00$$
$$\log\beta_4 = 0.66 \qquad \beta_3 = 0.0020$$

The QSAR for the vertebrate chicken DHFR was the reverse of the two bacterial enzymes: The most important effect was hydrophobic, in positions 3 and 4; the large negative MR_5 term (the larger the substituent, the weaker the binding) completely counterbalanced the π_5 term. Thus, given the option of binding in 3- or 5-space, a single meta substituent appeared to bind in hydrophobic 3-space regardless of whether or not the substituent is hydrophobic or hydrophilic. Electronic factors also came into play as evidenced by the Hammett constant σ in the QSAR: Increased electron density in the phenyl ring led to improved potency.

Hydrophobicity, or π, is a rough measure of the desolvation of the inhibitor. A coefficient of 1 suggests binding in a pocket, while a coefficient of 0.5 suggests binding of the substituent to a flat surface requiring only partial desolvation. Nonlinear functions of π usually indicate intermediate solvation, and suggest limited binding space at the enzyme active site. The initial slopes of π_3, π_4, and π_5 in Eq. (4.11) were about 0.5, suggesting a partial desolvation of the inhibitor on the enzyme surface. The bilinear equations for

TABLE 4.1. Potencies and Selectivity Indexes of 2,4-Diamino-5-(substituted-benzyl)pyrimidines

X	Potencies [$\log\left(\frac{1}{K_i}\right)$]		Selectivity Indexes	
	E. coli	Chicken	E. coli	L. casei
3,4,5-(OMe)$_3$ (**4.26**)	8.08[a]	3.98[a]	4.10[b]	2.90
3-C$_2$H$_5$, 4,5-(CH$_2$OH)$_2$	7.96[c]	3.58[d]	4.28[b]	3.13
3-C$_2$H$_2$, 4-CH$_2$OH, 5-NHCOMe (**4.28**)	8.02[c]	3.15[d]	4.87[d]	3.36

[a] Observed values.
[b] Calculated with $\log(1/K_i)$ of E. coli $-$ $\log(1/K_i)$ of chicken.
[c] Calculated with Eq. (4.8).
[d] Calculated with Eq. (4.11).

hydrophobicity of positions 3 and 4 are revealing: For π_3, the potency of the inhibitor first rose linearly to $\pi_3^0 = 2.45$ and then decreased linearly with a slope of -0.65; (0.39–1.04). The strongly negative slope implies that, after the optimum value of 2.45, the more hydrophobic substituents are subject to unfavorable steric effects. For π_4, the potency of the inhibitor first rose linearly to $\pi_4^0 \approx 3.00$, then leveled off with a slope of (0.4–0.32); indicating that the para substituent did not contact the enzyme active site after the optimum value of π. In conclusion, hydrophobic effects as given by π governed the potency of benzylpyrimidine inhibitors toward the vertebrate DHFR enzyme.

The QSAR results account for the selectivity of the useful drug TMP, **4.26**, for the bacterial enzymes. Positive steric effects modeled by MR were crucial for increasing inhibitory potency toward bacterial DHFR, while the hydrophobic properties of substituents governed potency toward vertebrate DHFR. Thus polar substituents with ideal MR values would increase binding to the polar bacterial DHFR, while the binding of these same substituents would not be favored at the hydrophobic surfaces of the vertebrate enzyme.

How QSAR might be used to design more selective antifolate drugs was illustrated by Selassie, et al. (1991). Based on QSAR Eq. (4.10) and (4.11), 2,4-diamino-5-(benzyl)pyrimidines substituted at the benzyl group with two meta substituents, one polar and one nonpolar, might be expected to be effective antibiotics. Two such molecules and their calculated potencies and selectivities, along with the corresponding values for the clinically important TMP (**4.26**) are given in Table 4.1. Chicken DHFR was assumed to be a reasonable surrogate for the human enzyme. Benzylpyrimidine **4.28** was calculated to be as potent as TMP (**4.26**) but was projected to be six times as selective.

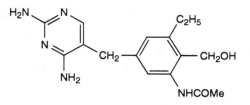

4.28

In summary, the QSAR analysis of DHFR by Selassie et al. (1991) showed the positive steric effect of substituents was critical in the inhibition of the two bacterial enzymes, whereas hydrophobic effects were most important for binding to the vertebrate enzyme.

The 3-D structures of DHFR from different species complexed with the antibiotic TMP (**4.26**) are known, and it is tempting to relate the QSAR findings to the X-ray results. X-ray analysis of DHFR containing bound TMP by Matthews et al. (1985a; 1985b) showed that the conformation of TMP bound to *E. coli* DHFR and chicken liver DHFR were distinctly different. In each mode the diaminopyridine ring of TMP was inserted in a similar position of the active site pocket, but the location of the trimethoxybenzyl group of TMP was different. In *E. coli* DHFR, the trimethoxybenzyl group of TMP was positioned low in the active site pocket, occupying a "lower" cleft, pointing down toward the empty NADPH-binding site; whereas in chicken DHFR (containing the bound NADPH cofactor), the benzyl group of TMP was accommodated in a side channel, positioning it in an "upper" cleft, away from the cofactor. Favorable van der Waals interactions between the active site and the trimethoxybenzyl group were observed for binding in both the upper and lower cleft. Polar interactions between TMP and the *E. coli* DHFR lower cleft active site were also found. Two methoxy side-chain groups of TMP in *E. coli* DHFR were positioned at the opening of the active site cleft where they were extensively solvated by fixed water molecules. These results lead to proposals explaining why different binding modes might give rise to different binding affinities. It is tempting to relate these structural results to the QSAR equations developed by Selassie et al. (1991). The QSAR method shows contact with a polar receptor surface is important for inhibition of bacterial DHFR, whereas hydrophobic effects are most important for binding to the chicken enzyme. Perhaps the polar active site deduced by QSAR in the bacterial enzyme corresponds to the solvent exposed region observed in the X-ray structure.

Unfortunately, this reasoning is not entirely proper. The structure of the *ternary* complex chicken DHFR–NADPH–TMP was compared to the *binary* complex *E. coli* DHFR–TMP, and as Matthews et al. note (1985a; 1985b), the environment of the solvated meta methoxy in the binary complex would undoubtedly be different in the biologically more relevant ternary complex containing NADPH. In fact, a direct interaction between one of the meta methoxy groups and the nicotinamide ring of the NADPH cofactor has been proposed as an important factor in specificity of TMP toward the bacterial enzyme (Baccanari et al., 1982; Champness et al., 1986). To make matters worse, the results of a recent X-ray structural determination of the ternary complex of TMP bound to mouse DHFR shows that the TMP binds in the bacterial mode (Groom et al., 1991)! These results illustrate the difficulties in using high-resolution crystal structures to design drugs: In the case of designing a species-specific antifolate, it may be more fruitful to use the almost old-fashioned, almost "low-tech," method of classical QSAR.

Nonetheless, there are several drawbacks and limitations of the QSAR method. These include the requirement for experimentally measured substituent parameters, and expertise in statistics. In addition, the simple organic systems used to derive the parameters are also imperfect models for biological systems. Because a structurally related series must be analyzed, a completely new type of active molecule will not be discovered by the QSAR technique. Another problem in QSAR analysis is parameter intercorrelation (collinearity). In the antifolate case study, π and MR may be collinear in the meta position for *E. coli*. Here collinearity between π and MR means a positive steric interaction is not always independent of hydrophobicity. Another drawback to QSAR is the lack of 3-D results.

Variations of traditional QSAR given by Eq. (4.2) include using other mathematical forms of the equation (Free and Wilson, 1964; Kubinyi, 1977), other descriptors (Hall and Kier, 1991; Bersuker et al., 1991; Hansch et al., 1975; Verloop et al., 1976), other statistical methods of analysis (Andrea and Kalayeh, 1991; Aoyama and Ichikawa, 1991; Jurs et al., 1979; Hansch et al., 1973; Dove et al., 1979), and strategies for the proper selection of substituent sets (Plummer, 1990). In recent years the classical QSAR approach has also been integrated with methods that explicitly consider the 3-D aspects of molecules. (Crippen, 1979; Hopfinger, 1980; Hansch et al., 1984; Hansch and Klein, 1986; Cramer et al., 1988; Doweyko, 1988).

4.5. PHARMACAPHORE MODELING

4.5.1 Background

The concept of pharmacaphore, a 3-D structural model defining the critical functional groups and bioactive conformation essential for binding to a receptor, is of major influence in rational drug design. A pharmacaphore model provides not only information to develop useful analogs, but may also yield some insight into the mechanism of drug action. The pharmacaphore approach can be considered a topographical or 3-D equivalent to the topological (''flat'' molecule) QSAR method. Like QSAR, it is used when the 3-D structure of a receptor site is unknown, and also like QSAR, it is developed indirectly by analyzing the structure–activity relationships of substances that interact with the receptor.

Deducing the pharmacaphore geometry from sets of molecules is accomplished by generating low-energy conformations for each active compound and finding a geometrical arrangement of groups common to all molecules. Two popular methods used to generate conformations are the active analog approach (Marshall, 1985) and distance geometry (Crippen, 1979). The active analog approach employs a systematic torsional search to generate conformations, whereas distance geometry randomly samples conformational space within distance upper and lower bounds.

4.5.2 Central Nervous System Drugs

In this section, we will examine a paper published by Lloyd and Andrews (1986) of the Victorian College of Pharmacy in Australia in which the active analog approach was used to find a common pharmacaphore for a diverse set of CNS drugs. The nervous system is traditionally divided into two parts: the CNS and peripheral nervous system. The CNS consists of the nerves of the brain and spinal cord, and the peripheral nervous system consists of the sensory and motor nerves in the rest of the body. Central nervous system drugs comprise the most widely used group of pharmacologically active agents. These include drugs with analgesic, hallucinogenic, anticonvulsant, stimulant, antidepressant, and antipsychotic activities. Because direct structural information on CNS drug receptors is difficult to obtain (Evans et al., 1992), many pharmacaphore models have been developed for specific classes of CNS drugs in order to determine the atom arrangements required to bind to the corresponding drug receptors (see, e.g., Rognan et al., 1992; Dorville et al., 1992). The majority of CNS active drugs contain both a phenyl group and a nitrogen atom. This observation forms the starting point for development of a general CNS drug pharmacaphore.

METHOD

The general CNS pharmacaphore was built beginning with the assumption of a four-point model, see Figure 4.13. Three receptor points were presumed critical for pharmaco-

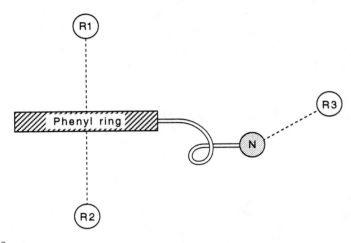

FIGURE 4.13.

Schematic of the R1, R2, and R3 receptor points of a CNS drug. View is perpendicular to plane of drug aromatic groups. [Reproduced with permission from Lloyd E. J.; Andrews, P. R. (1986), *J. Med. Chem.* **29**, 453. Copyright © 1986 American Chemical Society.]

logical effect: R1 and R2, representing hydrophobic bonding to a phenyl ring of the drug; and R3 representing hydrogen bonding with an electronegative atom of the receptor. The fourth point was a nitrogen atom of the drug. Lloyd and Andrews (1986) explain that although the key nitrogen atoms in these drugs vary from basic to acidic, all should participate in some form of hydrogen bonding in this model—either as donors or acceptors—with the same group attached to the receptor. The primary pharmacaphore was generated by using rigid or semirigid reference molecules, followed by fitting flexible analogs to this model.

The conformations of 14 CNS active compounds representing different classes of CNS drugs were analyzed by molecular mechanics. They were morphine (**4.29**), an analgesic; strychnine (**4.30**), a glycine antagonist; LSD (**4.31**), a hallucinogen; apomorphine (**4.32**), a dopamine agonist; mianserin (**4.33**), an antidepressant; phenobarbitone (**4.34**), a hypnotic; clonidine (**4.35**), an α-adrenergeric; diazepam (**4.36**), an anxiolytic; bicuculline (**4.37**), a γ-aminobutyric (GABA) antagonist; diphenylhydantoin (**4.38**), an anticonvulsant; amphetamine (**4.39**), a stimulant; imipramine (**4.40**), an antidepressant; chlorpromazine (**4.41**), an antipsychotic; and procyclidine (**4.42**), an anticholingeric, all given in Figure 4.14.

FIGURE 4.14.   *(facing page)*

Structures of the 14 CNS drugs used in the study. [Adapted with permission from Lloyd, E. J.; Andrews, P. R. (1986), *J. Med. Chem.* **29**, 453. Copyright © 1986 American Chemical Society.]

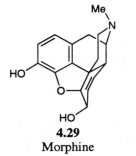

4.29
Morphine

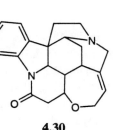

4.30
Strychnine

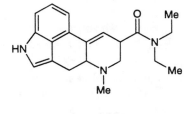

4.31
LSD

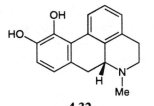

4.32
Apomorphine

4.33
Mianserin

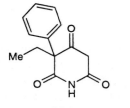

4.34
Phenobarbitone

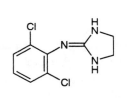

4.35
Clonidine

4.36
Diazepam

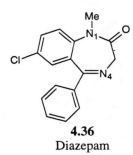

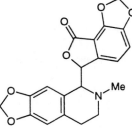

4.37
Bicuculline

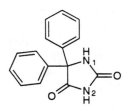

4.38
Diphenylhydantoin

4.39
Amphetamine

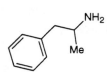

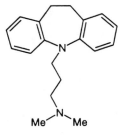

4.40
Imipramine

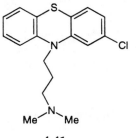

4.41
Chlorpromazine

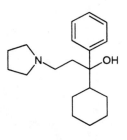

4.42
Procyclidine

The starting structures for the conformational searches were obtained from the corresponding X-ray crystal structures. Calculations were carried out using a simple force field that included only pairwise summation of van der Waals interactions. Electrostatic charges were ignored, and fixed values of bond distances and angles were used. The parameters were those of Giglio (1969). The torsional angles were varied in intervals of 10° for each variable to find the "global" minimum. Lloyd and Andrews (1986) pointed out that the quantitative energy differences between conformers determined in this manner were not of physical significance due to the limited nature of the force field. In addition, no matter how accurate the force field is, the receptor bound conformation of a drug is not necessarily related to the lowest energy conformations of the free drug (see also Section 4.3). Some of this difficulty can be circumvented by generating the primary pharmacaphore model using rigid or semirigid reference molecules. A rigid drug has only one possible low-energy conformation, and so by default this conformation must also be the biologically active receptor-bound conformation.

Molecular comparisons and superpostions of the CNS active drugs were performed using the molecular modeling system MORPHEUS, developed in house (Andrews and Lloyd, 1982).

RESULTS

Construction of the primary pharmacaphore is illustrated in Figure 4.15 for strychnine (**4.30**). The structures of the CNS active drugs morphine (**4.29**), strychnine (**4.30**), LSD (**4.31**), apomorphine (**4.32**), and mianserin (**4.33**) are semirigid, and these were used to define the arrangement of R1–R2 and N–R3 vectors common to all five molecules. There were 13 conformations of these molecules to consider, from which the hydrogen-bonded vectors could be superimposed and the least similar vectors eliminated. By using the relative energies of the conformers of LSD and morphine, the vectors were then reduced to five, one for each semirigid drug molecule (see Fig. 4.16).

The coordinates of the receptor points of the five chosen structures were then averaged to give the coordinates of the common drug–receptor molecule (Fig. 4.17). Both R1 and R2 were located 3.5 Å away from the drug aromatic group, R3 was situated 2.8 Å tetra-

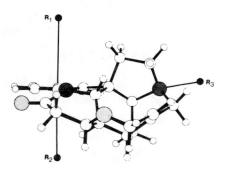

FIGURE 4.15.
Perspective drawing of strychnine (**4.30**) showing the receptor points. [Reproduced with permission from Lloyd, E. J.; Andrews, P. R. (1986), *J. Med. Chem.* **29**, 453. Copyright © 1986 American Chemical Society.]

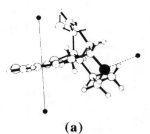

(a)

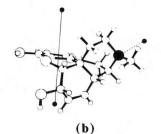

(b)

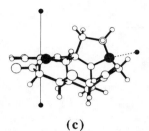

(c)

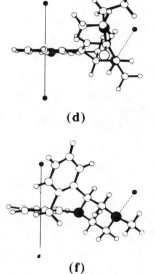

(d)

(e)

(f)

FIGURE 4.16.
Perspective drawing of the semirigid CNS active drugs showing the receptor points. (a) Morphine, axial hydrogen bond, N–R3; (b) morphine, equatorial hydrogen bond, N–R3; (c) strychnine; (d) LSD; (e) apomorphine (f) mianserin. The arrangements R1, R2, and R3 are represented by dark circles, and nitrogen and oxygen are indicated by dark and light shadings, respectively. [Reproduced with permission from Lloyd, E. J.; Andrews, P. R. (1986), *J. Med. Chem.* **29**, 453. Copyright © 1986 American Chemical Society.]

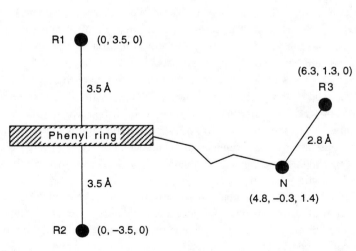

FIGURE 4.17.
Dimensions of the common model viewed from a point perpendicular to the plane of the drug aromatic group. [Reproduced with permission from Lloyd, E. J.; Andrews, P. R. (1986), *J. Med. Chem.* **29**, 453. Copyright © 1986 American Chemical Society.]

202

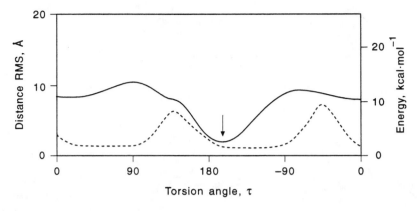

FIGURE 4.18.
Plot of potential energy (dashed line), and distance between receptor points (solid line) versus the phenyl ring torsional angle for phenobarbitone (**4.34**). [Reproduced with permission from Lloyd, E. J.; Andrews, P. R. (1986), *J. Med. Chem.* **29**, 453. Copyright © 1986 American Chemical Society.]

hedrally from the drug nitrogen, and the distance between the center of the phenyl ring and N atom was 5.0 Å. The drugs used to define the pharmacaphore were chiral, thus the drug–receptor model defined by them is also chiral, with the chirality determined by the orientation of the N–R3 vector relative to the R1–R2 vector.

Two sets of calculations were performed on the conformations of the nine remaining flexible drugs. First, the energies of the molecule were calculated; then a measure of the best fit between the four points R1, R2, R3, and N for the molecule under consideration and the same points in the common model was determined. Comparison of the energy data with the rms fit information led to the selection of conformations, which combined low energy (within 5.0 kcal/mol of the calculated minimum) and acceptable (rms < 1.0 Å) fit to the common model. The analysis used is illustrated for the phenyl ring torsion in phenobarbitone (**4.34**) (Fig. 4.18). Here a plot of energy and distance between receptor points versus torsion angle showed only one conformation ($\pi = -160°$) that allowed both low energy and best fit to the common model.

The best conformers for all 14 drugs are shown superimposed on each other in Figure 4.19. The figure shows the common placement of the aromatic rings and the nearly identical position of the corresponding receptor points R3 for this diverse group of CNS active drugs. The molecules share the common primary binding groups of an aromatic ring and a nitrogen in a precise spatial arrangement. The specific activity of individual classes of CNS drugs (e.g., hallucinogenic vs. stimulant) is likely determined by the position and placement of secondary binding groups.

How the general CNS pharmacaphore developed by Lloyd and Andrews could be applied to drug design was engagingly illustrated by Andrews, et al. (1986) of the Victorian College of Pharmacy. In one example, the researchers described a thought experiment in which morphine and LSD were ''spliced'' to give a composite structure, with the common pharmacaphore model defining how the two structures must be overlaped (Fig. 4.20). This new composite molecule might be expected to have hallucinogenic or analgesic activities.

(a)

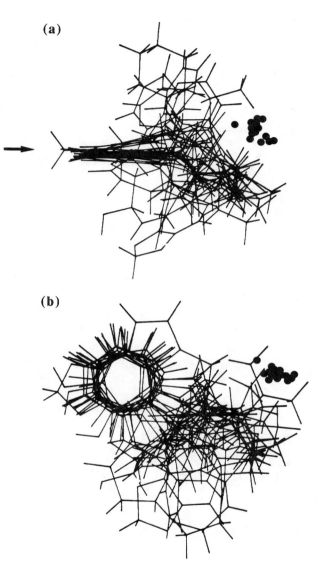

(b)

FIGURE 4.19.
Two views of line drawing superpositions of all 14 molecules in their lowest energy best-fit conformations to the common model. Here R3 is represented by the solid black circle. The phenyl rings are arrowed. (a) View perpendicular to the plane of the phenyl groups. (b) View above the phenyl groups. [Reproduced with permission from Lloyd, E. J.; Andrews, P. R. (1986), *J. Med. Chem.* **29**, 453. Copyright © 1986 American Chemical Society.]

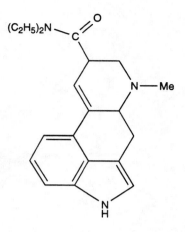

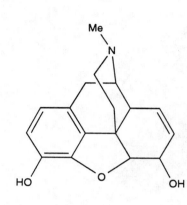

4.29

4.31

SPLICE

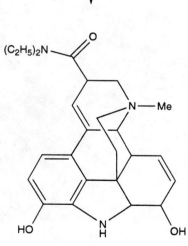

FIGURE 4.20.
Splicing morphine (**4.29**) with LSD (**4.31**).

In summary, a general pharmacaphore model for CNS drugs by the active analog approach was generated. Hydrophobic bonding between receptor and a phenyl ring of the drug, and a hydrogen bond between receptor and an N atom of the drug were presumed critical for pharmacological effect. Molecular mechanics was used to carry out conformational searches on the CNS active reference molecules. How the general CNS pharmacaphore might yield new CNS active drugs was described.

A practical problem in the development of a receptor pharmacaphore is that the receptor bound conformation of a drug is not necessarily the same as the lowest energy conformation of the free drug. This requires that the initial pharmacaphore model be generated using rigid or semirigid reference molecules. Limiting reference molecules to relatively small organics, rather than flexible peptides, for example, then limits the kinds of receptors that can be modeled. The greatest weakness of the pharmacaphore model is that it does not take into account the relative receptor affinities for different compounds. That is, each reference molecule is given equal weight in the development of the pharmacaphore, when, in fact, each reference molecule is not equally effective in binding to the receptor. To correct this deficiency, methods incorporating potency information in the pharmacaphore model, or 3-D structure-directed QSAR, have been proposed and applied: These methods include distance geometry QSAR (Ghose and Crippen, 1985), molecular shape analysis (Hopfinger, 1980), comparative molecular field analysis (CoMFA) (Cramer et al., 1988), and the Voroni binding site model (Boulu et al., 1990).

Homework

4.1. Extract the X-ray structure of HPR bound to a hydroxyethylene-based inhibitor from the Brookhaven Protein Data Bank (File 8HVP). Find the most important hydrogen-bonding interactions of the inhibitor with the protease, and compare it to the hydrogen-bonding schematic of Figure 4.4. Find the internal water bound to the inhibitor and to Ile50 and Ile150 (Ile150 is also labeled Ile50 in the X-ray structure). Convince yourself of the C_2 symmetry of the enzyme active site. How would you modify the inhibitor to increase water solubility without reducing potency? Delete the inhibitor, and create the artificially symmetric molecule **4.43** in the extended conformation below,

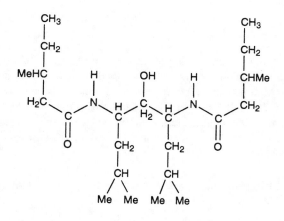

4.43

Minimize the structure, and then dock to the HPR active site. Visually maximize hydrogen bonds and minimize steric clashes. Compare the structure of this modeled inhibitor–HPR complex to 8HVP. Why might your designed drug be an effective and selective HPR inhibitor? Create the symmetric molecule **4.44** below,

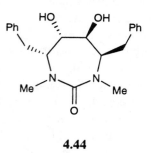

4.44

Minimize and then dock to the HPR active site. What is the importance of the urea carbonyl oxygen?

4.2. Create the cyclic peptide (**4.45**) drawn below.

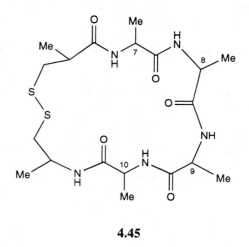

4.45

It is a simplified all L-alanine analog of Sandostatin (**4.21**). Using the dihedral values for a Type II′ β turn in Table 3.1, fix the φ, ψ dihedral angles of residues 7–10 to make a Type II′ β turn. Minimize the rest of the molecule. Find the intramolecular hydrogen bond between residue 7 and 10. Replace the L-alanine with D-alanine at position 8. Compare the two structures. Why might the turn structure be stabilized by the unnatural amino acid at this position? Replace Ala10 by proline. Why would a β-turn structure be destabilized by this replacement?

4.3. The $\log\left(\dfrac{1}{K_i}\right)$ of two substituted phenyl-based inhibitors was determined and expected to be a simple linear function of hydrophobicity: $\log\left(\dfrac{1}{K_i}\right) = a\pi + c$. Use the data below to develop the QSAR equation.

Substituent	$\log\left(\dfrac{1}{K_i}\right)$	π
n-Butyl	7.75	2.13
F	6.57	0.14

4.4. Calculate the $\log\left(\dfrac{1}{K_i}\right)$ for *E. coli* DHFR using the QSAR Eq. (4.8) for the following three compounds: the parent unsubstituted 2,4-diamino-5-benzylpyrimidine, and 2,4-diamino-5-(**3-CH₃**-benzyl)pyrimidine, and 2,4-diamino-5-[**3,4-(OH)₂**-benzyl]pyrimidine.

	MR	π
H	0.10	0.0
CH$_3$	0.57	0.52
OH	0.29	−0.78

Hint: Recall that hydrophobic substituents (nonpolar) will prefer the 3 position; hydrophilic (polar) will prefer the 5 position.

4.5. How is the idea of using rigid reference molecules to generate a pharmacaphore related conceptually to zeroing in on a biologically active form of a peptide by cyclizing it?

4.6. Carry out a conformational search on the antipsychotic AIDS lead haloperidol (**4.11**). Use either a torsional energy grid or annealed dynamics. Find several of the lowest energy conformers and see if any conform to the general CNS pharmacaphore developed by Lloyd and Andrews. If so, which phenyl group of haloperidol is involved in binding to R1 and R2 of the receptor?

Given your understanding of the CNS pharmacaphore, and the important parts of haloperidol as a CNS drug, splice haloperidol with one of the drugs listed in Figure 4.14 using pencil and paper. Try to guess what CNS effects your new designer drug might have.

References

Andrea, T. A.; Kalayeh, H. (1991), Applications of neural networks in quantitative structure–activity relationships of dihydrofolate reductase inhibitors, *J. Med. Chem.* **34**, 2824.

Andrews, P. R.; Lloyd, E. J. (1982), Molecular conformation and biological activity of central nervous system active drugs, *Med. Res. Rev.* **2**, 355.

Andrews, P. R.; Lloyd, E. J.; Martin, J. L.; Munro, S. L. A. (1986), Central nervous system drug design, *J. Mol. Graphics* **4**, 41.

Aoyama, T.; Ichikawa, H. (1991), Obtaining the correlation indices between drug activity and structural parameters using a neural network, *Chem. Pharm. Bull.* **39**, 372.

Baccanari, D. P.; Daluge, S.; King, R. W. (1982), Inhibition of dihydrofolate reductase: Effect of reduced nicotinamide adenine dinucleotide phosphate on the selectivity and affinity of diaminobenzylpyrimidines, *Biochemistry* **21**, 5068.

Bauer, W.; Briner, U.; Doepfner, W.; Haller, R.; Huguenin, R.; Marbach, P.; Petcher, T. J.; Pless, J. (1982), SMS 201–995: A very potent and selective octapeptide analogue of somatostatin with prolonged action, *Life Sci.* **31**, 1133.

Berman, E. M. Werbel, L. M. (1991), The renewed potential for folate antagonists in contemporary cancer chemotherapy, *J. Med. Chem.* **34**, 479.

Bersucker, I. B. and Dimoglo, A. S. (1991), The electron-topological approach to the QSAR problem. In Lipkowitz, K. B. and Boyd, D. B., Eds., *Reviews in Computational Chemistry*, Vol. II, VCH, New York, pp. 423–460.

Blaney, J. M.; Hansch, C.; Silipo, C.; Vittoria, A. (1984), Structure–activity relationships of dihydrofolate reductase inhibitors, *Chem. Rev.* **84**, 333.

Boulu, L. G.; Crippen, G. M.; Barton, H. A.; Kwon, H.; Marletta, M. A. (1990), Voronoi binding site model of a polycyclic aromatic hydrocarbon binding protein, *J. Med. Chem.* **33**, 771.

Brazeau, P.; Vale, W.; Burgus, R.; Ling, N.; Butcher, M.; Rivier, J.; Guillemin, R. (1973), Hypothalamic polypeptide that inhibits the secretion of immunoreactive pituitary growth hormone, *Science* **179**, 77.

Brown, M.; Rivier, J.; Vale, W. (1977), Somatostatin: Analogs with selected biological activities, *Science* **196**, 1467.

Buffington, L. A.; Garsky, V.; Rivier, J.; Gibbons, W. A. (1983), Conformation of somatostatin using scalar coupling constants from 270 and 600 MHz simulated proton magnetic resonance spectra, *Biophys. J.* **41**, 299.

Champness, J. N.; Stammers, D. K.; Beddell, C. R. (1986), Crystallographic investigation of the cooperative interaction between trimethoprim, reduced cofactor and dihydrofolate reductase, *FEBS Lett.* **199**, 61.

Cramer III, R. D.; Patterson, D. E.; Bunce, J. D. (1988), Comparative molecular field analysis (CoMFA). 1. Effect of shape on binding of steroids to carrier proteins, *J. Am. Chem. Soc.* **110**, 5959.

Crippen, G. M. (1979), Distance geometry approach to rationalizing binding data, *J. Med. Chem.* **22**, 988.

Deleuze, C.; Hull, W. E. (1982), Assignment and analysis of the 500 MHz ^{1}H NMR spectra of somatostatin and the acyclic precursor ($S^{3,14}$–acm)-somatostatin in dimethyl sulphoxide, *Org. Magn. Res.* **18**, 112.

DesJarlais, R. L.; Seibel, G. L.; Kuntz, I. D.; Furth, P. S.; Alvarez, J. C.; Ortiz de Montellano, P. R.; DeCamp, D. L.; Babe, L. M.; Craik, C. S. (1990), Structure-based design of nonpeptide inhibitors specific for the human immunodeficiency virus 1 protease, *Proc. Natl. Acad. Sci. USA* **87**, 6644.

Dorville, A.; McCort-Tranchepain, I.; Vichard, D.; Sather, W.; Maroun, R.; Ascher, P.; Roques, B. P. (1992), Preferred antagonist binding state of the NMDA receptor: Synthesis, pharmacology, and computer modeling of (phosphonomethyl)phenylalanine derivatives, *J. Med. Chem.* **35**, 2551.

Dove, S.; Franke, R.; Mndshojan, O. L.; Schkuljev, W. A.; Chashakjan, L. W. (1979), Discriminant-analytical investigation on the structural dependence of hyperglycemic and hypoglycemic activity in a series of substituted *o*-toluenesulfonylthioureas and *o*-toluenesulfonylureas, *J. Med. Chem.* **22**, 90.

Doweyko, A. M. (1988), The hypothetical active site lattice. An approach to modelling active sites from data on inhibitor molecules, *J. Med. Chem.* **31**, 1396.

Erickson, J.; Neidhart, D. J.; VanDrie, J.; Kempf, D. J.; Wang, X. C.; Norbeck, D. W.; Plattner, J. J.; Rittenhouse, J. W.; Turon, M.; Wideburg, N.; Kohlbrenner, W. E.; Simmer, R.; Helfrich, R.; Paul, D. A.; Knigge, M. (1990), Design, activity, and 2.8 Å crystal structure of a C_2 symmetric inhibitor complexed to HIV-1 protease, *Science* **249**, 527.

Evans, C. J.; Keith Jr., D. E.; Morrison, H.; Magendzo, K.; Edwards, R. H. (1992), Cloning of a delta opioid receptor by functional expression, *Science* **258**, 1952.

Foundling, S. I.; Cooper, J.; Watson, F. E.; Cleasby, A.; Pearl, L. H.; Sibanda, B. L.; Hemmings, A.; Wood, S. P.; Blundell, T. L.; Valler, M. J.; Norey, C. G.; Kay, J.; Boger, J.; Dunn, B. M.; Leckie, B. J.; Jones, D. M.; Atrash, B.; Hallett, A.; Szelke, M. (1987), High resolution X-ray analyses of renin inhibitor–aspartic proteinase complexes, *Nature (London)* **327**, 349.

Free Jr., S. M.; Wilson, J. W. (1964), A mathematical contribution to structure–activity studies. *J. Med. Chem.* **7**, 395.

Ghose, A. K.; Crippen, G. M. (1985), Use of physicochemical parameters in distance geometry and related three-dimensional quantitative structure-activity relationships: A demonstration using *Escherichia coli* dihydrofolate reductase inhibitors, *J. Med. Chem.* **28**, 333.

Giglio, E. (1969), Calculation of van der Waals interactions and hydrogen bonding in crystals, *Nature (London)* **222**, 339.

Gorden, P.; Comi, R. J.; Maton, P. N.; Go, V. L. W. (1989), Somatostatin and somatostatin analogue (SMS 201–995) in treatment of hormone-secreting tumors of the pituitary and gastrointestinal tract and non-neoplastic diseases of the gut, *Ann. Int. Med.* **110**, 35.

Groom, C. R.; Thillet, J.; North, A. C. T.; Pictet, R.; Geddes, A. J. (1991), Trimethoprim binds in a bacterial mode to the wild-type and E30D mutant of mouse dihydrofolate reductase, *J. Biol. Chem.* **266**, 19890.

Hall, L. H.; Kier, L. B. (1991), The molecular connectivity chi indexes and kappa shape indexes in structure-property modeling. In Lipkowitz, K. B. and Boyd, D. B., Eds., *Reviews in Computational Chemistry*, Vol. II, VCH, New York, pp. 367–422.

Hansch, C.; Fujita, T. (1964), ρ-σ-π Analysis. A method for the correlation of biological activity and chemical structure, *J. Am. Chem. Soc.* **86**, 1616.

Hansch, C.; Hathaway, B. A.; Guo, Z.-R.; Selassie, C. D.; Dietrich, S. W.; Blaney, J. M.; Langridge, R.; Volz, K. W.; Kaufman, B. T. (1984), Crystallography, quantitative structure–activity relationships, and molecular graphics in a comparative analysis of the inhibition of dihydrofolate reductase from chicken liver and *Lactobacillus casei* by 4,6-diamino-1,2-dihydro-2,2-dimethyl-1-(substituted-phenyl)-*s*-triazines, *J. Med. Chem.* **27**, 129.

Hansch, C.; Klein, T. E. (1986), Molecular graphics and QSAR in the study of enzyme–ligand interactions. On the definition of bioreceptors, *Acc. Chem. Res.* **19**, 392.

Hansch, C.; Maloney, P. P.; Fujita, T.; Muir, R. M. (1962), Correlation of biological activity of phenoxyacetic acids with Hammett substituent constants and partition coefficients, *Nature (London)* **194**, 178.

Hansch, C.; Silipo, C.; Steller, E. E. (1975), Formulation of *de novo* substituent constants in correlation analysis: Inhibition of dihydrofolate reductase by 2,4-diamino-5-(3,4-dichlorophenyl)-6-substituted pyrimidines, *J. Pharm. Sci.* **64**, 1186.

Hansch, C.; Unger, S. H.; Forsythe, A. B. (1973), Strategy in drug design. Cluster analysis as an aid in the selection of substituents, *J. Med. Chem.* **16**, 1217.

Hitchings Jr., G. H. (1989), Selective inhibitors of dihydrofolate reductase, *Angew. Chem. Int. Ed. Engl.* **28**, 879.

Holladay, L. A.; Puett, D. (1976), Somatostatin conformation: Evidence for a stable intramolecular structure from circular dichroism, diffusion, and sedimentation equilibrium, *Proc. Nat. Acad. Sci. USA* **73**, 1199.

Holladay, L. A.; Rivier, J.; Puett, D. (1977), Conformational studies on somatostatin and analogues, *Biochemistry* **16**, 4895.

Hopfinger, A. J. (1980), A QSAR investigation of dihydrofolate reductase inhibition by Baker triazines based upon molecular shape analysis, *J. Am. Chem. Soc.* **102**, 7196.

Hosur, M. V.; Bhat, T. N.; Kempf, D. J.; Baldwin, E. T.; Liu, B.; Gulnik, S.; Wideburg, N. E.; Norbeck, D. W.; Appelt, K.; Erickson, J. W. (1994), Influence on stereochemistry on activity and binding modes for C_2 symmetry-based diol inhibitors of HIV-1 protease, *J.Am. Chem. Soc.* **116**, 847.

Huff, J. R. (1990), HIV protease: A novel chemotherapeutic target for AIDS, *J. Med. Chem.* **34**, 2305.

Jaskolski, M.; Tomasselli, A. G.; Sawyer, T. K.; Staples, D. G.; Heinrikson, R. L.; Schneider, J.; Kent, S. B. H.; Wlodawer, A. (1991), *Biochemistry* **30**, 1600.

Jurs, P. C.; Chou, J. T.; Yuan, M. (1979), Studies of chemical structure-biological activity relations using pattern recognition. In Olson, E.C and Christoffersen, R. E., Eds., *Computer Assisted Drug Design,* American Chemical Society Symposium Series 112, American Chemical Society, Washington, DC, pp. 103–130.

Kempf, D. J.; Norbeck, D. W.; Codacovi, L.; Wang, X. C.; Kohlbrenner, W. E.; Wideburg, N. E.; Paul, D. A.; Knigge, M. F.; Vasavanonda, S.; Craig-Kennard, A.; Saldivar, A.; Rosenbrook Jr., W.; Clement, J. J.; Plattner, J. J.; Erickson, J. (1990), Structure-based, C_2 symmetric inhibitors of HIV protease, *J. Med. Chem.* **33**, 2687.

Kleinert, H. D.; Baker, W. R.; Stein, H. H. (1991), Renin inhibitors. *Adv. Phar.* **22**, 207.

Koshland Jr., D. E. (1958), Application of a theory of enzyme specificity to protein synthesis, *Proc. Natl. Acad. Sci. USA* **44**, 98.

Kubinyi, H. (1977), Quantitative structure–activity relationships. 7. The bilinear model, a new model for nonlinear dependence of biological activity on hydrophobic character, *J. Med. Chem.* **20**, 625.

Lamberts, S. W. J. (1987), A guide to the clinical use of the somatostatin analogue SMS 201-995 (Sandostatin), *Acta Endocrinologica (Copenh.) Suppl.* **286**, 54.

Lamberts, S. W. J.; Krenning, E. P.; Reubi, J.-C. (1991), The role of somatostatin and its analogs in the diagnosis and treatment of tumors, *Endocrine Rev.* **12**, 450.

Lapatto, R.; Blundell, T.; Hemmings, A.; Overington, J.; Wilderspin, A.; Wood, S.; Merson, J. R.; Whittle, P. J.; Danley, D. E.; Geoghegan, K. F.; Hawrylik, S. J.; Lee, S. E.; Scheld, K. G.; Hobart, P. M. (1989), X-ray analysis of HIV-1 proteinase at 2.7 Å resolution confirms structural homology among retroviral enzymes, *Nature (London)* **342**, 299.

Lloyd, E. J.; Andrews, P. R. (1986), A common structural model for central nervous system drugs and their receptors, *J. Med. Chem.* **29**, 453.

Lyle, T. A.; Wiscount, C. M.; Guare, J. P.; Thompson, W. J.; Anderson, P. S.; Darke, P. L.; Zugay, J. A.; Emini, E. A.; Schleif, W. A.; Quintero, J. C.; Dixon, R. A. F.; Sigal, I. S.; Huff, J. R. (1991), Benzocycloalkyl amines as novel C-termini for HIV protease inhibitors, *J. Med. Chem.* **34**, 1228.

Marshall, G. R. (1985), Structure–activity studies: A three-dimensional probe of receptor specificity, *Ann. N.Y. Acad. Sci.* **439**, 162.

Martin, Y. C. (1981), A practitioner's perspective of the role of quantitative structure–activity analysis in medicinal chemistry, *J. Med. Chem.* **24**, 229.

Matthews, D. A.; Bolin, J. T.; Burridge, J. M.; Filman, D. J.; Volz, K. W.; Kaufman, B. T.; Beddell, C. R.; Champness, J. N.; Stammers, D. K.; Kraut, J. (1985a), Refined crystal structures of *Escherichia coli* and chicken liver dihydrofolate reductase containing bound trimethoprim, *J. Biol. Chem.* **260**, 381.

Matthews, D. A.; Bolin, J. T.; Burridge, J. M.; Filman, D. J.; Volz, K. W.; Kraut, J. (1985b), Dihydrofolate reductase. The stereochemistry of inhibitor selectivity, *J. Biol. Chem.* **260**, 392.

Miller, M.; Jaskolski, M.; Rao, J. K. M.; Leis, J.; Wlodawer, A. (1989b), Crystal structure of a retroviral protease proves relationship to aspartic protease family, *Nature (London)* **337**, 576.

Miller, M.; Schneider, J.; Sathyanarayana, B. K.; Toth, M. V.; Marshall, G. R.; Clawson, L.; Selk, L.; Kent, S. B. H.; Wlodawer, A. (1989a), Structure of complex of synthetic HIV-1 protease with a substrate-based inhibitor at 2.3 Å resolution, *Science* **246**, 1149.

Momany, F. A. (1976), Conformational energy analysis of the molecule, luteinizing hormone-releasing hormone. 1. Native decapeptide, *J. Am. Chem. Soc.* **98**, 2990.

Moreau, J. P.; DeFeudis, F. V. (1987), Minireview. Pharmacological studies of somatostatin and somatostatin-analogues: Therapeutic advances and perspectives, *Life Sci.* **40**, 419.

Navia, M. A.; Fitzgerald, P. M. D.; McKeever, B. M.; Leu, C.-T.; Heimbach, J. C.; Herber, W. K.; Sigal, I. S.; Darke, P. L.; Springer, J. P. (1989), Three-dimensional structure of aspartyl protease from human immunodeficiency virus HIV-1, *Nature (London)* **337**, 615.

Osman, R.; Weinstein, H.; Green, J. P. (1979), Parameters and methods in quantitative structure-actvity relationships. In Olson, E.C and Christoffersen, R. E., Eds., *Computer Assisted Drug Design,* American Chemical Society Symposium Series 112, American Chemical Society, Washington, DC, pp. 21–78.

Pauling, L. (1948), Nature of forces between large molecules of biological interest, *Nature (London)* **161**, 707.

Pearl, L. H.; Taylor, W. R. (1987), A structural model for the retroviral proteases, *Nature (London)* **329**, 351.

Plummer, E. L. (1990), The application of quantitative design strtegies in pesticide discovery. In Lipkowitz, K. B. and Boyd, D. B., Eds., *Reviews in Computational Chemistry,* Vol. I, VCH, New York, pp. 119–168.

Rivier, J.; Brazeau, P.; Vale, W.; Guillemin, R. (1975), Somatostatin analogs. Relative importance of the disulfide bridge and of the Ala-Gly side chain for biological activity, *J. Med. Chem.* **18**, 123.

Rizo, J.; Gierasch, L. M. (1992), Constrained peptides: Models of bioactive peptides and protein substructures, *Annu. Rev. Biochem.* **61**, 387.

Rognan, D.; Boulanger, T.; Hoffmann, R.; Vercauteren, D. P.; Andre, J.-M.; Durant, F.; Wermuth, C.-G. (1992), Structure and molecular modeling of GABA$_A$ receptor antagonists, *J. Med. Chem.* **35**, 1969.

Rose, G.D.; Gierasch, L. M.; Smith, J. A. (1985), Turns in peptides and proteins, *Adv. Protein Chem.* **37**, 1.

Rutenber, E.; Fauman, E. B.; Keenan, R. J.; Fong, S.; Furth, P. S.; Ortiz de Montellano, P. R.; Meng, E.; Kuntz, I. D.; DeCamp, D. L.; Salto, R.; Rosé, J. R.; Craik, C. S.; Stroud, R. M. (1993), Structure of a nonpeptide inhibitor complexed with HIV-1 protease, *J. Biol. Chem.* **268**, 15343.

Schechter, I.; Berger, A. (1967), On the size of the active site in proteases. I. Papain. *Biochem. Biophy. Res. Commun.* **27**, 157.

Selassie, C. D. ; Fang, Z.-X.; Li, R.-L.; Hansch, C.; Debnath, G.; Klein, T. E.; Langridge, R.; Kaufman, B. T. (1989), On the structure selectivity problem in drug design. A comparative study of benzylpyrimidine inhibition of vertebrate and bacterial dihydrofolate reductase via molecular graphics and quantitative structure–activity relationships, *J. Med. Chem.* **32**, 1895.

Selassie, C. D.; Li, R.-L.; Poe, M.; Hansch, C. (1991), On the optimization of hydrophobic and hydrophilic substituent interactions of 2,4-diamino-5 -(substituted-benzyl)pyrimidines with dihydrofolate reductase, *J. Med. Chem.* **34**, 46.

Suguna, K.; Padlan, E. A.; Smith, C. W.; Carlson, W. D.; Davies, D. R. (1987), Binding of a reduced peptide inhibitor to the aspartic proteinase from *Rhizopus chinensis*: Implications for a mechanism of action, *Proc. Natl. Acad. Sci. USA* **84**, 7009.

Swain, A. L.; Gustchina, A.; Wlodawer, A. (1991), Comparison of three inhibitor complexes of human immunodeficiency virus protease. In Dunn, B. M., Ed., *Structure and Function of the Aspartic Proteinases*, Plenum, New York, pp. 433–442.

Thompson, W. J.; Fitzgerald, P. M. D.; Holloway, M. K.; Emini, E. A.; Darke, P. L.; McKeever, B. M.; Schleif, W. A.; Quintero, J. C.; Zugay, J. A.; Tucker, T. J.; Schwering, J. E.; Homnick, C. F.; Nunberg, J.; Springer, J. P.; Huff, J. R. (1992), Synthesis and antiviral activity of a series of HIV-1 protease inhibitors with functionality tethered to the P$_1$ or P$_{1'}$ phenyl substituents: X-ray crystal structure assisted design, *J. Med. Chem.* **35**, 1685.

Vacca, J. P.; Guare, J. P.; deSolms, S. J.; Sanders, W. M.; Giuliani, E. A.; Young, S. D.; Darke, P. L.; Zugay, J.; Sigal, I. S.; Schleif, W. A.; Quintero, J. C.; Emini, E. A.; Anderson, P. S.; Huff, J. R. (1991), L-687,908, a potent hydroxyethylene-containing HIV protease inhibitor, *J. Med. Chem.* **34**, 1225.

Veber, D. F.; Holly, F. W.; Nutt, R. F.; Bergstrand, S. J.; Brady, S. F.; Hirschmann, R.; Glitzer, M. S.; Saperstein, R. (1979), Highly active cyclic and bicyclic somatostatin analogues of reduced ring size, *Nature (London)* **280**, 512.

Veber, D. F.; Holly, F. W.; Paleveda, W. J.; Nutt, R. F.; Bergstrand, S. J.; Torchiana, M.; Glitzer, M. S.; Saperstein, R.; Hirschmann, R. (1978), Conformationally restricted bicyclic analogs of somatostatin, *Proc Natl. Acad. Sci. USA* **75**, 2636.

Veber, D. F.; Saperstein, R. (1979), Somatostatin, *Annu. Rep. Med. Chem.* **14**, 209.

Verloop, A.; Hoogenstraaten, W.; Tipker, J. (1976), Development and application of new steric substituent parameters in drug design. In Ariens, E. J., Ed., *Drug Design*, Vol. 7, Academic, New York, pp. 165–206.

Williams, R. J. P. (1977), Flexible drug molecules and dynamic receptors, *Angew. Chem. Int. Ed. Engl.* **16**, 766.

Wilmot, C. M.; Thornton, J. M. (1988), Analysis and prediction of the different types of β-turn in proteins. *J. Mol. Biol.* **203**, 221.

Wlodawer, A.; Erickson, J. W. (1993), Structure-based inhibitors of HIV-1 protease, *Annu. Rev. Biochem.* **62**, 543.

Wynants, C.; Van Binst, G.; Loosli, H. R. (1985), SMS 201–995, a very potent analogue of somatostatin, *Int. J. Peptide Protein Res.* **25**, 608.

Wynants, C.; Coy, D. H.; Van Binst, G. (1988), Conformational study of super-active analogues of somatostatin with reduced ring size by ¹H NMR, *Tetrahedron* **44**, 941.

Further Reading

Organic Medicinal Chemistry and Therapeutics

Ganellin, C. R.; Roberts, S. M., Eds. (1993), Medicinal Chemistry. *The Role of Organic Chemistry in Drug Research*, second edition, Academic, London.

Gilman, A. G.; Rall, T. W.; Nies, A. S.; Taylor, P., Eds. (1993), *Goodman and Gilman's The Pharmacological Basis of Therapeutics*, eighth edition, McGraw-Hill, New York.

Silverman, R. B. (1992), *The Organic Chemistry of Drug Design and Drug Action*, Academic, San Diego, CA.

Drug Modeling and Design

Andrews, P. R. (1979), The design of transition state analogs. In Olson, E.C and Christoffersen, R. E., Eds., *Computer Assisted Drug Design,* American Chemical Society Symposium Series 112, American Chemical Society, Washington, DC, pp. 149–160.

Bryan, W. M. (1991), Design of minimum active fragments of biologically active peptides, *Methods Enzymol.* **202**, 436.

Bugg, C. E.; Ealick, S. E., Eds. (1990), *Crystallographic and Modeling Methods in Molecular Design*, Springer-Verlag, New York.

Burt, S. K.; Greer, J. (1988), Search strategies for determining bioactive conformers of peptides and small molecules, *Annu. Rep. Med. Chem.* **23**, 285.

Cohen, N. C.; Blaney, J. M.; Humblet, C.; Gund, P.; Barry, D. C. (1990), Molecular modeling software and methods for medicinal chemistry, *J. Med. Chem.* **33**, 883.

Erickson, J. W.; Fesik, S. W. (1992), Macromolecular X-ray crystallography and NMR as tools for structure-based drug design, *Annu. Rep. Med. Chem.* **27**, 271.

Giannis, A.; Kolter, T. (1993), Peptidomimetics for receptor ligands—discovery, development, and medical perspectives, *Angew. Chem. Int. Ed. Engl.* **32**, 1244.

Greer, J.; Erickson, J. W.; Baldwin, J. J.; Varney, M. D. (1994), Application of the three-dimensional structures of protein molecules in structure-based drug design, *J. Med. Chem.* **37**, 1035.

Gupta, S. P. (1987), QSAR studies on enzyme inhibitors, *Chem. Rev.* **87**, 1183.

Humblet, C.; Dunbar Jr., J. B.(1993), 3D database searching and docking strategies, *Annu. Rep. Med. Chem.* **28**, 275.

Marshall, G. R. (1987), Computer-aided drug design, *Annu. Rev. Pharmacol. Toxicol.* **27**, 193.

Marshall, G. R.; Barry, C. D.; Bosshard, H. E.; Dammkoehler, R. A.; Dunn, D. A. (1979), The conformation parameter in drug design: the active analog approach. In Olson, E.C and Christoffersen, R. E., Eds., *Computer Assisted Drug Design,* American Chemical Society Symposium Series 112, American Chemical Society, Washington, DC, pp. 205–226.

Martin, Y. C. (1992), 3D database searching in drug design, *J. Med. Chem.* **35**, 2145.

Perun, T. J.; Propst, C. L., Eds. (1989), *Computer-Aided Drug Design*, Marcel-Dekker, New York.

Wearth, B. (1994), *The Billion Dollar Molecule. One Company's Quest for the Perfect Drug*, Simon & Schuster, New York.

Whittle, P. J; Blundell, T. L. (1994), Protein structure-based drug design, *Annu. Rev. Biophys. Biomol. Struct.* **23**, 349.

Supplemental Case Studies

Al-Obeidi, F.; Hadley, M. E.; Pettitt, B. M.; and Hruby, V. J. (1989), Design of a new class of superpotent cyclic α-melanotropins based on quenched dynamic simulations, *J. Am. Chem. Soc.* **111**, 3413.

Appelt, K.; Bacquet, R. J.; Bartlett, C. A.; Booth, C. L. J.; Freer, S. T.; Fuhry, M. A. M.; Gehring, M. R.; Herrmann, S. M.; Howland, E. F.; Janson, C. A.; Jones, T. R.; Kan, C.-C.; Kathardekar, V.; Lewis, K. K.; Marzoni, G. P.; Matthews, D. A.; Mohr, C.; Moomaw, E. W.; Morse, C. A.; Oatley, S. J.; Ogden, R. C.; Reddy, M. R.; Reich, S. H.; Schoettlin, W. S.; Smith, W. W.; Varney, M. D.; Villafranca, J. E.; Ward, R. W.; Webber, S.; Webber, S. E.; Welsh, K. M.; and White, J. (1991), Design of enzyme inhibitors using iterative protein crystallographic analysis, *J. Med. Chem.* **34**, 1925.

Culberson, J. C. and Walters, D. E. (1991), Three-dimensional model for the sweet taste receptor. In Walters, D. E.; Orthoefer, F. T. and DuBois, G. E., Eds., *Sweeteners. Discovery, Molecular Design, and Chemoreception*, American Chemical Society Symposium Series 450, Boston, p. 214.

Goto, S.; Guo, Z.; Futatsuishi, Y.; Hori, H.; Taira, Z.; and Terada, H. (1992), Quantitative Structure–Activity Relationships of benzamide derivatives for anti-leukotriene activities, *J. Med. Chem.* **35**, 2440.

Ring, C. S.; Sun, E.; McKerrow, J. H.; Lee, G. K.; Rosenthal, P. J.; Kuntz, I. D.; Cohen, F. E. (1993), Structure-based inhibitor design by using protein models for the development of antiparasitic agents, *Proc. Natl. Acad. Sci. USA* **90**, 3583.

Sheridan, R. P. and Venkataraghavan, R. (1987), Designing novel nicotinic agonists by searching a database of molecular shapes, *J. Computer-Aided Molecular Design* **1**, 243.

CHAPTER 5

DNA

5.1. INTRODUCTION

Deoxyribonucleic acid, or DNA, is essential to life. It contains the information necessary for protein synthesis and transmits the genetic information of heredity from one generation to the next. A great deal is now known about the structures of crystalline DNAs from X-ray work, and optical and NMR studies have provided complimentary information about DNAs in solution.

The double-stranded DNA polymer consists of the hydrogen-bonded purine-pyrimidine base pairs guanine-cytosine (G-C) and adenine-thymine (A-T) just like rungs on a ladder (Fig. 5.1). Adenine and guanine are purine bases, cytosine and thymine are pyrimidine bases.

The uprights of the ladder are formed by alternating deoxyribose sugars and phosphates linked by 3′, 5′-phosphodiester bonds. The sugars are linked to the bases by glycosyl bonds. The structure of a short segment of a single-stranded DNA chain is shown in

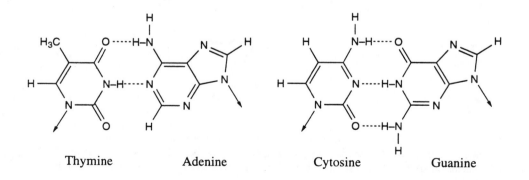

| Thymine | Adenine | Cytosine | Guanine |

FIGURE 5.1.
Watson–Crick hydrogen-bonded base pairs in DNA.

215

216

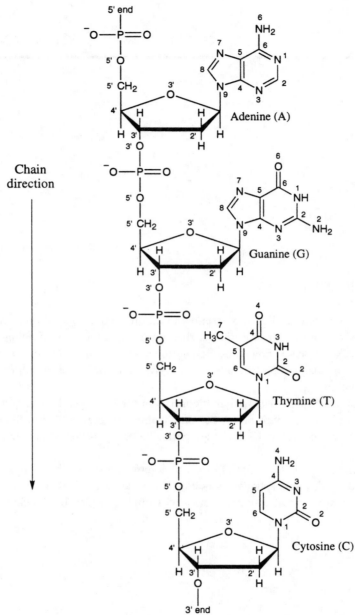

FIGURE 5.2.
Schematic of the covalent structure of a polynucleotide chain of DNA. In shorthand notation, this fragment is d(pApGpTpC), where d means it is DNA (deoxyribonucleic acid) rather than RNA (ribonucleic acid), and the p represents phosphate. In even shorter shorthand this fragment is d(pAGTC).

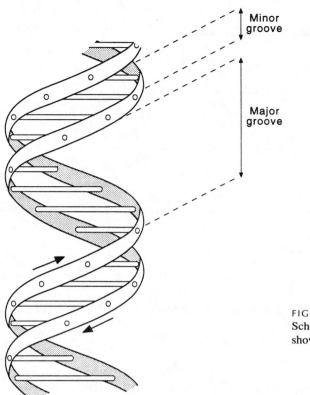

FIGURE 5.3.
Schematic of B-DNA. The arrows
show 5′ to 3′ chain direction.

Figure 5.2. By convention the forward direction of a polynucleotide chain is from 5′ → 3′; that is, one counts from P → O5′ → C5′ → C4′ → C3′ → O3′ → P, where sugar atoms are distinguished from base atoms by a prime.

The ladder of double-stranded DNA is twisted, adopting a helical conformation. The phosphodiester bonds of the DNA chains run in opposite directions; they are antiparallel. The most common form of DNA, the one discovered by Watson and Crick, B-DNA, is illustrated in Figure 5.3. Note that the negatively charged phosphate groups and polar sugar residues of DNA are on the outside of the double helix, exposed to aqueous solution. The strong electrostatic repulsion between the phosphate groups is screened by water and cations. The relatively hydrophobic bases are inside the helix, shielded from water, and stacked nearly perpendicular to the helix axis. In comparison to globular proteins, a larger fraction of the atoms of the double helix are accessible to water. The double helix itself is stabilized by the hydrogen bonds between complimentary base pairs; and by van der Waals nonbonded, dipole–dipole electrostatic, and dipole–induced dipole polarization interactions between the stacked bases. The binding of polyvalent ions, and site-specific hydration also contribute to the stability of DNA.

A notable and biologically important structural feature of B-DNA is the presence of the two grooves in the outer envelope of the double helix (Fig. 5.3). The grooves, called the major and minor grooves, are of different width and depth, and are a consequence of

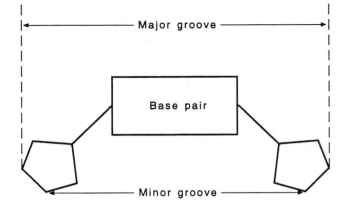

Base pair

Minor groove

FIGURE 5.4.
Schematic of a cross section of DNA showing the origin of the major and minor grooves.
The rectangle represents the base pair, the pentagons represent sugars.

the glycosyl bonds between the sugars and bases of a given base pair not being directly opposite one another (see Fig. 5.4).

5.1.1 Structural Nomenclature

In contrast to proteins, there is only one secondary structural type for DNA: the double helix. The helix structure is not perfectly regular, however. The DNA conformation is dependent on base sequence and environmental conditions. In order to begin to understand the basic principles of DNA helix dynamics, the molecular basis for the recognition of particular base sequences by other molecules, and the structural changes induced by drug binding to DNA, we must first define both the quantitative and qualitative terms commonly used to describe the geometry of DNA. This nomenclature will be used throughout this chapter.

BACKBONE

The standard International Union of Pure and Applied Chemistry (IUPAC) nomenclature for the DNA backbone torsional parameters is defined in Figure 5.5. The backbone dihedral angles are given by α, β, γ, δ, ϵ, and ζ. Actual angle values are often reported, but frequently an angle range terminology is reported. The *gauche* and *trans* terminology is used by crystallographers; here the three possible staggered conformations are defined by *gauche*$^+$ or g^+ (about $+60°$), *trans* or t (about $\pm 180°$) and *gauche*$^-$ or g^- (about $-60°$). The sign is positive if the far bond is rotated clockwise with respect to the near bond. The Klyne–Prelog notation from organic stereochemistry is also commonly used in which the conformation ranges are given by syn (about $0°$), anti (about $180°$), synclinal (about $\pm 60°$), and anticlinal (about $\pm 120°$)

Torsional rotations about five of these backbone bonds would be expected to be relatively unrestricted; only the $C3'-C4'$ bond in the deoxyribose ring (δ), would be torsionally restricted. The backbone torsional angles are not independent; conformational changes in the DNA helix are associated with concerted motions of other torsion angles.

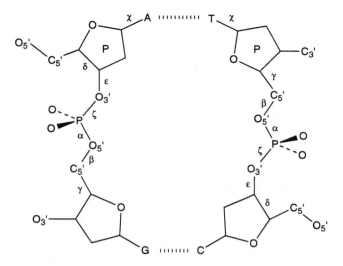

FIGURE 5.5.
Atomic numbering scheme and definition of backbone torsional angles for DNA. [Reproduced with permission from Withka, J.; Swaminathan, S.; Beveridge, D. L.; Bolton, P. H. (1991), *J. Am. Chem. Soc.* **113**, 5041. Copyright © 1991 American Chemical Society.]

SUGAR

The exocyclic sugar-base (glycosyl) torsion angle χ determines the orientation of the base relative to the C1′ ring of the deoxyribose. The syn and anti terminology (anti corresponds to antiperiplanar in Klyne–Prelog notation; syn is synperiplanar) is often used to describe the conformations of the glycosyl angle. In anti, the bulk of the heterocycle (the pyrimidine ring in purines, and O2 in pyrimidines) is pointing away from the sugar, and in syn the heterocycle is over or toward the sugar (Fig. 5.6). The term "high anti" is also

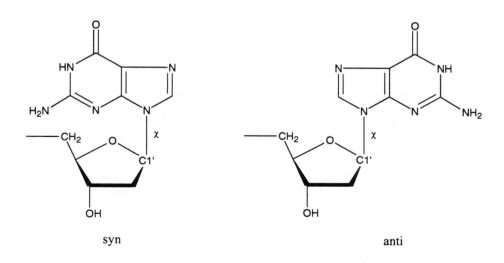

syn   anti

FIGURE 5.6.
Schematic of syn and anti orientations of the sugar-base link in guanine.

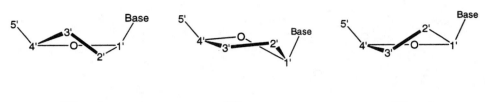

C3'-*endo*   C1'-*exo*   C2'-*endo*

FIGURE 5.7.
Schematic representation of common sugar puckering modes.

used to describe the glycosyl angle. High anti is actually a part of syn, with the N1−C6 bonds of the pryimidine or N9−C8 bonds of the purine nucleosides nearly eclipsed with the C1'−C2' bond of the sugar.

The five-membered sugar ring is not flat. It is commonly puckered in an envelope form where one atom is out of the plane to one side or other. The value of the pucker is given by P, a pseudorotation phase angle, and it is intrinsically related to the δ torsion angle. The corresponding qualitative terms describing the sugar pucker are exo and endo. Xn'-*endo* or Xn'-*exo* describes whether the out of planeness of the atom X is on the same side of the ring as the C5' atom or the opposite side (Fig. 5.7). Another set of terms sometimes used to define the common DNA sugar conformations are N, (C3'-*endo*, where $0° \leq P \leq 36°$) and S, (C2'-*endo*, where $144° \leq P \leq 190°$).

HELIX

The definition of the helicoidal coordinates describing the relationship of the nucelotide bases and base pairs to each other and to the helix axis using the Cambridge convention (Dickerson, 1989) are given in Figure 5.8. The helicoidal coordinates can be conveniently divided into four groups: intrabase pair, interbase pair, axis-base pair, and axis-junction parameters.

The intrabase pair parameters are shear, stretch, stagger, buckle, propeller twist, and opening and involve two bases of a pair. Shear and stretch are linear in-plane displacements of one base pair with respect to the other. Stagger is the out-of-plane linear displacement. Angular displacements are given by propeller twist, buckle, and opening. Propeller twist is defined as the rotation of the bases along their longitudinal axis, and buckle is the dihedral angle of the bases along their short axis. Both relate to the noncoplanarity of the base pairs. Opening is the base pair opening parameter; opening toward the minor groove is defined as negative.

The interbase pair parameters are shift, slide, rise, tilt, roll, and twist, and involve two successive base pairs. Both shift and slide describe linear shearing displacements. A positive shift corresponds to a displacement of the lower base pair in the direction of the minor groove. Rise is the vertical displacement of one base pair with respect to another. Twist is the rotation between successive base pairs about the axis perpendicular to the base plane. Positive twist corresponds to a right-handed helix. Roll is a rotation of a base pair about its long axis in the direction of the major or minor groove, and the sign of roll is positive if the major groove is compressed. Tilt is the rotation around the short axis towards either sugar phosphate backbone.

The axis-base parameters are x displacement, y displacement, inclination, and tip. The x and y displacement specify the displacement of a nucleotide base pair from the helical axis, with a positive x displacement corresponding to a displacement of the base pair towards the major groove. The parameters inclination and tip specify the orientation of the base pairs with respect to the helical axis, and are both positive for a right-hand rotation.

The axis-junction parameters are axis x displacement, axis y displacement, axis inclination, and axis tip, and are used to measure DNA curvature at each position. The local displacement and orientation of the helix axis parameters are defined with respect to the straight, canonical (idealized) B-DNA reference axis.

DNA HELIX TYPES

Three different conformations of the DNA double helix have been discovered: Two forms of right-handed DNA and a left-handed DNA. The major structural features of the three helical types are compared in Table 5.1, and schematic models of the DNA types

TABLE 5.1. Structural Features of the Major DNA Conformations

Property	A-DNA	B-DNA	Z-DNA
Helix handedness	Right	Right	Left
Repeating helix unit	1bp	1bp	2bp
Rotation per base pair (°)	32.7	36.0	−30
Base pairs per turn	11	10	12
Rise per base pair (Å)	2.9	3.4	3.8
Inclination of base pairs (°)	12	0	−7
Diameter (nm)	2.6	2.4	1.8
Glycosyl angle conformation	anti	anti	anti at C; syn at G
Sugar pucker	C3′-endo	O1′-endo to C2′-endo	C2′-endo at C; C2′-exo to C1′-exo at G

are illustrated in Figure 5.9: A-DNA is wide and stubby compared to B-DNA. The major groove of A-DNA is also narrower and deeper than B-DNA, and the minor groove wider and shallower, with bases canted sharply to the helix axis. The left-handed Z-DNA helix is even more slender than B-DNA. The major groove has become a convex surface, and the minor groove a deep cleft. The sugar-phosphate skeleton has a zigzag structure (thus the name Z-DNA), rather than the smooth spiral found in B-DNA.

The conformation of DNA is dependent on humidity, salt content, and base pair composition. The most important biological form of DNA is B, but Z-DNA has been observed in the cell. A-DNA is not found under physiological conditions.

222

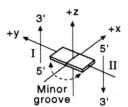

Coordinate frame

Intrabase parameters

 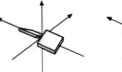

Shear (SHR) Stretch (STR) Stagger (STG)

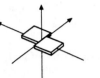 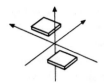

Buckle (BKL) Propeller twist (PRP) Opening (OPN)

Interbase parameters

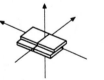 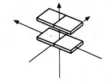

Shift (SHF) Slide (SLD) Rise (RIS)

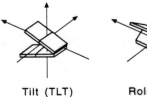 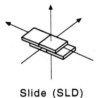 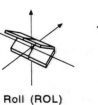

Tilt (TLT) Roll (ROL) Twist (TWS)

FIGURE 5.8.
Definitions of helicoidal parameters. The standard coordinate frame is defined in the upper left. The y direction points along the long axis of the base pair, x along the short axis, and the z direction is perpendicular to the plane of the pair. The rectangles represent the base pairs. [Adapted with

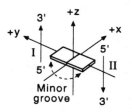

Coordinate frame

Axis-base pair parameters

x displacement (XDP) y displacement (YDP)

Inclination (INC) Tip (TIP)

Axis-junction parameters

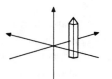
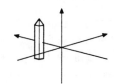

Axis *x*
displacement (AXD) Axis *y*
displacement (AYD)

Axis inclination (AIN) Axis tip (ATP)

permission from Dickerson, R. E. (1989) *J. Biomolecular Structure Dynamics*, **6**, 627 and Srinivasan, J.; Withka, J. M.; Beveridge, D. L. (1990), *Biophys. J.* **58**, 533.]

A-DNA   **B-DNA**   **Z-DNA**

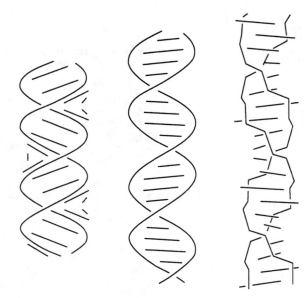

FIGURE 5.9.
The three major structural forms of DNA. [Reproduced with permission from the Annual Review of Biochemistry. Vol. 51, © 1982, by Annual Reviews Inc.]

5.1.2. DNA Motion

The B-DNA double helix is a mobile structure, and its dynamic activity is intrinsic to its function. Individual atoms have rapid vibrational motions at room temperature (on the time scale of ~ 0.01 ps). Groups of atoms move more slowly. Short-range, local motions such as the single-bond torsions of the DNA phosphate backbone, sugar pucker deformations, and motions of the bases occur on the picosecond time scale. The collective movement of large groups of atoms, such as bending of the DNA helix, occur over longer time scales.

5.1.3. DNA Modeling

In this chapter we will discuss the application of molecular modeling to DNA. First, we will examine two MD studies of the DNA double helix, with and without solvent. In the next sections, we will describe molecular mechanics and dynamics investigations of the interaction of drugs or drug models with DNA. Case studies on netropsin binding to the minor groove of DNA, on proflavine intercalation, and on the covalent binding of cisplatin to DNA will be discussed.

Nucleic acid force fields evolved from protein force fields, and so unsurprisingly, the major programs currently being used for DNA simulations are also popular for protein simulation. These are AMBER, GROMOS, and CHARMM, and are described in more detail in Section 3.2.

5.2. DYNAMICS OF DNA

The molecular basis for the dynamic conformational properties of double helical DNA is of enormous interest because the 3-D form and local flexibility of DNA is thought to play an important role in the recognition of binding sites by proteins or other biologically important molecules. The contribution of MD to this science is still in its infancy. The first MD studies of DNA were reported in 1983 (Tidor et al., 1983; Levitt, 1983), nearly 5 years after the first papers on protein dynamics. Since then several MD studies on nucleic acids have appeared. Solvent effects would be expected to be even more important for DNA than for proteins because DNA is charged at the surface, and nucleic acids have a larger surface/volume ratio than globular proteins. Most researchers in the field agree that high-quality descriptions of DNA must include counterions and a large number of water molecules. Such calculations have rarely been carried out, due to the enormous computational requirements. Most simulations have been made in vacuum, and the effect of solvent approximated by reduced charges on phosphate groups, or a distance dependent dielectric parameter. Apart from solving the challenging problems of correctly describing DNA solvation and electrostatics, there is another major difficulty: understanding the results. The huge amount of data generated in a dynamics run requires a compact and complete characterization of the structural changes during DNA dynamics.

Below we shall discuss two DNA dynamics case studies. First, we will briefly examine a pioneering DNA dynamics investigation carried out in 1983. Second, we will describe an exhaustive dynamics simulation that includes an elegant method of data presentation.

5.2.1 DNA *in Vacuo*

One of the first dynamics simulations of the DNA double helix was carried out by Levitt (1983) at the Weizmann Institute of Science. The results demonstrated a high degree of conformational flexibility of the DNA double helix at room temperature. Below we will discuss Levitt's findings for a $(A)_{24}(T)_{24}$ duplex in a vacuum.

METHODS

The starting structure of the heavy atoms of the $(A)_{24}(T)_{24}$ duplex was taken from an X-ray fiber diffraction study; all hydrogen atoms were added, giving 1530 atoms. The force field included bond stretch, bend, and torsions. van der Waals and hydrogen bonding were included. Electrostatic interactions were neglected. Before starting the dynamics trajectory, the X-ray conformation was relaxed by energy minimization. The dynamics simulation was carried out *in vacuo* for 96 ps. The temperature of the system was increased to 300 K and the time step was 0.002 ps.

RESULTS

Stereoscopic drawings of three of the $(A)_{24}(T)_{24}$ conformations during the dynamics simulation are illustrated in Figure 5.10. The hydrogen bonds between base pairs were all found to be stable during the dynamics run. The figure does, however, show dramatic changes in DNA conformation. Most of the motion of the DNA double helix was smooth

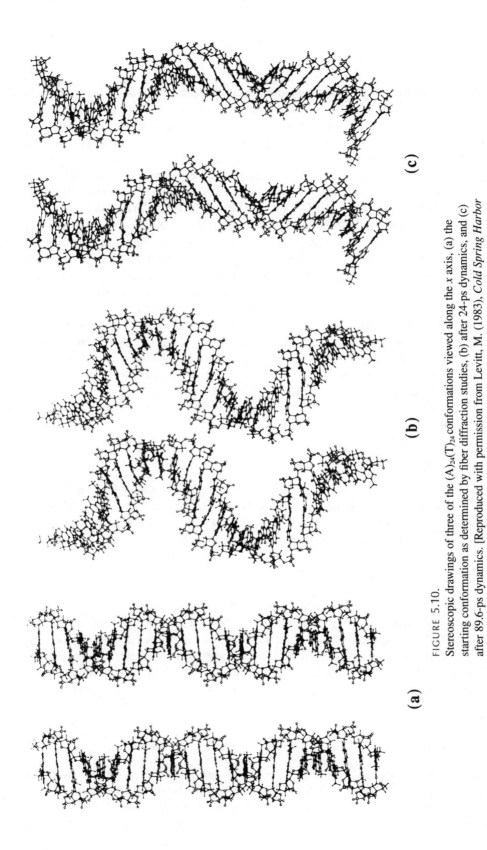

FIGURE 5.10.

Stereoscopic drawings of three of the $(A)_{24}(T)_{24}$ conformations viewed along the x axis, (a) the starting conformation as determined by fiber diffraction studies, (b) after 24-ps dynamics, and (c) after 89.6-ps dynamics. [Reproduced with permission from Levitt, M. (1983), *Cold Spring Harbor Symp. Quant. Biol.* **47**, 251. Copyright © 1983 Cold Springs Harbor Laboratory Press.]

bending into the major groove with an estimated bending vibrational frequency less than
1 cm^{-1} (lifetime or period of $\sim$ 30 ps). Including waters in the calculation would be
expected to dampen this motion. The large-scale and low-energy conformational changes
were not reflected by substantial changes in torsional angles (see Table 5.2). Careful analysis showed four single-bond torsion angles of the central four base pairs were different in
the straight and bent conformations: α changed by $-4°$, γ by $4°$, ϵ by $2°$, and χ by $6°$.

Base pair kinking (local unstacking of base pairs) occurred once during the simulation
at about 80 ps. A detailed stereoscopic molecular drawing of the kinking is illustrated in
Figure 5.11, and the torsional and helical parameters of the kink given in Table 5.2. The
kinking was caused as the helix bent out of the major groove (and into the minor groove).
The angle between the base pairs at the kink was $90°$, although there were no major
torsional angle changes (see Table 5.2) in the kink region. This result again illustrates that
major changes in the global structure did not require major changes in the torsional angles.

In an interesting experiment, Levitt repeated the MD calculations on $(A)_{24}(T)_{24}$ in a
vacuum, but this time allowing for electrostatic interactions. Partial atomic charges were
derived from ab initio quantum calculation, and the dielectric constant was set to 1. Minimization with this potential gave a normal double helical structure, but the structure was
unstable and rapidly denatured during the dynamics. The base pair hydrogen bonds remained intact, but the helix unwound and then re-formed to a very distorted structure,
indicating a very serious defect in the calculation. Levitt suggested the unwinding was
caused by using a dielectric constant of 1, which allows far too strong of a repulsion
between partially charged base pairs.

The MD investigations *in vacuo* by Levitt show that DNA can exhibit dramatic
changes in conformation. A large-scale bending into the minor groove was found, which

TABLE 5.2. Torsion Angles and Other Calculated DNA Properties[a]

	Torsion Angle (in degrees)							Base		Helix	
	α	β	γ	δ	ϵ	ζ	χ	INC[b]	PRP[b]	RIS[b]	TWS[b]
$(A)_{24}(T)_{24}$											
X-ray	313	171	36	156	155	265	218	6	4	3.4	36.0
0 ps	294	172	64	107	173	274	250	16	23	3.4	35.6
24 ps	296	174	57	90	179	286	267	23	12	3.4	27.7
48 ps	294	171	61	98	177	282	261	19	17	3.4	31.3
72 ps	294	173	57	94	180	284	258	21	15	3.1	28.8
96 ps	294	172	64	99	175	281	260	26	17	3.6	32.5
$(A)_{24}(T)_{24}$kink[c]											
78 ps (8–11)	298	171	59	94	171	281	267	17	12	3.4	31.0
(34–41)	294	172	65	99	178	281	269	16	12	3.4	31.0
84 ps (8–11)	303	176	63	106	173	277	266	30	20	4.0	36.5
(38–41)	295	172	75	106	176	275	263	31	20	4.0	36.5

[a] The first and last base pairs were not included when calculating mean values.
[b] The angle between the normal to the base and the overall helix axis is INC; the propeller twist is PRP;
the rise is RIS; the twist is TWS. See Figure 5.8 for definitions of PRP, RIS, and TWS.
[c] Only the four nucleotides indicated were included in the calculation of mean values.

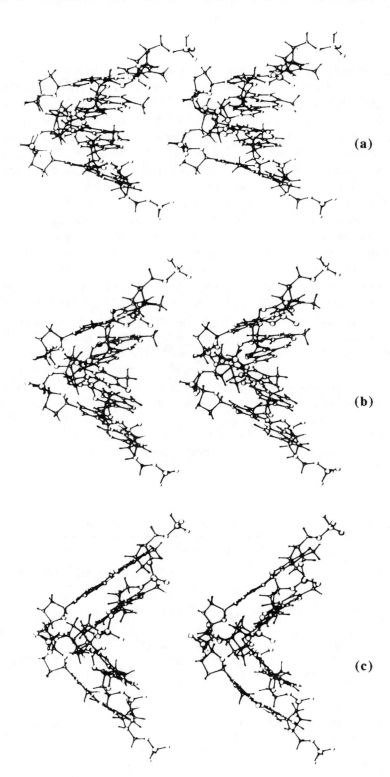

Stereoscopic drawings of the kinking that occurs at about 80 ps. View along the x-axis of the fragment $(A_8A_9A_{10}A_{11})(T_{41}T_{40}T_{39}T_{38})$. (a) At 78 ps, before kinking, (b) at 81 ps, during kink formation, (c) at 84 ps, after the double helix has kinked. [Reproduced with permission from Levitt, M. (1983), *Cold Spring Harbor Symp. Quant. Biol.* **47**, 251. Copyright © 1983 Cold Springs Harbor Laboratory Press.]

was a consequence of only small changes in backbone torsional angle. Correctly accounting for electrostatic interactions is vital to modeling DNA. Here, in this early work, reasonable results were observed when electrostatic interactions were ignored altogether. Including charges and using a dielectric of 1 gave a drastically distorted DNA double helix.

5.2.2 DNA in Water

A serious problem in quantifying the structure and dynamics in DNA simulations is the comprehensive representation of the conformational and helicoidal parameters and their time evolution. Just taking snapshots of the changing molecule or tabulating average parameters may not be enough. Ravishanker et al. (1989) presented a rigorous and original method for the analysis of the dynamical evolution of structural parameters on a DNA double helix. This method is termed a "Curves," dials, and windows (CDW) approach. The procedure is based on the IUPAC conformational backbone coordinates, and "Curves" (Lavery and Sklenar, 1988; Lavery and Sklenar, 1989), a mathematically complete set of helicoidal coordinates. "Dials and windows" is the compact graphical display of the structural and dynamic results of the conformational and helicoidal parameters, respectively. The complete time evolution of the parameters of a DNA double helix can be depicted in a set of six composite figures.

Using a CDW analysis, Swaminathan et al. (1991) of Wesleyan University carried out a detailed and elegant analysis of a 140-ps MD study of solvated d(CGCGAATTCGCG) double helix. The model reproduced several of the features observed in the crystal structure of the dodecamer, such as local axis deformations and the large propeller twist in the base pairs. We will examine this 1991 work below.

METHODS

An idealized or canonical form of B-DNA (Arnott et al., 1980) was used as the starting configuration of double helical d(CGCGAATTCGCG). To generate the initial DNA-counterion complex, a sodium cation was placed 6 Å from each nucleotide phosphorus atom along the OPO bisector. With the DNA structure fixed, and a restraint potential applied to the Na−P distance, a Monte Carlo (MC) simulation was carried out, see Appendix D for a discussion of Monte Carlo selection procedures. The addition of 1927 water molecules was enough to provide more than two complete hydration shells around the DNA. Still keeping the DNA fixed, the Na cation–P potential was adjusted to allow for some movement of the sodium, and another MC equilibration was carried out. Finally, all sodium restraints were removed and the water and sodium cations were equilibrated by MC. Minimization on the full system was carried out. The GROMOS force field, with a harmonic constraint function for hydrogen bonds involved in base pairing added, was used. The simple point charge (SPC) model was used for the water. Switching functions were used for all nonbonded interactions. The system was equilibrated by MD for 35 ps at 300 K. A MD simulation for 100 ps was then carried out.

RESULTS

A sequence of structures of double helical d(CGCGAATTCGCG) over the course of the dynamics simulation is shown in Figure 5.12. The DNA double helix remained stable during the course of over 140 ps of MD simulation. The rms deviations of the dynamical structure from the idealized forms of A- and B-DNA at the end of the dynamics simulation showed that the calculated structure was in the B-DNA family. Nonetheless, a number of significant irregularities developed over the course of the simulation as compared with canonical B. In order to understand the nature of helicoidal variations, sugar pucker changes, and other conformational transitions observed in the dynamics trajectory, Swaminathan et al. (1991) undertook a CDW analysis of the dynamics results. First, we will examine the time evolution of the conformational parameters by dials, followed by analysis of the helicoidal parameters for duplex d(CGCGAATTCGCG) by windows.

The behavior of the conformational parameters of the dodecamer with time is shown by Dials in Figure 5.13 for strand 1, and in Figure 5.14 for strand 2. The figures show conformation wheels, or "dials," for each of the 206 backbone dihedral angles, α, β, γ, δ, ϵ, and ζ, the exocyclic sugar base torsion angle χ, and for the sugar pucker P, a pseudorotation wheel (see Section 5.1.1 for a definition of these parameters). The vertical position of the wheel, corresponding to north on a compass wheel, is taken as 0°. The layout of the dials in the graphic representation is isomorphic with the structure, 5' to 3', top to bottom on strand 1, and vice versa for strand 2. The order of the dials is based on a sequence that places the phosphodiester torsion angles (α and ζ) in interior positions within the dial system for a given nucleotide. The radial coordinate of the dial is treated as the time axis, extending from $t = 0$ at the center to $t = 140$ ps at the circumference. The value of each parameter as a function of time is then displayed as an additional entry on each dial. Thus each dial contains a compact record of the dynamical trajectory of the corresponding structural parameter in the course of the simulation. The dotted lines in each of the dials are the corresponding crystallographic values for the native dodecamer. The shaded areas indicate the sterically forbidden regions of conformational space predicted by Olsen (1982). Dials for canonical A (Arnott and Hukins, 1972) and B (Arnott et al., 1980) forms of DNA are given at the bottom of the figures for orientation and reference. Note that the structural differences between the native, crystallographic dodecamer, and the canonical A and B forms of DNA are readily apparent in Figures 5.13 and 5.14.

Dials analysis of the MD trajectory for duplex d(CGCGAATTCGCG) showed the parameter δ, the internal sugar torsion angle, was relatively stable. The backbone parameter ϵ was also generally stable with the exception of an interesting transient conformational transition at A17. Over the time course of the simulation, ϵ made a transition of t to (g^-), where the parentheses indicate a transitory form. This transition was correlated (dependent or interelated) with a multiple transition g^- to (t) to g^- in the A17 phosphodiester torsion ζ, which was otherwise reasonably stable in the region of a high g^-. The behavior observed is essentially a transition from a B_I to a B_{II} state. The B_I and B_{II} states are the two fundamental backbone conformations observed in crystalline B-DNA, with the B_I conformer the normal, more common form (Fratini et al., 1982).

The phosphodiester torsions α and γ, leading to an oscillation of the position of the phosphorus atom, showed the most dynamical activity in the trajectory. The transitions in

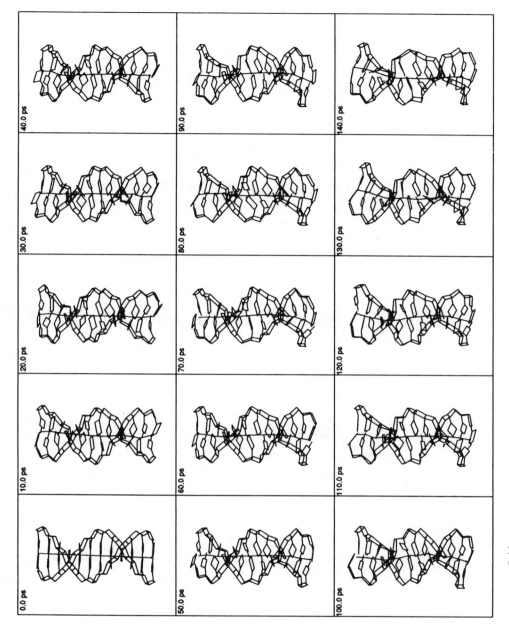

FIGURE 5.12.

Molecular dynamics trajectory for double helical d(CGCGAATTCGCG). [Reproduced with permission from Swaminathan, S.; Ravishanker, G.; Beveridge, D. L. (1991), *J. Am. Chem. Soc.* **113**, 5027. Copyright © (1991) American Chemical Society.]

231

232

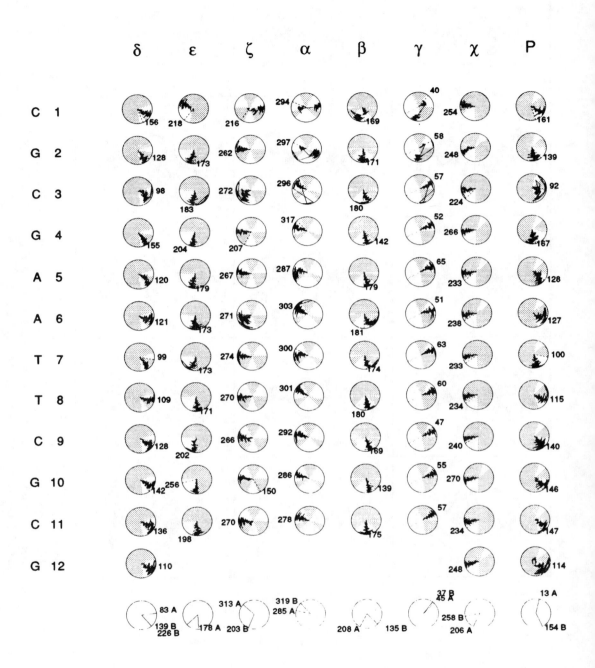

FIGURE 5.13.
Conformational dials for strand 1 of duplex d(CGCGAATTCGCG). [Reproduced with permission from Swaminathan, S.; Ravishanker, G.; Beveridge, D. L. (1991), *J. Am. Chem. Soc.* **113**, 5027. Copyright © (1991) American Chemical Society.]

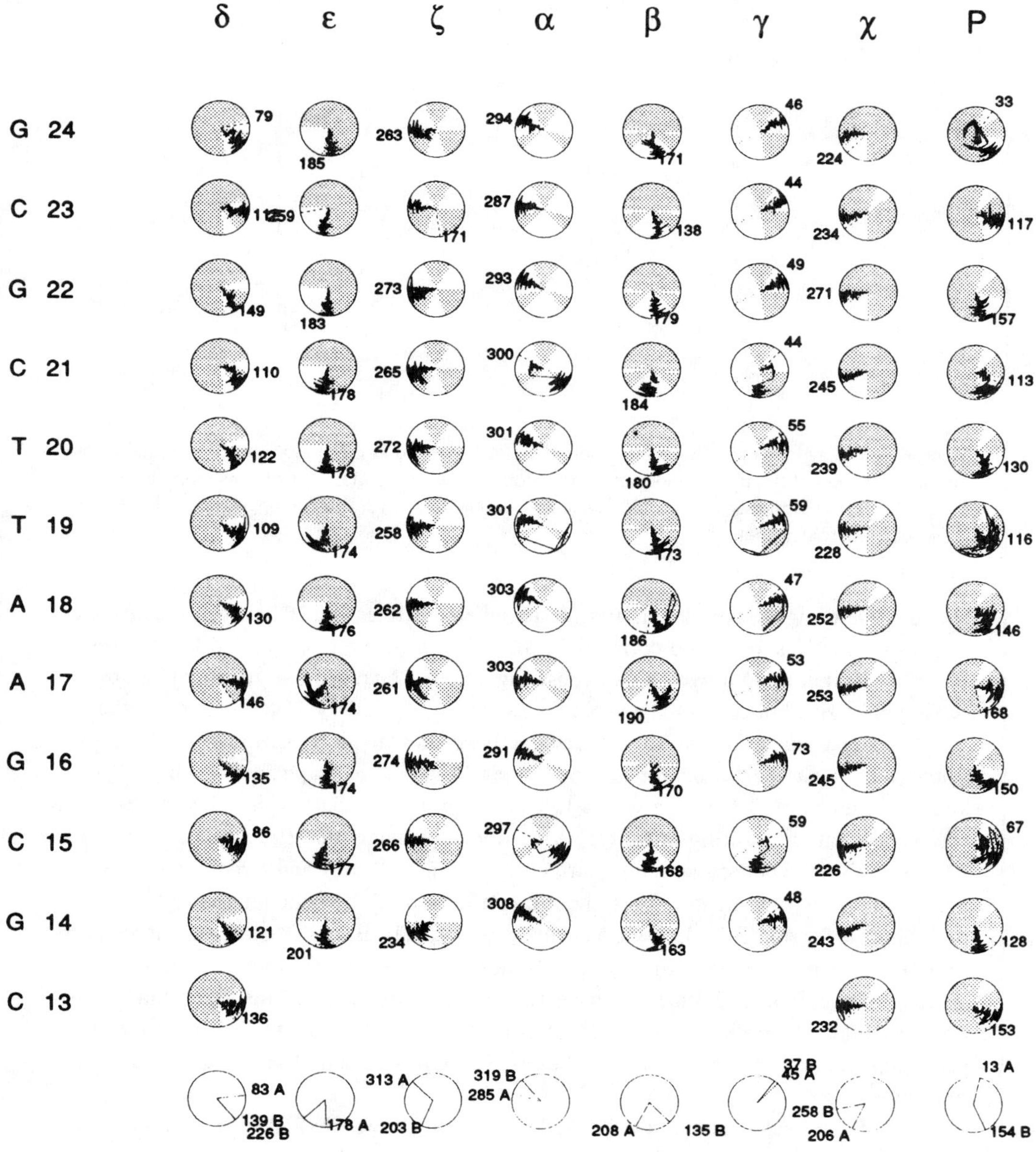

FIGURE 5.14.
Conformational dials for strand 2 of duplex d(CGCGAATTCGCG). [Reproduced with permission from Swaminathan, S.; Ravishanker, G.; Beveridge, D. L. (1991), *J. Am. Chem. Soc.* **113**, 5027. Copyright © (1991) American Chemical Society.]

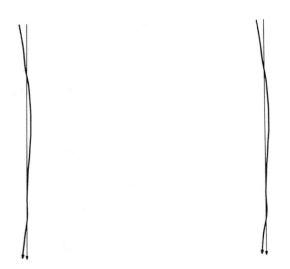

FIGURE 5.15.
Stereoscopic drawings of the calculated helix axis of canonical B-DNA (straight line) superimposed on the helix axis of the native dodecamer. [Reproduced with permission from Swaminathan, S.; Ravishanker, G.; Beveridge, D. L. (1991), *J. Am. Chem. Soc.* **113**, 5027. Copyright © (1991) American Chemical Society.]

α and γ were highly correlated, and the parameter β intervening between α and γ was conformationally stable. The correlated transitions in α and γ, separated by β, define a "crankshaft motion," in which compensating changes occur in two torsional parameters separated by one bond, in such a way that the helical secondary structure is conserved.

The sugar puckering was dynamically active even at the termination of the trajectory. There were a number of significant departures from C2'-*endo* puckering; the C2'-*endo* structure typical of B-DNA gave way to C1'-*exo* in the simulation. In a few cases, there was a transient repuckering to the C3'-*endo* value characteristic of A-DNA.

The DNA secondary structure variables, such as roll, tilt, and twist are sensitive to the way in which the helix axis is chosen. (Definition of the helicoidal coordinates was presented in Section 5.1.) In a Curves analysis, the helical axis is obtained uniquely by minimizing a function that simultaneously describes the change in orientation between successive nucleotides and the nonlinearity of the helicoidal axis. This algorithm is especially useful for describing irregular DNA structures. The calculated helix axes for canonical B duplex d(CGCGAATTCGCG) is compared to the native dodecamer crystal structure in Figure 5.15.

The dynamical behavior of the helical axis is shown in Figure 5.16. The model showed axis deformation in the vicinity of C3 and G10; the bends are localized at the junction between the CGCG end and the AATT center. These axis deformations were oscillatory with an estimated vibrational frequency of about 10 cm^{-1} (lifetime of 3 ps). Local deformations, termed "roll points" have also been observed in the experimental X-ray structure of the dodecamer crystal structure at C3pG4 and C9pG10. The close accord between the positions of the calculated and experimental deformations suggested the roll

FIGURE 5.16.
Stereoscopic drawings of superimposed helix axes from the MD snapshots for duplex d(CGCGAATTCGCG). [Reproduced with permission from Swaminathan, S.; Ravishanker, G.; Beveridge, D. L. (1991), *J. Am. Chem. Soc.* **113**, 5027. Copyright © (1991) American Chemical Society.]

points were intrinsic to the DNA rather than a consequence of crystal packing effects.

The behavior of the helicoidal parameters of the system over time is described using windows. A windows graphic display is composed of one window for each helicoidal parameter, giving 230 total windows. The time axis is on the vertical, increasing from bottom to top, and the value of each parameter as a function of time is then plotted on a vertical axis. The dotted lines in each of the windows are the corresponding crystallographic values for the native dodecamer. The shaded areas indicate the sterically forbidden regions of conformational space delineated by Olsen (1982). Windows for canonical A and B forms of DNA are given at the bottom of the figures for orientation and reference.

The dynamical behavior of the axis parameters x displacement (XDP), y displacement (YDP), inclination (INC), and tip (TIP) is shown in Figure 5.17. The x displacement is a parameter critical in differentiating the canonical B form of DNA (-0.71 Å) from the A form (-5.43 Å). On the basis of XDP, the dodecamer remained in the B range over the course of the dynamics trajectory. The parameters YDP and TIP were dynamically stable. The parameter base pair INC specifies the orientation of the base pairs with respect to the helical axis. The inclination changed to a form in between canonical B and A forms of DNA during the simulation.

The dynamical behavior of the intrabase parameters shear (SHR), stretch (STR), stagger (STG), and propeller twist (PRP) is shown in Figure 5.18: The distance parameters SHR, STR, and STG were relatively stable. A large propeller twist (PRP) in the base pairs was observed in the dynamical model, and stabilized at values of about 30°. A large amplitude propeller twist is also a prominent feature of the crystallographic form of the dodecamer.

236

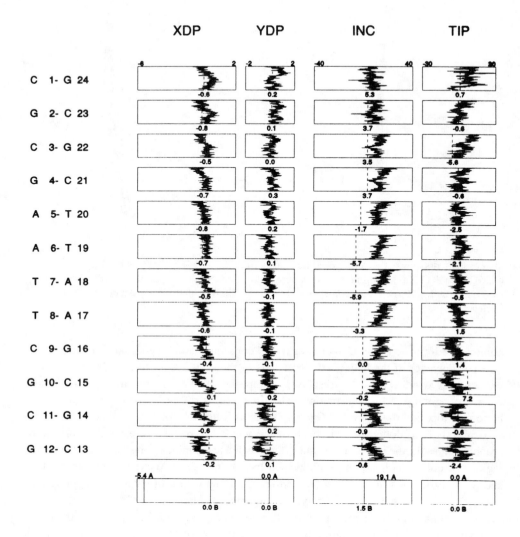

FIGURE 5.17.
Helicoidal windows for axis-base parameters for duplex d(CGCGAATTCGCG). The values assumed in the native dodecamer crystal structure are indicated by dashed lines. [Reproduced with permission from Swaminathan, S.; Ravishanker, G.; Beveridge, D. L. (1991), *J. Am. Chem. Soc.* **113**, 5027. Copyright © (1991) American Chemical Society.]

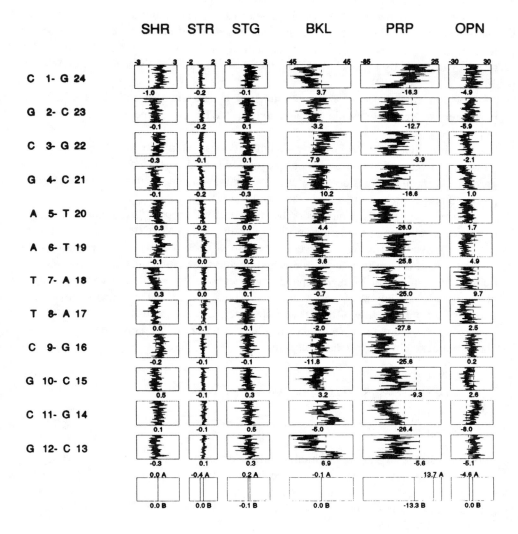

FIGURE 5.18.
Helicoidal windows for intrabase parameters for duplex d(CGCGAATTCGCG). The values assumed in the native dodecamer crystal structure are indicated by dashed lines. [Reproduced with permission from Swaminathan, S.; Ravishanker, G.; Beveridge, D. L. (1991), *J. Am. Chem. Soc.* **113**, 5027. Copyright © (1991) American Chemical Society.]

The results for the interbase pair parameters shift (SHF), slide (SLD), rise (RIS), tilt (TLT), roll (ROL), and twist (TWS) are shown in Figure 5.19: The parameters SHF, SLD, and RIS remained fairly stable. The base pair RIS and TWS values remained at or near the value for canonical B-DNA. The parameter ROL showed evidence of a positive displacement in the region of the axis deformations (C3 and G10), indicating an opening toward the minor grove, that is, a bending into the major groove, coupled with a slight tendency of the helix to underwind from the idealized value of TWS = 36°, to TWS = 30°.

The dynamical behavior of the axis–junction parameters is given in Figure 5.20. The parameters are generally oscillatory, with axis tip (ATP) displacements at the C3 and C11 steps.

Swaminathan et al. (1991) also carried out a corresponding simulation *in vacuo*, using reduced charges and found a pronounced narrowing, almost a collapse, of the minor groove. The explicit inclusion of water in the simulation appears to be important in supporting a proper groove structure in the simulated DNA.

In summary, the dynamics work by Swaminathan et al. (1991) demonstrates that dynamical evolution of the structure of duplex d(CGCGAATTCGCG) in water can be graphically displayed using a Curves, dials, and windows analysis. Several large-scale amplitude motions were observed in this high-quality simulation, including axis deformations, crankshaft motion of the phosphodiester backbone, and sugar repuckering. Analysis of the dynamics study shows good agreement with several of the features observed in the X-ray crystal structure of the DNA dodecamer. Explicit inclusion of solvation in the calculation was found to be necessary; the effect of water on DNA groove structure could not be mimicked by using reduced charges.

5.3. BINDING TO THE MINOR GROOVE

5.3.1 Background

Netropsin (**5.1**) is an antiviral antitumor antibiotic that, although too toxic for clinical use, has received intense experimental study as the archetype of a molecule that binds to the minor groove of B-DNA (Zimmer and Wähnert, 1986; Neidle et al., 1987).

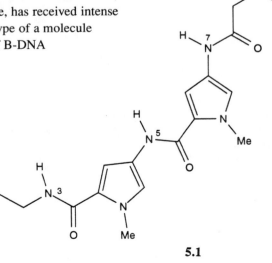

5.1

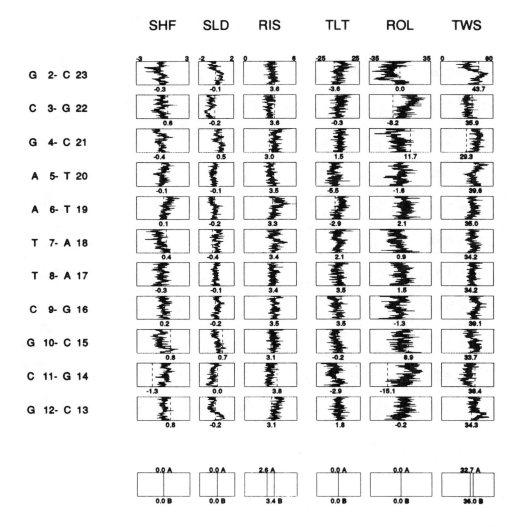

FIGURE 5.19.
Helicoidal windows for interbase parameters for duplex d(CGCGAATTCGCG). The values assumed in the native dodecamer crystal structure are indicated by dashed lines. [Reproduced with permission from Swaminathan, S.; Ravishanker, G.; Beveridge, D. L. (1991), *J. Am. Chem. Soc.* **113**, 5027. Copyright © (1991) American Chemical Society.]

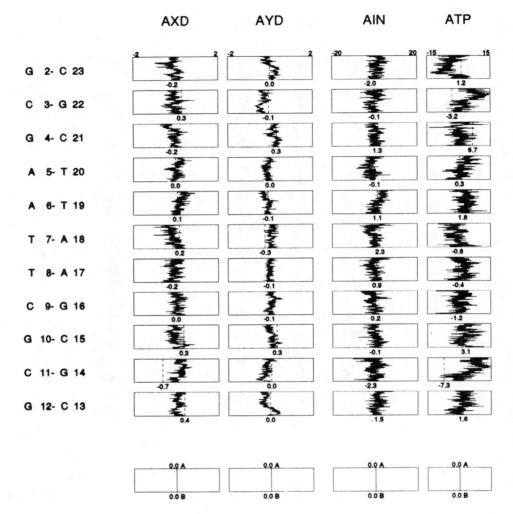

FIGURE 5.20.
Helicoidal windows for axis junction parameters for duplex d(CGCGAATTCGCG). The values assumed in the native dodecamer crystal structure are indicated by dashed lines. [Reproduced with permission from Swaminathan, S.; Ravishanker, G.; Beveridge, D. L. (1991), *J. Am. Chem. Soc.* **113**, 5027. Copyright © (1991) American Chemical Society.]

The natural curvature of this crescent-shaped dication is complimentary to the minor groove, and this drug fits snugly into the double helix. It is generally believed that the DNA binding property of netropsin is responsible for the drug's ability to inhibit DNA synthesis. The X-ray structure of netropsin bound to d(CGCGATATCGCG)$_2$ is illustrated in Figure 5.21 (Coll et al., 1989).

Footprinting experiments, a DNA cleaving method that maps drug-binding location, have shown that netropsin preferentially binds to runs of A-T base pairs over G-C pairs (Van Dyke et al., 1982). The primary difference between A-T and G-C pairs in the minor groove is the presence of a guanine NH$_2$ group at C2 (see Fig. 5.22). Thus G-C pairs may prevent binding of netropsin because of steric hindrance between the drug and DNA (see Fig. 5.23).

The relative importance of other possible factors responsible for A-T base specificity, such as hydrogen bonds, van der Waals interactions, electrostatic interactions, and curvature of the ligand, is still a subject of active debate by both experimentalists and theorists (Rao et al., 1988; Dasgupta et al., 1990; Zakrzewska et al., 1987). This work has inspired

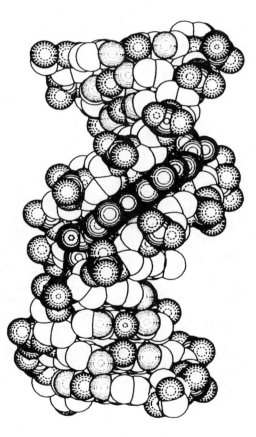

FIGURE 5.21.
van der Waals diagram of a complex of netropsin and d(CGCGATATCGCG)$_2$ looking into the minor groove. [Reproduced with permission from Coll, M.; Aymami, J.; van der Marel, G. A.; van Boom, J. H.; Rich, A.; Wang, A. H.-J. (1989), *Biochemistry* **28**, 310. Copyright © (1989) American Chemical Society.]

242

(a)   **(b)**

Major groove   Major groove

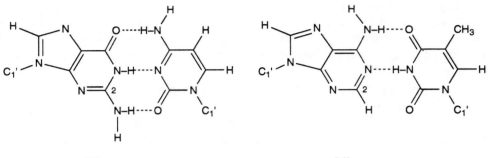

Minor groove   Minor groove

FIGURE 5.22.
Major and minor groove of Watson–Crick base pairs (a) G-C and (b) A-T. The C2 positions of the purines G and A are labeled.

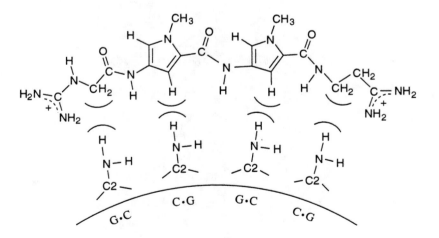

FIGURE 5.23.
Schematic of repulsive steric interactions in a netropsin-d(GCGC)₂ complex; the floor of the minor groove is represented by the curved line at the bottom of the figure. [Adapted with permission from Kopka, M. L.; Yoon, C.; Goodsell, D.; Pjura, P.; Dickerson, R. E. (1985), *Proc. Natl. Acad. Sci. USA* **82**, 1376.]

attempts to design synthetic analogs of netropsin as possible gene control agents (see, e.g., Debart et al., 1989).

5.3.2 Netropsin Binding

In this section, we will examine a molecular mechanics study by Gago et al. (1989) of Oxford University. This work modeled the binding of netropsin to the B-DNA conformation of the DNA dodecamers d(ATATATATATAT)$_2$, abbreviated as ATAT, and d(GCGCG-CGCGCGC)$_2$, abbreviated as GCGC *in vacuo*. Based on the molecular mechanics results, Gago et al. proposed that the width of the minor groove in the DNA–netropsin complexes was a determinant of the A-T base selectivity of netropsin.

METHODS

The starting double helical B-DNA structures were created by restricting all bond lengths and angles to those established experimentally for fiber DNA. Hydrogens on carbon atoms in DNA were not explicitly included. The starting netropsin structure was created by adding hydrogen atoms to the known crystal structure of the unbound drug using standard bond lengths and angles.

The AMBER suite of programs was used for the calculations. The missing AMBER parameters of netropsin were interpolated. Charges on the netropsin were obtained by the semiempirical molecular orbital (MO) method AM1. The DNA charges were those given in AMBER. Counterions were included. A sodium ion was placed in the plane of each phosphate group and given an enlarged van der Waals radius to mimic a hexahydrated sodium. The charge on the sodium was set to 1. A distance-dependent dielectric function was used to dampen the electrostatic interactions. Water was not included.

The B-DNA structure was minimized *in vacuo* first by allowing only the counterions to move, then the whole macromolecule was minimized. Netropsin was minimized separately. The docking of netropsin to DNA was accomplished by using the interactive molecular graphics program HYDRA. The drug was docked to the minor grove of DNA, optimizing the hydrogen-bond contacts and minimizing the van der Waals contacts. The best complexes from the preliminary survey of the conformational space were then fully optimized.

RESULTS

The structure of the ATAT netropsin complex after optimization showed the drug molecule was anchored in the center of the minor groove of ATAT, with its pyrrole rings twisted in a manner that matched the natural curvature of the minor groove of B-DNA. A comparative illustration of the hydrogen bonds between netropsin and the electronegative atoms of the minor groove of the ATAT tract determined by molecular mechanics and X-ray crystallography (Coll et al., 1989) is shown in Figure 5.24. The number and positions of hydrogen bonds predicted by mechanics are in poor agreement with the X-ray results for a related complex.

The binding energy of the netropsin to ATAT was calculated to be more negative than the binding to GCGC, in agreement with experiment. The relative importance of non-

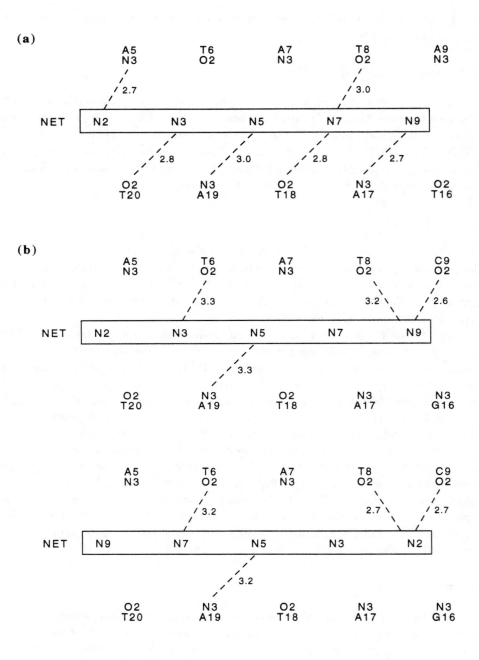

FIGURE 5.24.
Schematic representation of the hydrogen bonds between the electronegative atoms of the ATAT tract of the minor groove in (a) netropsin and d(ATATATATAT)$_2$, calculated by molecular mechanics, and (b) two orientations of netropsin bound to the related DNA d(CGCGATATCGCG)$_2$, determined by X-ray crystallography (Coll et al., 1989). Dashed lines indicate hydrogen-bonding interactions and numbers indicate the length of the hydrogen bonds in angstroms. [Schematic adapted with permission from Gago, F.; Reynolds, C. A.; Richards, W. G. (1989) The binding of nonintercalative drugs to alternating DNA sequences, *Mol. Pharmacol.* **35**, 232–241.]

bonded van der Waals, electrostatic, and hydrogen-bonded interactions to the differential binding energy was also examined by Gago et al. (1989). The calculated interaction energy of the DNA–netropsin complex was broken down into its nonbonded, electrostatic, and hydrogen-bonded contributions (Table 5.3). In this table, E_{el} is the electrostatic contri-

TABLE 5.3. Interaction Energy in kilocalories per mole between Netrospin and Tetranucleotides

Complex	E_{vdw}	E_{el}	E_{HB}	Drug–DNA
ATAT–Netrospin	−63.4	−176.7	−4.1	−244.2
GCGC–Netropsin	−55.9	−160.0	−4.4	−220.3

bution, E_{vdw} corresponds to the 6–12 nonbonded dispersion–repulsion (van der Waals) term, E_{HB} is the 10–12 term for hydrogen bonding. The molecular mechanics results showed that the difference in the hydrogen-bond energy contribution to the interaction energy between the two dodecanucleotides was small, but there was a substantial differential in nonbonding interaction between the ATAT and GCGC netropsin complexes, with the ATAT–netropsin complex being favored. These results are in agreement with Dickerson and co-worker's (Kopka et al., 1985) suggestion that, although hydrogen bonds help to position netropsin correctly along the minor grove of ATAT, the actual recognition of the base sequences results from close van der Waals contacts between netropsin pyrrole CH hydrogen atoms and adenine C2 hydrogen atoms. Comparison of the nonbonding van der Waals interactions in netropsin–ATAT and netropsin–GCGC complexes is illustrated schematically in Figure 5.25. A large difference in electrostatic interaction between the ATAT and GCGC complexes was also calculated by Gago et al. (1989). This result is not in accord with the experimental study by Marky and Breslauer (1987) on the salt dependencies of the binding constants of netropsin with ATAT and GCGC. Marky and Breslauer showed the salt dependencies were nearly identical, leading to the conclusion that the electrostatic contribution was essentially equal for netropsin binding to both ATAT and GCGC. In a related study, Gago and Richards (1990) point out that, because water was not included in the molecular mechanics calculation, the electrostatic term was likely overestimated in the 1989 work.

Gago et al. (1989) discovered an interesting structural difference between the ATAT–netropsin and GCGC–netropsin complexes. The minor groove was narrower in the central part of the region covered by netropsin in the ATAT complexes than in the GCGC counterpart. The drug appears to induce the collapse of the minor groove in AT rich DNA. Distinctively narrower minor grooves in uncomplexed AT-rich DNA oligomers as compared to GC-rich DNA oligomers have been observed crystallographically (Bhattacharyya and Bansal, 1992), and this has lead to the proposal that the minor groove width of the uncomplexed DNA helix is a dominant factor in the specificity of groove-binding drugs such as netropsin (Kopka et al., 1985). The X-ray findings, taken together with the molecular mechanics results, suggest that netropsin can induce the *further* collapse of the minor groove in AT-rich DNA.

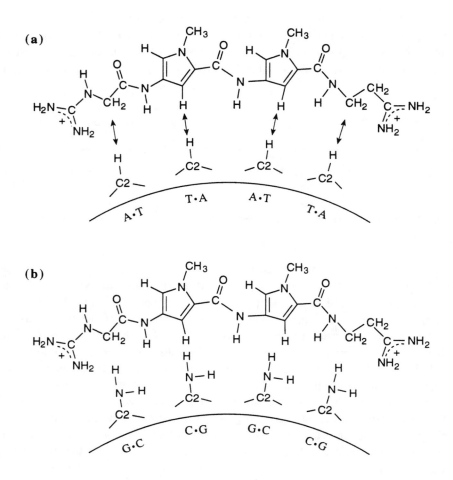

FIGURE 5.25.
Schematic of nonbonding interactions in netropsin–DNA complexes. Attractive nonbonded interactions are indicated by double headed arrows; the floor of the minor groove is represented by a curved line at the bottom of the figure. (a) netropsin–ATAT complex and (b) netropsin–GCGC complex. [Adapted with permission from Kopka, M. L.; Yoon, C.; Goodsell, D.; Pjura, P.; Dickerson, R. E. (1985), *Proc. Natl. Acad. Sci. USA* **82**, 1376.]

The possibility of drug-induced compression of the minor groove has also been noted by Wang and co-workers (Coll et al., 1989). The X-ray structures of several netropsin and netropsin-like minor groove-binding drugs complexed with the dodecamer d(CGCGA-TATCGCG)$_2$ have been solved, and the ATAT tract of the complexes have narrow grooves, in agreement with the mechanics results. Unfortunately, the widths of the minor grooves of the drug–DNA complexes cannot be compared to the minor groove width of the uncom-plexed dodecamer, because the free d(CGCGATATCGCG)$_2$ could not be crystallized. Wang and co-workers used crystal packing arguments to support the idea that the drug could induce minor groove compression. The drug–DNA complexes all crystallize in an orthorhombic lattice, where there are close intermolecular contacts between two phosphate groups of two symmetry-related dodecamers. One set of phosphate groups is in the ATAT stretch. Thus, DNAs with a wide minor groove would be severely crowded in the crystal, and so are unable to pack into this lattice.

In summary, the paper by Gago, et al. (1989) describes the *in vacuo* minimization of duplexes d(ATATATATATAT)$_2$, abbreviated as ATAT, and d(GCGCGCGCGCGC)$_2$, abbreviated as GCGC, with and without complexed netropsin, in order to determine the energetic factors responsible for a minor groove drug binding. In agreement with experiment, the netropsin binding to ATAT was calculated to be more negative than the binding to GCGC. However, the calculated relative importance (and nature) of the electrostatic, non-bonding and hydrogen-bonding interactions was not in accord with thermodynamic and X-ray crystallography experiments, probably because waters were not included in the simulation.

5.4. DYNAMICS OF INTERCALATION

5.4.1 Background

The concept of intercalation, in which a rigidly planar polycyclic molecule inserts or intercalates between the adjacent stacked base pairs of double helical DNA as if it was itself another base pair, was first introduced by Lerman (1961) to explain the binding of acridines to DNA, see Figure 5.26.

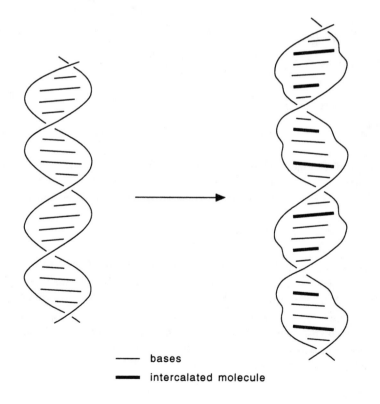

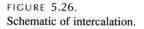

—— bases

━━ intercalated molecule

FIGURE 5.26.
Schematic of intercalation.

The biological consequences of intercalation can be significant. Some intercalators are carcinogens and mutagens, while others are important in clinical pharmacology as antibiotics or antitumor drugs (Hurley, 1989). The cell-killing effect of intercalators is ascribed not just to the intercalator's DNA-binding ability, but also to its effect on a complex involving the intercalator, DNA, and an enzyme called DNA topoisomerase II. Topoisomerase II can induce double-stranded breaks by binding to DNA in a state called a "cleavable complex." Intercalators can stabilize the cleavable complex somehow, and stimulate the irreversible breakage of DNA strands leading to cell death (Yang et al., 1985; Pommier et al., 1985). Baguley has proposed that intercalators with antitumor activity have two binding domains, one for specific DNA–intercalator interaction and the other for specific topoisomerase II–intercalator interaction (Baguley, 1990).

One of the simplest intercalators is the acridine dye proflavine (3,6-diaminoacridine). Although proflavine (**5.2**) is too toxic to be used as a systemic antibacterial agent, it was used as a topical disinfectant during World War I. The related compound mepacrine (**5.3**) has clinical antiparasitic activity, and the acridine amsacrine (**5.4**) is used in the treatment of acute leukemia and other cancers.

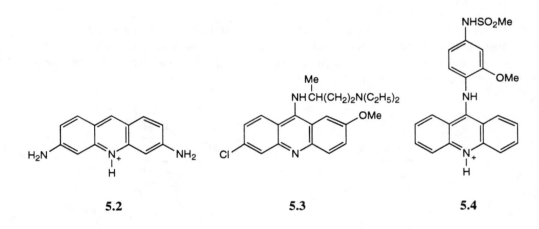

The structural change in DNA that allows the interbase spacing to essentially double to accommodate the intercalator has been investigated by a variety of experimental (Hurley, 1989; Long and Barton, 1990) and theoretical techniques (Neidle and Jenkins, 1991).

5.4.2 Proflavine Intercalation

In this section, we will examine a MD study of DNA intercalation by Herzyk et al. (1992) of the Institute of Cancer Research and Birkbeck College, both located in England. This work investigated the conformation and dynamics of the binding of proflavine (**5.2**) to a dodecamer of double helical DNA in water. The results showed that the structural consequences of intercalation were not confined to the binding site. In addition, significant widening of the minor and major groove widths in the proflavine complex as compared to the native DNA was also observed.

METHODS

Herzyk et al. (1992) carried out a dynamics study of the unintercalated dodecamer d(CGATACGATACG)$_2$, and on the same sequence with a proflavine intercalated between the central G-C pairs. Calculations included counterions and water. The starting structure of the native, unintercalated dodecamer was that of idealized B-DNA (Arnott and Hukins, 1972). In order to minimize end effects during the simulation, optimizations were carried out *in vacuo* on a 20-*mer*, and then the extra end base pairs removed. Counterions were added to the resulting structure. A sodium ion was placed 3 Å from each phosphorus atom along the OPO bisector and 1474 water molecules were added, giving a total of 5204 atoms. First, the water molecules were minimized, keeping the nucleotide and associated ions fixed, followed by further minimization of the full system except for the first and last base pairs, which were kept fixed.

The starting model of the intercalated dodecamer was generated from a crystal structure (Fig. 5.27) of a duplex dinucleoside monophosphate CpG DNA fragment intercalated with proflavine (Neidle et al., 1980). Nine base pair fragments in the canonical (idealized) B conformation were added on either end of the central section of the dinucleoside mono-phosphate–proflavine intercalation complex giving a 20-*mer* with a proflavine molecule intercalated between the central C-G pair. This structure was minimized *in vacuo* keeping the central dCpG-proflavine fragment and all phosphorus atoms, except those adjacent to the dCpG-proflavine duplex, fixed. The central dodecamer was excised, surrounded by 22 sodium ions and 1509 water molecules. This system, containing 5337 atoms, was optimized in two stages. First, only the water molecules were minimized, keeping the solute fixed, followed by optimization of the full system except for the first and last base pairs, which were kept fixed.

The AMBER 3.1 program (all atom) was used for the calculations. van der Waals and electrostatic interactions between pairs of atoms more than 9.5 Å apart were ignored. The

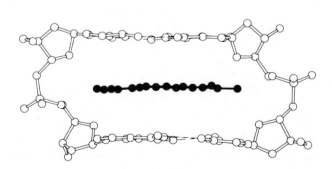

FIGURE 5.27.
Molecular structure of a 2:2 dCpG–proflavin complex in the crystalline solid. View parallel to the base pairs. [Adapted with permission from Swaminathan, S.; Beveridge, D. L.; Berman, H. M. (1990), *J. Phys. Chem.* **94**, 4660. Copyright © 1990 American Chemical Society.]

DNA charges were those given in AMBER, and charges on the proflavine were obtained by the QUEST program (Singh and Kollman, 1984). A distance-dependent dielectric function was applied to dampen the electrostatic interactions in the *in vacuo* calculations. The dielectric constant was set to 1 in calculations that included water. The SHAKE algorithm was applied to constrain the bond lengths of bonds containing hydrogen atoms only.

The dynamics simulation on the dodecanucleotide alone was carried out for 90 ps, using a time step of 0.001 ps. The rms deviations of the structure and the total energy showed the system had equilibrated after 10 ps. The last 80 ps of the dynamics run was used for analysis. The dynamics simulation on the dodecanucleotide–drug complex was also continued for 90 ps, using a time step of 0.001 ps. Inspection of the rms deviations and the total energy showed the system had equilibrated only after 20 ps. The analysis was carried out over the last 70 ps of the simulation.

RESULTS

The minimized conformations of the dodecamer and **5.2**-dodecamer complex are illustrated in Figure 5.28. Analysis of the conformation of the dodecanucleotide alone during the dynamics simulation showed the helical structure was in the B-DNA family. Groove widths were also B-like although the minor groove was consistently wider than classical canonical B-DNA. Percentage of different torsion conformations for the native DNA duplex is collected in Table 5.4. Correlations in backbone torsion angles, especially the pairs α/γ and ϵ/ζ were observed during this study. Angle β was conformationally stable over almost all of the nucleotides, and uncorrelated with changes in α or γ. The backbone conformational results parallel the findings of Swaminathan et al. (1991) on the dynamics of d(CGCGAATTCGCG)$_2$, described in Section 5.2. Sugar pucker angle changed during the simulation of solvated d(CGATACGATACG)$_2$: the C2'-*endo* structure typical of B-DNA changed to a C3'-*endo* value characteristic of A-DNA, especially in

TABLE 5.4. Percentage of Different Backbone Torsion Conformations for the Native Dodecanucleotide Structure

	AVER1[a]			AVER2[b]			OPT[c]		
	g^+	t	g^-	g^+	t	g^-	g^+	t	g^-
α	5	7	88	6	9	85			100
β	5	93	2	3	95	2		100	
γ	89	11		86	14		100		
δ	56	44		56	44		33	67	
ϵ		89	11		89	11		100	
ζ	1	18	81	1	16	83			100
χ		65[d]	35[e]		64[d]	36[e]		33[d]	67[e]

[a] Molecular dynamics averages calculated using all base pairs.
[b] Molecular dynamics averages calculated using the central hexamer only.
[c] Values adopted in the energy-minimized structure.
[d] χ–anti.
[e] χ–high anti.

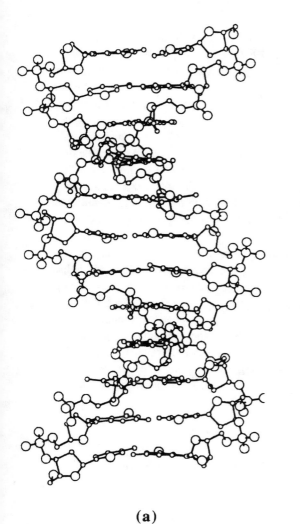 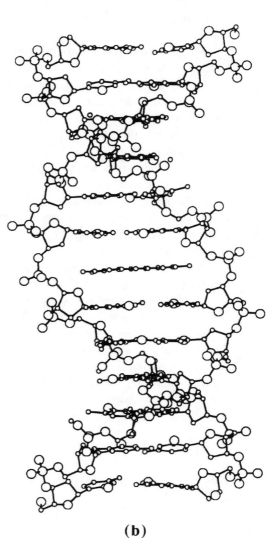

(a) **(b)**

FIGURE 5.28.
Structures of the (a) dodecamer and (b) proflavine–dodecamer complex after energy refinement.
[Reproduced with permission from Herzyk, P.; Neidle, S.; Goodfellow, J. M. (1992,) *J. Biomol. Structure Dyn.* **10**, 97.]

the central region of strand 1. This result was somewhat different than for the duplex simulated by Swaminathan et al. (1991): Here the C2'-*endo* structure gave way to C1'-*exo* in the simulation, with only a transient repuckering to the C3'-*endo* structure. The hydrogen bonding between base pairs was fairly stable with only occasional and transient changes.

Dynamics simulation of the proflavine drug–DNA complex showed abrupt conformational transitions at 20 ps, leading Herzyk et al. (1992) to believe that the structure was a stable one, relevant to the global minimum. Percentage of different torsion conformations for the proflavine complex is collected in Table 5.5. Correlations in the backbone torsion angle pairs α/γ and ϵ/ζ were observed, but there were fewer correlated transitions compared with the drug free structure. Angle β was conformationally stable. Sugar pucker angles changed from the mainly C2'-*endo* conformation towards O4'-*endo* and C3'-*endo* puckering. More sugars adopted the C3'-*endo* structure in the drug complex than the drug free form. This finding is at odds with the results of solid state NMR on proflavine-fiber DNA (Tang et al., 1990), which ruled out any significant fraction of 3'-*endo* sugars. Most of the base pair hydrogen bonds were stable during the course of the dynamics simulation. There were, however, base pair hydrogen-bond disruptions in the A-T base pairs flanking the intercalation site.

The structural consequences of intercalation were not restricted to the binding site. The backbone α and γ angles, which were correlated in each simulation, showed distinct changes between uncomplexed and complexed DNA. The percentage of α angles in the *gauche*⁻ conformation was 85 and 58% for the dodecamer duplex and drug complex, respectively. A similar difference was observed in the percentage of γ angles in the *gauche*⁺ (g^+) conformation (cf. Table 5.4 to Table 5.5). The sugar pucker parameters all showed an increase in the amount of C3'-*endo* conformations relative to C2'-*endo* conformations in the drug complex. The δ torsion angle showed increases in the percentage of g^+ conformations, and the χ angle showed increases in the percentage of *t* compared with g^- conformations (cf. Table 5.4 to Table 5.5).

TABLE 5.5. Percentage of Different Backbone Torsion Conformations for the Proflavine Complex

	AVER1[a]			AVER2[b]			OPT[c]		
	g^+	t	g^-	g^+	t	g^-	g^+	t	g^-
α	4	31	65	4	39	58	0	5	95
β	6	94		5	95		−100		
γ	66	34		58	42		100		
δ	64	36		70	30		33	67	
ϵ		89	11		92	8	5	90	5
ζ		14	86		12	88	5	10	85
χ		74	26		78	22		33	67

[a] Molecular dynamics averages calculated using all base pairs.
[b] Molecular dynamics averages calculated using the central hexamer only.
[c] Values adopted in the energy-minimized structure.

Both minor and major groove widths were significantly widened in the proflavine complex compared to the uncomplexed dodecamer. The widening of the minor groove by intercalating drugs may have specific biological significance; for example, it could explain the experimentally observed enhancement of cutting ability by the enzyme DNase I near daunomycin (also an intercalator) binding sites (Chaires et al., 1987; Suck et al., 1988). The calculated changes in DNA groove structure upon proflavine intercalation may also be important in understanding the basis for cleavable complex stabilization by antitumor agents.

To summarize, this extensive computer simulation by Herzyk et al. (1992) investigated the binding of proflavine (**5.2**) to double helical d(CGCGAATTCGCG)$_2$ in water. Both the conformational features and dynamic properties of the double helix were effected by the intercalation. Significant widening of the minor and major groove widths in the proflavine complex as compared to the uncomplexed DNA was observed. This result may be relevant to the mechanism of action of intercalating antitumor agents.

5.5. COVALENT BONDING

5.5.1 Background

Cisplatin [*cis*-diamminedichloroplatinum(II)] **5.5**, a planar molecule, is one of the very few anticancer drugs with significant clinical activity against a wide range of solid tumors and is widely used despite its severe side effects (Loehrer and Einhorn, 1984). *trans*-Diamminedichloroplatinum(II) **5.6**, the geometric isomer of cisplatin, has no antitumor activity.

5.5   **5.6**

The primary mechanism of action of cisplatin involves covalent binding to DNA, although cisplatin's binding to DNA does not, in itself, account for the drug's anticancer activity. Among the proposals for the mechanism of cisplatin's cytotoxicity is the binding of a protein HMG1 to cisplatin-modified double helical DNA (Pil and Lippard, 1992), which may prevent recognition of the damaged nucleic acid by cellular repair enzymes.

Detailed *in vitro* studies of the interaction of cisplatin with DNA or fragments of DNA show that the cisplatin binds to DNA primarily through 1,2-intrastrand cross-links between the adjacent N7 atoms of the nucleobases guanine and adenine (Reedijk et al., 1987; Sherman and Lippard, 1987). The structure of a single-stranded cisplatin-d(GpG) model is illustrated schematically in Figure 5.29.

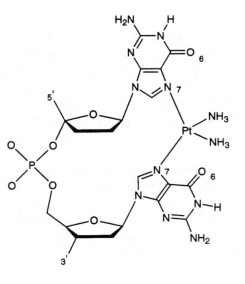

FIGURE 5.29.
Schematic representation of the structure of d(GpG) coordinated to cisplatin. [Adapted from den Hartog, J. H. J.; Altona, C.; Chottard, J.-C.; Girault, J.-P.; Lallemand, J.-Y.; de Leeuw, F. A. A. M.; Marcelis, A. T. M.; Reedijk, J. (1982), *Nucleic Acids Res.* **10**, 4715, by permission of Oxford University Press.]

The kinetically most favored binding sites of duplex DNA with cisplatin are adjacent guanine base GpG sequences (**5.7**), with less frequent attachment of cisplatin to adjacent adenine and guanine residues of ApG sequences (**5.8**), where the adenine is on the 5′ terminus of the complex (Fichtinger-Schepman et al., 1985):

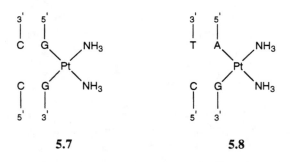

5.7   **5.8**

Cisplatin binding does not occur to d(GpA) (van der Veer et al., 1986), where the adenine is on the 3′ terminus of the complex, (**5.9**):

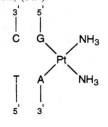

5.9

X-ray work (Sherman et al., 1985; Admiraal et al., 1987; Sherman et al., 1988) shows that the bases are destacked in complexes of cisplatin bound to single-stranded dinucleotides, allowing the N7 atoms to be in nearly the proper position for square planar coordination. The X-ray structure of the single-stranded cisplatin dinucleotide *cis*-[Pt(NH$_3$)$_2${d(pGpG}}] is illustrated in Figure 5.30 (Sherman et al., 1985). The coordinated bases of the single-stranded dinucleotide tilt toward each other with dihedral angles between ring planes of 76°–87°. Such a local disruption of base pairs could induce local distortions in double helical DNA. A kink in the double helix axis might occur if the

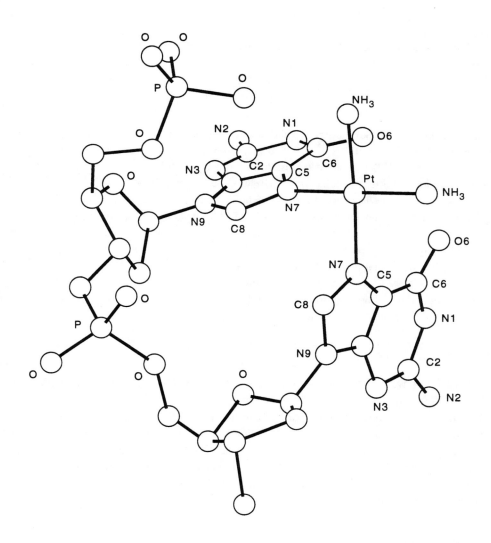

FIGURE 5.30.
The X-ray structure of *cis*-[Pt(NH$_3$)$_2${d(pGpG}}]. [Adapted with permission from Sherman, S. E.; Gibson, D.; Wang, A. H.-J.; Lippard, S. J. (1985), *Science* **230**, 412. Copyright © 1985 by the AAAS.]

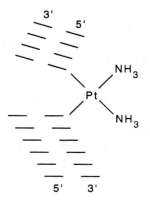

FIGURE 5.31.
Schematic model of a kinked platinated DNA double helix. [Adapted with permission from Kozelka, J.; Chottard, J.-C. (1990), *Biophys. Chem.* **35**, 165.]

surrounding bases stack on the perturbed nucleotides (Fig. 5.31). The anomalous gel electrophoresis mobility shifts (Leng, 1990) of platinated double-stranded oligonucleotides are consistent with the bending or kinking of the DNA helix by 34° toward the major groove, and unwinding the helix by 13° (Bellon and Lippard, 1990).

The geometric constraints of the double helix are different than that of single-stranded DNA, making detailed predictions of duplex DNA cisplatin adduct structures based on the X-ray structures of short, single-stranded DNA-cisplatin adducts, such as the one shown in Figure 5.30, problematic. The NMR studies also cannot completely determine the structure of cisplatin complexed to double helical DNA (den Hartog et al., 1985a; Herman et al., 1990); for example, the geometry of the kink cannot be directly related to NMR parameters. Molecular modeling might fill this knowledge gap, providing useful structural understanding of the biologically relevant double-stranded cisplatin adducts.

5.5.2 Cisplatin Binding

In this section, we will describe a molecular mechanics study of cisplatin (**5.5**) binding to duplex DNA tetramers. Hambley (1991) of the University of Sydney, Australia has modeled the monofunctional and bifunctional binding of cisplatin to sequences d(GpApGpG)$_2$ and d(GpGpApG)$_2$ in order to investigate the factors that might stereochemically influence the binding of cisplatin to adenine-containing sequences. The modeling work suggests that the experimentally observed binding specificity of cisplatin for ApG over GpA sequences is a consequence of secondary interactions between an ammine (NH$_2$) ligand and groups on the purine base.

METHODS

The starting structure of the uncomplexed tetranucleotide duplexes were constructed using a helix building program. Both A and B forms of DNA were constructed. A B-DNA model with sugar conformations other than the normal C2′-*endo* conformation was also constructed, and called B'-DNA. The starting structures were minimized prior to complexation with cisplatin.

The AMBER force field was used to model the DNA. The angle deformation parameters were modified, and the centers of repulsion for hydrogen atoms were moved toward the heavy atoms to which they were bonded. A distance-dependent dielectric function was applied to dampen the electrostatic interactions in the *in vacuo* calculations. Electrostatic charges on DNA were taken from the AMBER force field, and the charges on the complexes were estimated to be $+0.05$ on Pt, and $+0.20$ on the ammine (NH_2) nitrogen atoms.

RESULTS

cis-Diammineplatinum(II) complexes [Pt $(NH_3)_2]^{2+}$ bifunctionally bound to the sequences d(GpApGpG)$_2$, abbreviated as GAGG, and d(GpGpApG)$_2$, abbreviated as GGAG, were modeled in order to understand binding specificity.

Cisplatin binding to the ApG sequence was examined by molecular mechanics. The minimized A-DNA type structure of [Pt $(NH_3)_2]^{2+}$ bound to the ApG sequence of GAGG is illustrated in Figure 5.32. Hydrogen bonds between one ammine and a terminal oxygen of the 5'-phosphate group, and another between the other ammine ligand and the O6 of the coordinated guanine (3' side) were observed, schematically illustrated below for one strand (Fig. 5.33): The existence of similar hydrogen bonds have been established for single-stranded DNA cisplatin adducts by solution NMR (den Hartog et al., 1985b; Fouts et al., 1988) and X-ray crystallography (Reedijk et al., 1987; Sherman et al., 1988), and may play an important role in stabilizing the conformation of the Pt-coordinated bases.

In the minimized B-DNA type structure of [Pt $(NH_3)_2]^{2+}$ bound to GAGG, the hydrogen bond between the ammine ligand and O6 of the coordinated guanine was also observed, but the orientation of the backbone was such that hydrogen bonding between the ammine and terminal oxygen of the 5'-phosphate group was not possible. A conforma-

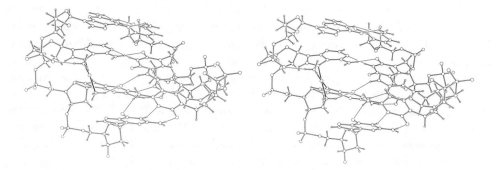

FIGURE 5.32.
Stereoview of [Pt(NH$_3$)$_2$]$^{2+}$ bound to d(GpApGpG)$_2$. Hydrogen bonds are indicated by thin lines. [Reproduced with permission from Hambley, T. W. (1991), *Inorg. Chem.* **30**, 937. Copyright © (1991) American Chemical Society.]

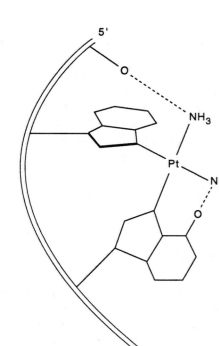

FIGURE 5.33.
Schematic of hydrogen bonds in Pt $(NH_3)_2{}^{2+}$ bound to the d(ApG) sequence. View of one strand; ribbon represents the sugar phosphate backbone.

tional rearrangement of the 5'-sugar ring from C2'-*endo* to C3'-*endo* giving a B'-DNA structure, brought the phosphate into a position where the hydrogen bond could form. The C3'-*endo* sugar puckers are more characteristic of A-DNA than B-DNA, and changes in sugar geometry of the 5'-nucleotide from C2'-*endo* to C3'-*endo* have also been observed in X-ray (Reedijk et al., 1987; Sherman et al., 1988) and solution NMR studies (den Hartog et al., 1985a; Caradonna and Lippard, 1988) of single-stranded DNA–cisplatin complexes.

Cisplatin binding to the GpA sequence was also investigated by molecular mechanics. The minimized A-DNA type structure of $[Pt (NH_3)_2]^{2+}$ bound to the GpA sequence of GGAG is given in Figure 5.34. Both the minimized A-DNA or B'-DNA GGAG-cisplatin adducts showed a hydrogen bond between one ammine ligand and a terminal oxygen of the 5'-phosphate group, analogous to the one observed for GAGG. However, since the purine on the 3'-side was an adenine rather than guanine, with an NH_2 instead of an O6, hydrogen bonding between the ammine and purine was impossible. Rather, there were repulsive H···H contacts in the range 2.59–2.64 Å. These hydrogen bonds are schematically illustrated for one strand in Figure 5.35.

These results suggested that the preference of cisplatin binding to ApG over GpA sequences was a consequence of interactions between NH_3 ligands and groups in the 6

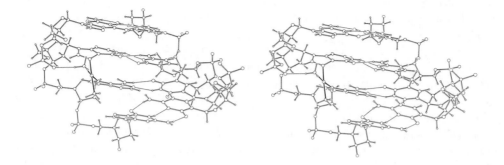

FIGURE 5.34.
Stereoview of Pt(NH₃)₂²⁺ bound to d(GpGpApG)₂. Hydrogen bonds are indicated by thin lines. [Reproduced with permission from Hambley, T. W. (1991), *Inorg. Chem.* **30**, 937. Copyright © (1991) American Chemical Society.]

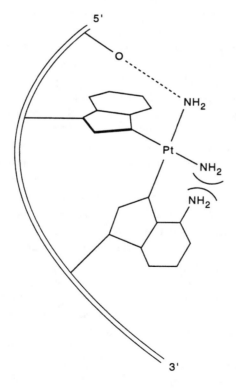

FIGURE 5.35.
Schematic of hydrogen bonds in Pt (NH₃)₂²⁺ bound to the d(GpA) sequence. View of one strand; ribbon represents the sugar phosphate backbone.

position of the 3'-purine. However, because the binding of cisplatin to DNA is thought to be kinetically controlled (Johnson et al., 1985), five-coordinate transition states were also modeled in order to determine whether the same interactions would occur during the binding process.

The transition state of cisplatin binding was also modeled by mechanics. Platinum coordinated to two purine bases and three NH_3 ligands, with five-coordinate geometries intermediate between idealized square pyramidal and trigonal bipyramidal were modeled. Interactions between ammine ligand and groups in the 6 position of the 3'-purine similar to those observed in the bifunctional model were seen, but the increased bulk about the Pt atom (five-coordinate vs. four-coordinate) resulted in these contacts becoming shorter. Thus for cisplatin binding to GAGG, the $O6 \cdots H_3N$ hydrogen bond shortened from 2.91–2.93 Å to 2.72–2.76 Å indicating increased stabilization; conversely in binding to GGAG, the $NH_2 \cdots H_3N$ contact shortened from 2.59–2.64 Å to 2.45–2.47 Å, suggesting increased destabilization. The proposal that the experimentally observed binding specificity of cisplatin for ApG over GpA sequences was a consequence of secondary interactions between an ammine ligand and groups on the purine base was also supported by models of five-coordinate transition state complexes.

The binding specificity model developed by Hambley has implications for the design of new analogs of cisplatin. Hambley (1991) proposed that a cisplatin analog with one ammine hydrogen-bond donor group and another group able to accept a hydrogen bond from the adenine NH_2 might bind to GpA sequences of DNA. A complex with one ammine and one sulfoxide ligand would be stereochemically ideal, and such compounds might have considerably different anticancer properties than cisplatin. Ling et al. (1993) reported the preparation and characterisation of a series of these platinum amminesulfoxide complexes. The compounds were tested for anticancer activity and found to be modestly active. Preliminary DNA cleavage results indicated the compounds bound preferentially to GA sequences, in agreement with the molecular mechanics model.

To summarize, this molecular mechanics study by Hambley (1991) investigates the binding of cisplatin to GAGG and GGAG *in vacuo* to determine the stereochemical factors involved in binding specificity. The corresponding five-coordinate transition states were also examined. The results suggested that the preference of cisplatin binding to ApG over GpA sequences was a consequence of interactions between NH_3 ligands and groups in the 6 position of the 3'-purine. In the structure of Pt $(NH_3)_2^{2+}$ bound to GAGG, a hydrogen bond between the ammine ligand and O6 of the coordinated guanine was observed, whereas in GGAG, repulsive $H \cdots H$ contacts between the ammine ligand and the NH_2 of coordinated adenine was found.

It is informative to compare the Hambley cisplatin study to the netropsin investigation by Gago et al. (1989) described in Section 5.3. Both attempt to examine the origins of drug-binding specificity by molecular mechanics, and both calculations were carried out using the AMBER force field *in vacuo*. Given the results of Section 5.2, both dynamics and water are necessary to correctly describe DNA, and so the conclusions of both the netropsin and cisplatin studies are suspect. Nonethless, the cisplatin study by Hambley may be more accurate, because of a cancellation of errors. The modeling work compared the structures of two cisplatin complexes that were isomers of each other [Pt$(NH_3)_2^{2+}$ bound to GGAG vs. GAGG]. The solvation of these two complexes should be similar, and therefore, neglect of solvation might not alter the structural conclusions. In the case of the netropsin work, the thermodynamic and structural study was done by comparing reactions: DNA and netropsin reacting to form a DNA–netropsin complex. The reactions were examined for the DNA duplexes d(ATATATATATAT)$_2$ and d(GCGCGCGCGCGC)$_2$. The duplexes are not isomers. The differential effect of solvation on DNA as compared to complexed DNA, and the two sequences of DNA, is likely to be significant.

Homework

5.1. Using pencil and paper to draw a schematic of the structure of d(pGpCpA) as in Figure 5.2.

5.2. Sketch a deoxyribose ring. Minimize. Do a 1-ps dynamics run at 600 K. Make a movie of the trajectory and examine the sugar pucker. Assign Xn'-*exo* and Xn'-*endo* conformations to the puckering modes observed in the trajectory.

5.3. Create the B-DNA double helix d(GGGGGGGGGG)$_2$. Find the sugars. Find the phosphate groups. Find the bases. Find the hydrogen-bonds between the base pairs. Locate the major and minor groove.

5.4. Is the left-handed Z-DNA helix the mirror image of the right-handed B-DNA helix? Explain.

5.5. Extract the X-ray structure of a small B-DNA duplex from the Brookhaven Protein Data Bank (File 1DN9). Locate the waters of crystallization. Examine the rise per base pair, diameter of the helix, glycosyl angle conformation, and sugar pucker. How does your analysis of this structure compare to the idealized B-DNA structure of Table 5.1?

5.6. Use paper and pencil to calculate the electrostatic energy of two partial charges, both negative 0.27, separated by a distance of 3.5 Å using Coulomb's law (Eq. 3.16). Assume ϵ is 80 (water). Repeat the calculation using a distance of 6 Å and then 2 Å. Repeat the first calculation using an ϵ of 1 (vacuum). Repeat the calculation using a distance dependent dielectric (Eq. 3.22), using an ϵ^* of 1. The commonly used cutoff distance is 9.5 Å. Calculate the electrostatic energy of the two charges using a distance of 9.5 Å. Can you ignore electrostatic energy at this distance?

5.7. Minimize the DNA double helix d(GGGGGGGGGG)$_2$ created in Problem 5.3 *in vacuo* using a distance dependent dielectric parameter. Carry out a 2-ps dynamics run at 300 K. Examine the trajectory movie. Why should you use a distance dependent dielectric parameter when doing an *in vacuo* calculation? What kind of atomic motion do you see in the trajectory? If the time scale of DNA helix bending is on the order of several picoseconds, is it possible to see this motion during a 2-ps calculation? If you have the resources, do a longer calculation.

5.8. Extract the X-ray structure of a small duplex of DNA complexed with netropsin from the Brookhaven Protein Data Bank (File 1DNE). Find the attractive VDW interactions between netropsin and the DNA. Find the attractive hydrogen bonds. Netropsin is nearly symmetrical; can you remove the netropsin from the DNA duplex, flip it over and dock it back into the DNA? Does the netropsin still fit, even without minimization?

5.9. The molecule illustrated below (**5.10**) is an analog of netropsin in which two methylpyrroles have been replaced by two methylimidazoles (CH groups are replaced by N groups in the heterocycles).

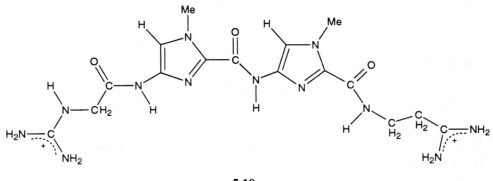

5.10

Draw a van der Waals interaction scheme as in Figure 5.23 for this molecule with the minor groove of duplex AGGA. Compare to Figure 5.25(a). Would you predict that this molecule would bind to G-C base pair runs of double helical DNA? Why or why not?

5.10. In Section 5.3, the *in vacuo* binding of DNA dodecamers to netropsin was simulated in order to determine the origins of A-T specificity. Can you redesign this theoretical experiment using paper and pencil to allow for a better cancellation of differential solvation errors? Your experiment must also be *in vacuo* on dodecamers. Assume that netropsin binds to DNA over four base pairs. *Hint:* See how the binding of cisplatin to DNA was simulated in Section 5.5.

5.11. What electronic factors stabilize the intercalator within the DNA double helix? Can all of these electronic factors can be simulated by the commonly used biological force fields CHARMM and AMBER (see Section 3.2)? Explain.

References

Admiraal, G.; van der Veer, J. L.; de Graaff, R. A. G.; den Hartog, J. H. J.; Reedijk, J. (1987), Intrastrand bis(guanine) chelation of d(CpGpG) to *cis*-Platinum: An X-ray single-crystal structure analysis, *J. Am. Chem. Soc.* **109**, 592.

Arnott, S.; Hukins, D. W. L. (1972), Optimised parameters for A-DNA and B-DNA. *Biochem. Biophys. Res. Commun.* **47**, 1504.

Arnott, S.; Chandrasekaran, R.; Birdsall, D. L.; Leslie, A. G. W.; Ratliff, R. L. (1980), Left-handed DNA helices, *Nature (London)* **283**, 743.

Baguley, B. C. (1990), The possible role of electron-transfer complexes in the antitumour action of amsacrine analogues, *Biophys. Chem.* **35**, 203.

Bellon, S. F.; Lippard, S. J. (1990), Bending studies of DNA site-specifically modified by cisplatin, *trans*-diamminedichloroplatinum(II) and *cis*-[Pt(NH₃)₂(*N*3-cytosine)Cl]⁺. *Biophy. Chem.* **35**, 179.

Bhattacharyya, D.; Bansal, M. (1992), Groove width and depth of B-DNA structures depend on local variation in slide, *J. Biomol. Structure Dyn.* **10**, 213.

Caradonna, J. P.; Lippard, S. J. (1988), Synthesis and characterization of [d(ApGpGpCpCpT)]₂ and its adduct with the anticancer drug *cis*-diamminedichloroplatinum(II). *Inorg. Chem.* **27**, 1454.

Chaires, J. B.; Fox, K. R.; Herrera, J. E.; Britt, M.; Waring, M. J. (1987), Site and sequence specificity of the daunomycin-DNA interaction, *Biochemistry* **26**, 8227.

Coll, M.; Aymami, J.; van der Marel, G. A.; van Boom, J. H.; Rich, A.; Wang, A. H.-J. (1989), Molecular structure of the netropsin-d(CGCGATATCGCG) complex: DNA conformation in an alternating AT segment, *Biochemistry* **28**, 310.

Dasgupta, D.; Howard, F. B.; Sasisekharan, V.; Miles, H. T. (1990), Drug-DNA binding specificity: Binding of netropsin and distamycin to poly(d2NH$_2$A-dT), *Biopolymers* **30**, 223.

Debart, F.; Perigaud, C.; Gosselin, G.; Mrani, D.; Rayner, B.; Le Ber, P.; Auclair, C.; Balzarini, J.; De Clercq, E.; Paoletti, C.; Imbach, J.-L. (1989), Synthesis, DNA binding, and biological evaluation of synthetic precursors and novel analogues of netropsin, *J. Med. Chem.* **32**, 1074.

den Hartog, J. H. J.; Altona, C.; Chottard, J.-C.; Girault, J.-P.; Lallemand, J.-Y.; de Leeuw, F. A. A. M.; Marcelis, A. T. M.; Reedijk, J. (1982), Conformational analysis of the adduct of *cis*-[Pt(NH$_3$)$_2${d(GpG)}]$^+$ in aqueous solution. A high field (500–300 MHz) nuclear magnetic resonance investigation, *Nucleic Acids Res.* **10**, 4715.

den Hartog, J. H. J.; Altona, C.; van Boom, J. H.; van der Marel, G. A.; Haasnoot, C. A. G.; Reedijk, J. (1985a), *Cis*-Diamminedichloroplatinum(II) induced distortion of a single and double stranded deoxydecanucleosidenonaphosphate studied by nuclear magnetic resonance, *J. Biomol. Structure Dyn.* **2**, 1137.

den Hartog, J. H. J.; Altona, C.; Van Der Marel, G. A.; Reedijk, J. (1985b), A ^{1}H and ^{31}P NMR study of *cis*-Pt(NH$_3$)$_2$[d(CpGpG)-*N*7(2),*N*7(3)]. The influence of a 5'-terminal cytosine on the structure of the *cis*-Pt(NH$_3$)$_2$[d(GpG)-*N*7,*N*7] intrastrand cross-link, *Eur. J. Biochem.* **147**, 371.

Dickerson, R. E. (1989), Definitions and nomenclature of nucleic acid structure parameters, *J. Biomol. Structure Dyn.*, **6**, 627.

Fichtinger-Schepman, A. M. J.; van der Veer, J. L.; den Hartog, J. H. J.; Lohman, P. H. M.; Reedijk, J. (1985), Adducts of the antitumor drug *cis*-diamminedichloroplatinum(II) with DNA: Formation, identification, and quantitation, *Biochemistry* **24**, 707.

Fouts, C. S.; Marzilli, L. G.; Byrd, R. A.; Summers, M. F.; Zon, G.; Shinozuka, K. (1988), HMQC and ^{1}H and ^{31}P NMR studies of platinum amine adducts of tetradeoxyribonucleotides. Relationship between ^{31}P shift and potential hydrogen-bonding interactions in pGpG moieties crosslinked by platinum, *Inorg. Chem.* **27**, 366.

Fratini, A. V.; Kopka, M. L.; Drew, H. R.; Dickerson, R. E. (1982), Reversible bending and helix geometry in a B-DNA dodecamer: CGCGAATT^{Br}CGCG, *J. Biol. Chem.* **257**, 14686.

Gago, F.; Reynolds, C. A.; Richards, W. G. (1989), The binding of nonintercalative drugs to alternating DNA sequences, *Mol. Pharm.* **35**, 232.

Gago, F.; Richards, W. G. (1990), Netropsin binding to poly[d(IC)]·poly[IC] and poly[d(GC)·poly[d(GC)]: A computer simulation, *Mol. Pharm.* **37**, 341.

Hambley, T. W. (1991), Molecular mechanics analysis of the stereochemical factors influencing monofunctional and bifunctional binding of *cis*-diamminedichloroplatinum(II) to adenine and guanine nucleobases in the sequences d(GpApGpG)·d(CpCpTpC) and d(GpGpApG)·d(CpTpCpC) of A- and B-DNA, *Inorg. Chem.* **30**, 937.

Herman, F.; Kozelka, J.; Stoven, V.; Guittet, E.; Girault, J.-P.; Huynh-Dinh, T.; Igolen, J., Lallemand, J.-Y.; Chottard, J.-C. (1990), A d(GpG)-platinated decanucleotide duplex is kinked. An extended NMR and molecular mechanics study, *Eur. J. Biochem.* **194**, 119.

Herzyk, P.; Neidle, S.; Goodfellow, J. M. (1992), Conformation and dynamics of drug-DNA intercalation. *J. Biomol. Structure Dyn.* **10**, 97.

Hurley, L. H. (1989), DNA and associated targets for drug design, *J. Med. Chem.* **32**, 2027.

Johnson, N. P.; Mazard, A. M.; Escalier, J.; Macquet, J. P. (1985), Mechanism of the reaction between *cis*-[PtCl$_2$(NH$_3$)$_2$] and DNA *in vitro*, *J. Am. Chem. Soc.* **107**, 6376.

Kopka, M. L.; Yoon, C.; Goodsell, D.; Pjura, P.; Dickerson, R. E. (1985), The molecular origin of DNA-drug specificity in netropsin and distamycin, *Proc. Natl. Acad. Sci. USA* **82**, 1376.

Kozelka, J.; Chottard, J.-C. (1990), How does cisplatin alter DNA structure? A molecular mechanics study on double-stranded oligonulceotides, *Biophys. Chem.* **35**, 165.

Lavery, R.; Sklenar, H. (1988), The definition of generalized helicoidal parameters and of axis curvature for irregular nucleic acids, *J. Biomol. Structure Dyn.* **6**, 63.

Lavery, R.; Sklenar, H. (1989), Defining the structure of irregular nucleic acids: Conventions and principles. *J. Biomol. Structure Dyn.* **6**, 655.

Leng, M. (1990), DNA bending induced by covalently bound drugs. Gel electrophoresis and chemical probe studies. *Biophys. Chem.* **35**, 155.

Lerman, L. S. (1961), Structural considerations in the interaction of DNA and acridines, *J. Mol. Biol.* **3**, 18.

Levitt, M. (1983), Computer simulation of DNA double-helix dynamics, *Cold Spring Harbor Symp. Quant. Biol.* **47**, 251.

Ling, E. C. H.; Allen, G. W.; Hambley, T. W. (1993), The preparation and charaterisation of some aminesulfoxidedichloroplatinum(II) complexes, *J. Chem. Soc. Dalton Trans.* 3705.

Loehrer, P. J.; Einhorn, L. H. (1984), Diagnosis and treatment. Drugs five years later. Cisplatin, *Ann. Int. Med.* **100**, 704.

Long, E. C.; Barton, J. K. (1990), On demonstrating DNA intercalation, *Acc. Chem. Res.* **23**, 271.

Marky, L. A.; Breslauer, K. J. (1987), Origins of netropsin binding affinity and specificity: Correlations of thermodynamic and structural data, *Proc. Natl. Acad. Sci. USA* **84**, 4359.

Neidle, S.; Berman, H. M.; Shieh, H. S. (1980), Highly structured water network in crystals of a deoxydinucleoside–drug complex. *Nature (London)* **288,** 129.

Neidle, S.; Jenkins, T. C. (1991), Molecular modeling to study DNA intercalation by anti-tumor drugs, *Methods Enzymol.* **203**, 433.

Neidle, S.; Pearl, L. H.; Skelly, J. V. (1987), DNA structure and perturbation by drug binding, *Biochem. J.* **243**, 1.

Olsen, W. (1982), Theoretical studies of nucleic acid conformation: Potential energies, chain statistics, and model building. In Neidle, S. Ed., *Topics in Nucleic Acid Structure,* Part 2, Macmillan, London. pp. 1–96.

Pil, P. M.; Lippard, S. J. (1992), Specific binding of chromosomal protein HMG1 to DNA damaged by the anticancer drug cisplatin, *Science* **256**, 234.

Pommier, Y.; Schwartz, R. E.; Zwelling, L. A.; Kohn, K. W. (1985), Effects of DNA intercalating agents on topoisomerase II induced DNA strand cleavage in isolated mammalian cell nuclei, *Biochemistry* **24**, 6406.

Rao, K. E.; Dasgupta, D.; Sasisekharan, V. (1988), Interaction of synthetic analogues of distamycin and netropsin with nucleic acids. Does curvature of ligand play a role in distamycin-DNA interactions? *Biochemistry* **27**, 3018.

Ravishanker, G.; Swaminathan, S.; Beveridge, D. L.; Lavery, R., and Sklenar, H. (1989), Conformational and helicoidal analysis of 30 ps of molecular dynamics on the d(CGCGAATTCGCG) double helix: "Curves", dials and windows, *J. Biomol. Structure Dyn.* **6**, 669.

Reedijk, J.; Fichtinger-Schepman, A. M. J.; van Oosterom, A. T.; van de Putte, P. (1987), Platinum amine coordination compounds as anti-tumor drugs. Molecular aspects of the mechanism of action. In Cantor, C. R., Ed., *Coordination Compounds: Synthesis and Medical Application*, Structure and Bonding, Vol. 67, Springer-Verlag, New York, pp. 53–90.

Sherman, S. E.; Gibson, D.; Wang, A. H.-J.; Lippard, S. J. (1985), X-ray structure of the major adduct of the anticancer drug cisplatin with DNA: *cis*-[Pt(NH$_3$)$_2${d(pGpG)}], *Science* **230**, 412.

Sherman, S. E.; Gibson, D.; Wang, A. H.-J.; Lippard, S. J. (1988), Crystal and molecular structure of *cis*-[Pt(NH$_3$)$_2${d(pGpG)}], the principal adduct formed by *cis*-diamminedichloroplatinum(II) with DNA, *J. Am. Chem. Soc.* **110**, 7368.

Sherman, S. E.; Lippard, S. J. (1987), Structural aspects of platinum anticancer drug interactions with DNA, *Chem. Rev.* **87**, 1153.

Singh, U. C.; Kollman, P. A. (1984), An approach to computing electrostatic charges for molecules, *J. Computational Chem.* **5**, 129.

Suck, D.; Lahm, A.; Oefner, C. (1988), Structure refined to 2 Å of a nicked DNA octanucleotide complex with DNase I, *Nature (London)* **332**, 464.

Swaminathan, S.; Beveridge, D. L.; Berman, H. M. (1990), Molecular dynamics simulation of a deoxynucleoside–drug intercalation complex: dCpG/proflavin, *J. Phys. Chem.* **94**, 4661.

Swaminathan, S.; Ravishanker, G.; Beveridge, D. L. (1991), Molecular dynamics of B-DNA including water and counterions: A 140-ps trajectory for d(CGCGAATTCGCG) based on the GROMOS force field, *J. Am. Chem. Soc.* **113**, 5027.

Tang, P.; Juang, C.-L.; Harbison, G. S. (1990), Intercalation complex of proflavine with DNA: Structure and dynamics by solid-state NMR, *Science* **249**, 70.

Tidor, B.; Irikura, K. K.; Brooks, B. R.; Karplus, M. (1983), Dynamics of DNA oligomers, *J. Biomol. Structure Dyn.* **1**, 231.

van der Veer, J. L.; van den Elst, H.; den Hartog, J. H. J.; Fichtinger-Schepman, A. M. J.; Reedijk, J. (1986), Reaction of the antitumor drug *cis*-diamminedichloroplatinum(II) with the trinucleotide d(GpApG): Identification of the two main products and kinetic aspects of their formation, *Inorg. Chem.* **25**, 4657.

Van Dyke, M. W.; Hertzberg, R. P.; Dervan, P. B. (1982), Map of distamycin, netropsin, and actinomycin binding sites on heterogeneous DNA: DNA cleavage-inhibition patterns with methidiumpropyl-EDTA·Fe(II), *Proc. Natl. Acad. Sci. USA* **79**, 5470.

Withka, J.; Swaminathan, S.; Beveridge, D. L.; Bolton, P. H. (1991), Time dependence of nuclear overhauser effects of duplex DNA from molecular dynamics trajectories, *J. Am. Chem. Soc.* **113**, 5041.

Yang, L.; Rowe, T. C.; Nelson, E. M.; Liu, L. F. (1985), *In vivo* mapping of DNA topoisomerase II-specific cleavage sites on SV40 chromatin, *Cell* **41**, 127.

Zakrzewska, KJ.; Lavery, R.; Pullman, B. (1987), A theoretical study of the sequence specificity in binding of lexitropsins to B-DNA, *J. Biomol. Structure Dyn.*, **4**, 833.

Zimmer, C.; Wähnert, U. (1986), Nonintercalating DNA-binding ligands: Specificity of the interaction and their use as tools in biophysical, biochemical and biological investigations of the genetic material, *Prog. Biophys. Mol. Biol.* **47**, 31.

Zimmerman, S. B. (1982), The three-dimensional structure of DNA, *Annu. Rev. Biochem.*, **51**, 395.

Further Reading

General DNA

Blackburn, G. M.; Gait, M. J., Eds. (1990), *Nucleic Acids in Chemistry and Biology,* Oxford University Press, Oxford, UK.

DNA Structure

Dickerson, R. E. (1992), DNA structure from A to Z, *Methods Enzymol.* **211**, 67.

Dickerson, R. E.; Drew, H. R.; Conner, B. N.; Wing, R. M.; Fratini, A. V.; Kopka, M. L. (1982), The anatomy of A-, B-, and Z-DNA, *Science* **216**, 475.

Palecek, E. (1991), Local supercoil-stabilized DNA structures, *Crit. Rev. Biochem. Mol. Biol.* **26**, 151.

Saenger, W. (1984), *Principles of Nucleic Acid Structure*, Springer-Verlag, Berlin.

Trifonov, E. N. (1991), DNA in profile, *Trends Biochem. Sci.* 16, 467.

Westhof, E. (1988), Water: An integral part of nucleic acid structure, *Annu. Rev. Biophys. Biophys. Chem.* **17**, 125.

DNA Dynamics

Alam, T. M.; Drobny, G. P. (1991), Solid state NMR studies of DNA structure and dynamics, *Chem. Rev* **91**, 1545.

Gorenstein, D. G. (1994), Conformation and dynamics of DNA and protein-DNA complexes by ^{31}P NMR, *Chem. Rev* **94**, 1315.

Hagerman, P. J. (1988), Flexibility of DNA, *Annu. Rev. Biophys. Biophys. Chem.* **17**, 265.

McCammon, J. A.; Harvey, S. C. (1987), *Dynamics of Proteins and Nucleic Acids*, Cambridge University Press, Cambridge, UK.

Patel, D. J.; Pardi, A.; Itakura, K. (1982), DNA conformation, dynamics, and interactions in solution, *Science* **216**, 581.

DNA-Drug Interaction

Franklin, T. J.; Snow, G. A. (1989), *Biochemistry of Antimicrobial Action*, fourth edition, Chapter 4, Chapman and Hall, London.

Shafer, R. H.; Brown, S. C. (1988), Nuclear magnetic resonance studies of drug nucleic acid interactions. In Kallenbach, N. R., Ed., *Chemistry and Physics of DNA-Ligand Interactions*, Adenine Press, Schenectady, NY, pp. 109–142.

Silverman, R. B. (1992), *The Organic Chemistry of Drug Design and Drug Action*, Chapter 6, Academic, San Diego, CA.

Wang, A. H.-J. (1990), Molecular recognition of DNA minor groove binding drugs. In Bugg, C. E. and Ealick, S. E., Eds, *Crystallographic and Modeling Methods in Molecular Design*, Springer-Verlag, New York, pp. 123–150.

Supplemental Case Studies

Boehncke, K.; Nonella, M.; Schulten, K.; and Wang, A. H.-J. (1991), Molecular dynamics investigation of the interaction between DNA and distamycin, *Biochemistry* **30**, 5465.

Laaksonen, A.; Nilsson, L. G.; Jönsson, B.; and Teleman, O. (1989), Molecular dynamics simulation of double helix Z-DNA in solution, *Chem. Phys.* **129**, 175.

Rao, S. N. and Kollman, P. (1990), Simulations of the B-DNA molecular dynamics of d(CGCGAAT-TCGCG)$_2$ and d(GCGCGCGCGC)$_2$: An analysis of the role of initial geometry and a comparison of united and all-atom models, *Biopolymers* **29**, 517.

Rao, S. N. and Remers, W. A. (1990), All atom molecular mechanics simulations on covalent complexes of anthramycin and neothramycin with deoxydecanucleotides, *J. Med. Chem.* **33**, 1701.

Veal, J. M. and Wilson, W. D. (1991), Modeling of nucleic acid complexes with cationic ligands: A specialized molecular mechanics force field and its application, *J. Biomol. Structure Dyn.* **8**, 1119.

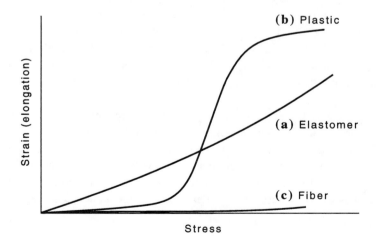

CHAPTER 6

Synthetic Polymers

6.1. INTRODUCTION

The continuing discovery of new polymeric materials and the increasing use of synthetic polymeric materials as replacements for metals, ceramics, and natural fibers has made the synthesis and characterization of polymers an active area of research. Molecular mechanics has played a role both in the determination of polymer structure and in the estimation and explanation of polymeric properties.

6.1.1 Polymer Types

Synthetic polymers can be classified as either elastomers, plastics, or fibers, based upon the behavior of the material under the influence of external forces, such as an applied stress, see Figure 6.1. Elastomers readily undergo deformation and exhibit large reversible elongations under small applied stresses, that is, they exhibit elasticity [see Fig. 6.1(a)].

FIGURE 6.1.
Stress–strain curves for several types of polymeric material.

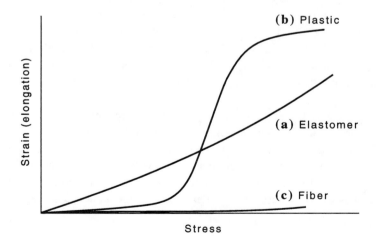

267

This elasticity can be attributed to a moderate degree of cross-linking. Cross-linking is a process wherein chemical bonds are formed between polymer chain strands (see Fig. 6.2). Examples of elastomeric polymers include silicones and carbon based polymers such as polyisoprene, polyisobutylene, polybutadiene, and poly(styrene-butadiene), where the cross-linking reaction involves the double bonds retained in the chains.

Plastic materials generally change their shape on the application of a significant force and, unlike elastomers, retain the distorted shape on removal of the force [note the bend in the curve of Fig. 6.1(b)]. In addition, plastic polymers can be classified in terms of the dependence of this distortional behavior on temperature. The two most common classifications are thermosetting and thermoplastic materials. Thermosetting materials are deformable but become permanently hard when heated above a particular temperature (representative of the kinetic onset of a chemical reaction resulting in cross-linking) and do not soften upon reheating, see Figure 6.2(b). Examples of thermosetting materials include phenolic resins, amino resins, polyester resins, and epoxy resins. A thermoplastic polymer will soften when heated above a particular temperature called the glass transition temperature, T_g. The material can then be shaped and upon cooling will harden in this new form. On reheating it will soften again and can be reshaped. Examples of thermoplastic polymers include polyethylene, polypropylene, polystyrene, poly(vinyl chloride), and poly(methyl methacrylate).

Synthetic fibers, as a rule, are thermoplastic polymers that are drawn to generate a fiber structure. Fibers are resistant to deformation [see Fig. 6.1(c)]. General classes of fibers include polyamides [nylons such as poly(hexamethylene adipamide)], polyesters [poly(ethylene terephthalate)], polyureas [poly(nonamethylene urea)], acrylics (polyacrylonitrile), hydrocarbons (polyethylene and isotactic polypropylene), halogen substituted alkenes [poly(vinyl chloride), poly(vinylidene chloride), and poly(tetrafluoroethylene)], and vinyl alkenes [polypropylene and poly(vinyl alcohol)].

6.1.2 Polymer Simulation

The molecular structures simulated and described in the other chapters may be large but they are finite, and the interactions are localized either within a discrete molecule, or between a discrete molecule and a binding or active site. When we discuss the properties of discrete molecules we, as chemists, tend to ascribe reactivity and structural properties to the shapes and functional group characteristics of individual molecules. For polymeric materials substantial knowledge can be gained from studies of individual polymer strands, but it is also important to consider the intermolecular influence of adjacent polymer strands.

Polymeric materials, in general, do not contain a single molecular species but rather consist of a distribution of chain lengths, and hence a distribution of molecular species. Even the simplest characterization of a chemical species, its mass, is not precisely defined for a polymeric material because different experiments yield different results due to the ensemble nature of the material. Techniques that measure the number of molecules in solution, such as osmotic pressure measurements, provide a number average molar mass, Eq. (6.1),

$$\langle M \rangle_n = \frac{\sum N_i M_i}{\sum N_i} \tag{6.1}$$

(a)

(b)

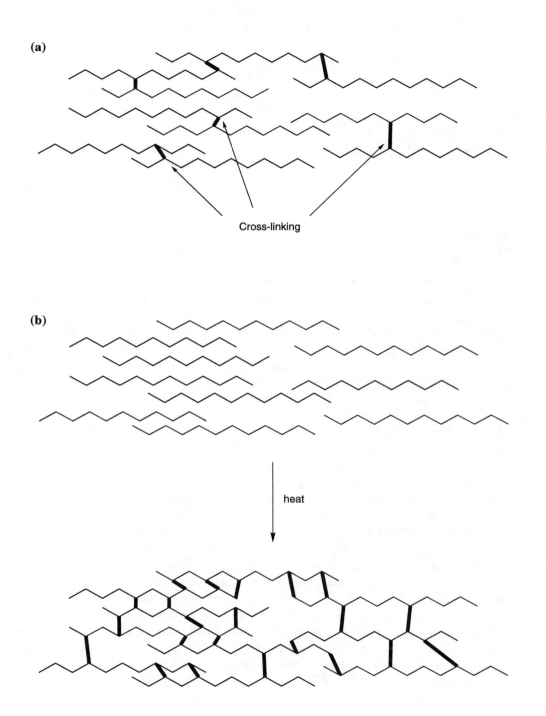

heat

Cross-linking

FIGURE 6.2.
Cross-linking in (a) elastomers and (b) thermosetting materials.

where N_i is the number of molecules of species i with molar mass M_i. Techniques that measure the size of molecules, such as light scattering, provide a weight averaged molar mass, Eq. (6.2).

$$\langle M \rangle_w = \frac{\sum N_i M_i^2}{\sum N_i M_i} \tag{6.2}$$

Additional, different, molar masses can be obtained from ultracentrifuge measurements on polymer solutions, Eq. (6.3).

$$\langle M \rangle_z = \frac{\sum N_i M_i^3}{\sum N_i M_i^2} \qquad \langle M \rangle_{z+1} = \frac{\sum N_i M_i^4}{\sum N_i M_i^3} \tag{6.3}$$

Because one is not examining a single molecular species and because each technique responds differently to mass distribution, each will yield a different estimate of the molar mass.

As with conventional molecules, developing a molecular level description of the physical (mechanical) and chemical properties of a polymeric material requires a knowledge of the molecular shape. For a given molecular composition we need to know if the individual polymer strands adopt a helical conformation, an extended conformation, or a random conformation. As discussed in Section 6.3, the response of a polymeric material to applied stress (stretching) depends on the chain conformation.

X-ray crystallography provides a detailed description of molecular shape for any material that is arranged in a regular, repeating pattern in the solid state. The diffraction pattern created by the scattering of X-ray radiation off this repeating pattern of electron densities (atoms) is analyzed in X-ray crystallography to obtain molecular structures. For small molecules, obtaining 3-D periodic samples is often possible. For proteins, it is more difficult but still possible. For polymeric materials, it is not possible due to conformational defects in the polymeric strands. Conformational defects introduce breaks in the 3-D periodicity of the lattice and cause a breakdown in the X-ray diffraction pattern.

Since polymeric materials consist of very long chains with relatively small intrachain torsional barriers, the probability of there being a defect in the idealized structure is rather high (and grows with increasing temperature and increasing chain length). For example, the lowest energy conformation for a strand or single chain of polyethylene is the fully extended or all-*trans* conformation [see Fig. 6.3(a)]. The barrier for rotation about any one of the sp^3-sp^3 C—C backbone torsions is at most a few kilocalories per mole (~ 3 kcal/mol), making structures with non-*trans* conformations, such as in Figure 6.3(b), kinetically accessible. Structures with a single *gauche* conformation are less than 1 kcal/mol in internal energy above the extended structure. For a strand of polyethylene made up from 20 monomer units there are 74 possible structures with a single *gauche* or defect conformation (20 intramonomer torsions, 19 intermonomer torsions, minus 2 methyl end groups, all times 2 for g^+ and g^- possibilities) and only one with an all-*trans* conformation. The Boltzmann equation, Eq. (6.4), can be used to find the population of the lowest energy *trans* conformation (see Appendix B).

$$P_i = \frac{f_i e^{\frac{-E_i}{RT}}}{\sum_j f_j e^{\frac{-E_j}{RT}}} \tag{6.4}$$

(a)

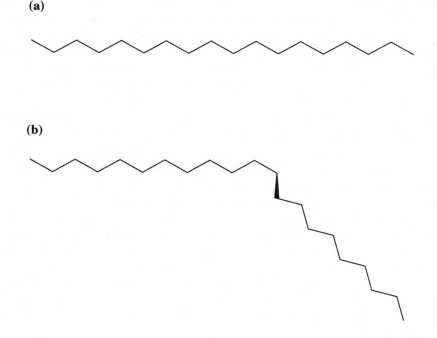

(b)

FIGURE 6.3.
Conformations of polyethylene. (a) the all-*trans* or extended conformation (b) a conformation with a single *gauche* defect.

In Eq. (6.4) P_i is the probability or population of conformation i, f_i is the number of conformations of energy E_i (or degeneracy of conformation i), and the j summation is over all the conformations. Assuming the *gauche* defect structures to each be 0.7 kcal/mol above the extended structure (derived from butane) and also assuming a single conformational defect in each chain, plugging our data in Eq. (6.4) at 150 K gives Eq. (6.5) and using a temperature of 300 K yields Eq. (6.6).

$$P_{trans,150\ K} = \frac{1 \times \exp(0)}{1 \times \exp(0) + 74 \times \exp[-0.7/(0.00198 \times 150)]} \quad (6.5)$$

$$P_{trans,300\ K} = \frac{1 \times \exp(0)}{1 \times \exp(0) + 74 \times \exp[-0.7/(0.00198 \times 300)]} \quad (6.6)$$

The *trans* structure is estimated to be 12.5% of the population at 150 K and 4.2% at 300 K! Even assuming that defect structures are 3 kcal/mol above the extended structure, more than four times higher in energy than normal, the Boltzmann population of the *trans* conformation at room temperature would still only be 67.8%, leaving a 32.2% population of defect structures. If the chain is stretched, the energetic preference for the *trans* conformation would be increased substantially, and this statistical conformational defect problem could be partially offset. In the laboratory, crystal structures of polymeric materials are almost always obtained on fiber samples of the material.

The presence of these conformational defects, along with the fact that a polymeric material consists of a distribution of chain lengths, lead to structural defects in samples of solid state polymeric materials. These defects cause polymeric solids to have differing degrees of crystallinity; in general they are not truly crystalline. This disorder makes X-ray diffraction structural determinations difficult. Molecular modeling has proved to be a useful aid in the determination of crystal structures for nearly crystalline polymers. Examples are provided in Section 6.2.

Because polymers are principally used as structural materials the molecular modeling of polymers has principally focused on the processibility and stiffness or rigidity of the materials. That is, molecular modeling questions are concerned with how easy it is to cause a material to adopt a particular shape at a given temperature, and on how well the material occupies space. The deformability and/or stiffness of homogenous, isotropic, elastic materials can be described in terms of their elastic constants or moduli, which are physical constants of a material analogous to the force constant of a harmonic bond stretch potential. The elastic moduli indicate the degree to which a polymeric material changes its size or shape in response to an applied force (and then relaxes back to the original size or shape). The magnitudes of the moduli of a material and the shapes of the stress–strain curves are related to both intramolecular and intermolecular attributes of the material, and thus are of intrinsic interest as well as being of practical importance (see Fig. 6.1). Examples of the estimation of moduli and discussion of the underlying physical models are provided in Section 6.3.

Given the dynamic nature of molecular structure in solution, the distribution of species present in a sample of a polymeric material, and the range of local environments present in a glassy or amorphous substance, it is only possible to discuss the physical properties of materials under these conditions "on average." As such, the ensemble representations of statistical mechanics provide the most plausible framework for discussing polymer solutions and polymers in a glassy state. The parameters of the resulting simplified models are often adjusted to reproduce experimental observation. This practice makes the extension to new materials problematic. Molecular mechanics has been used to obtain the data employed in the statistical mechanics analysis, potentially permitting application to new/unknown materials. The use of scalar coupling data from NMR spectroscopy and energetic and structural information from molecular mechanics to understand the degree of extension or compression of polymeric liquid crystals containing polyethers is described in Section 6.4. An overview of the rotational isomeric state model and parameterization for polyethers is provided in Section 6.4 as well. A molecular mechanics conformational searching procedure has been developed to obtain the average coordination number and differential binding energies of a Flory–Huggins analysis. The process is outlined in Section 6.5. Application of this technique to a liquid–liquid mixture (for calibration), a polymer solution, and a polymer blend is described in Section 6.5. Flory–Huggins theory is reviewed in Section 6.5 as well.

One property of particular importance in determining the processibility of plastic polymeric materials is the glass transition temperature, T_g. This temperature can be related to the flexibility of individual polymer chains (heights of torsional barriers) and the strengths of interstrand forces, particularly hydrogen bonding. As such, molecular mechanics, which is parameterized to reproduce the energetics of these forces, has proven useful in the estimation of T_g. A specific approach for estimating T_g is discussed in Section 6.6.

6.2. CRYSTALLINE STRUCTURE DETERMINATION

6.2.1 Background

Natta and Corradini (1959) presented a methodology for constructing a guess for the crystal structure of a polymeric material based on conformational analysis and chain packing. As discussed in Section 2.53 crystalline materials can be thought of as a set of unique atoms, molecules, or polymer chain fragments contained in a unit cell, which is translationally replicated to generate the macroscopic, crystalline sample. The crystal structure is obtained from the molecular coordinates of the atoms within the unit cell and the six unit cell coordinates. The unit cell or lattice distances are a, b, and c and the unit cell angles are α, β, and γ (see Fig. 2.23).

In the Natta and Corradini procedure, the conformational preferences of an individual polymer chain were obtained from organic conformational analysis (or more recently from molecular mechanics energy minimization). The arrangement of the individual strands in the unit cell, and hence the crystal packing arrangement and unit cell dimensions, was based on the assumption that all polymer chains within a crystal would be equivalent. This methodology, with the addition of nonbonded interactions (crystal packing forces), led to the determination of a number of polymer crystal structures in the 1960s and 1970s. Three increasingly more complete enhancements to this approach are discussed below.

6.2.2 Poly(ethylene oxybenzoate)

Difficulties with the Natta and Corradini approach surfaced in 1977 when Kusanagi et al., of Osaka University reported the solution to the crystal structure of the α form of poly(ethylene oxybenzoate) **6.1**.

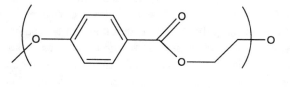

6.1

6.1 is an aromatic poly(ester ether), similar in chemical formula to but structurally quite distinct from the industrially important fiber poly(ethylene terephthalate) (terylene) **6.2**.

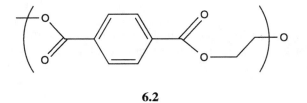

6.2

Terylene (**6.2**) is used in carpets and 2-L soft drink containers. In addition, terylene is more than 15 times stiffer than **6.1**. Kusanagi et al. (1977) sought a structural explanation for this stiffness difference by obtaining the crystal structure of **6.1**. Poly(ethylene oxybenzoate) is observed to have two crystalline forms, a β form, β-**6.1** with an extended chain conformation, analogous to the extended structure of terylene, and an α form, α-**6.1,** thought to adopt a more contracted structure. The X-ray diffraction structure of β-**6.1** had been solved, but the X-ray structure of α-**6.1** form had not, due to the lack of adequate structural models to use as starting guesses in the least-squares structure factor refinement. A satisfactory solution had not been obtained by starting from seven structural models built from optimized single-chain models, each consistent with the observed $P2_12_12_1$ space group. The best structure factor least-squares refinement from this set of models led to an R factor of 26%, which is large even for fiber structures. The parameter R is given by Eq. (6.7), where $(I_o)^{1/2}$ are the observed structure factors and $(I_c)^{1/2}$ are the calculated structure factors.

$$R = \frac{\sum | (I_o)^{1/2} - (I_c)^{1/2} |}{\sum (I_o)^{1/2}} \qquad (6.7)$$

This suggested that the polymer chains of **6.1** might not adopt the preferred isolated chain conformation when in a crystalline environment.

METHODS

In order to include the effects on the chain conformation of intermolecular interactions due to adjacent chains, an energy minimization of the backbone torsional coordinates of α-**6.1** was carried out in a crystalline environment (Kusanagi, 1975). The torsional potential consisted of threefold terms between the sp^3 centers and twofold terms for the aromatic-to-ester linkage and the aromatic-to-ether linkage. Intrastrand van der Waals interactions were described with a 6–12 potential and intrastrand electrostatics were represented as dipole–dipole interactions. The crystalline environment was simulated by placing the repeat unit of the polymer strand in a unit cell and adding in the nonbonded interactions between atoms in the unit cell and atoms in translationally equivalent cells (periodic boundary conditions as discussed in Section 2.5.3). Only the short-range repulsive portion of the potential was used to simplify the computation. The nonbond potential was made short range by representing it by a half harmonic potential, see Eq. (6.8).

$$V_{vdW} = \frac{1}{2}k(\rho - \rho_0)^2 \qquad \text{for} \qquad \rho \leq \rho_0 \qquad (6.8)$$

$$V_{vdW} = 0 \qquad \text{for} \qquad \rho > \rho_0$$

In Eq. (6.8) the force constants and natural nonbond distances were fit to 6–12 potential function data. Lacking an attractive term for the crystal potential, the unit cell parameters a, b, c, α, β, and γ were fixed.

RESULTS

Following optimization of the backbone torsional coordinates of α-**6.1** in the simulated crystalline environment, subsequent structure factor least-squares refinement from this improved model led to a dramatically enhanced X-ray structure solution ($R = 13\%$).

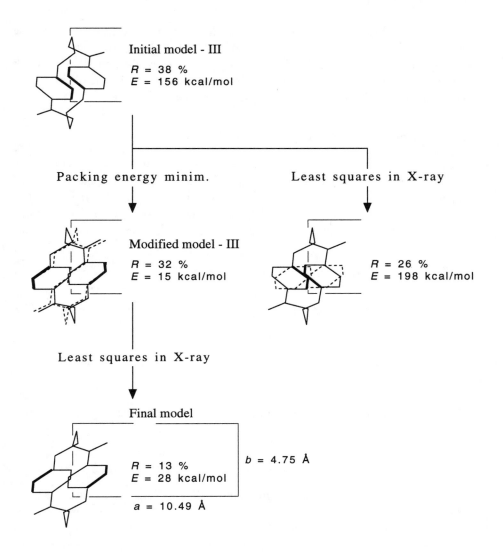

FIGURE 6.4.
Structural impact of energy minimization within the crystal lattice for **6.1**. [Reproduced with permission from Kusanazi, H.; Tadokoro, H.; Chatani, Y.; Suchiro, K. (1977), *Macromolecules* **10**, 405. Copyright © 1977 American Chemical Society.]

The original attempts at structure solution failed because the single-chain energy minimization had led to a conformation in which, looking down the chain axis, the benzene rings of the chain were canted towards the right. The effect of the crystal lattice during crystal energy minimization was predominantly on the benzene rings, causing the benzene rings to flip, canting towards the left. These results are summarized in Figure 6.4.

The Natta–Corradini procedure was quite useful as a means of obtaining trial polymer structures for X-ray structure determination because the conformational preferences due to intrastrand energetics generally are larger than interstrand energetics; however, when interstrand (crystal packing) forces are comparable to or larger than intrastrand forces the two effects must be treated in concert. Kusanagi et al. (1977) determined that interstrand effects were important for α-**6.1**. They also found that even a crude model of interstrand forces was enough, when used in conjunction with molecular mechanics, to obtain a good trial polymer structure of α-**6.1** for their X-ray structure determination.

6.2.3 Poly(oxymethylene): Use of Helical Coordinates

Sorensen et al. (1988) of the University of Utah used an approach that extended the procedure of Section 6.2.2 to include the attractive part of the van der Waals potential as well as interstrand electrostatic interactions. Their approach was applied to polyethylene, **6.3**, to test the methodology and force field, and applied to poly(oxymethylene), **6.4**, to explain the observation of two crystalline forms with substantially different backbone torsional angles.

6.3   **6.4**

Structure **6.4** is a highly crystalline material with good abrasion resistance and high thermal stability.

METHODS

To provide a simplified framework for optimization and given the helical nature of polymer chains, the polymer chain coordinates were expanded in helical coordinates, see Figure 6.5. The Cartesian coordinates of the repeat unit, the cell lattice coordinates, the helical coordinates, and any displacement coordinates associated with more than one polymer strand in the unit cell were simultaneously optimized using a Newton–Raphson optimization procedure. Periodic boundary conditions were used and the nonbond summations were carried out over two complete packing shells (18 chains).

RESULTS

Sorensen et al. (1988) were able to reproduce the experimental structure and packing energy of polyethylene, indicating that their approach was viable. For poly(oxymethylene), two crystalline forms of comparable packing energy were found. One form had the backbone stretched in comparison to a single-strand calculation (O$-$C$-$O$-$C ϕ = 78°, compared to 71.5° for an isolated single strand). The other form had the backbone compressed in comparison to the single-strand system (O$-$C$-$O$-$C ϕ = 63°), see Figure 6.6. The similarity in calculated energy between the two forms was consistent with the experimental observation of two crystalline forms. Computationally, the two forms only differed by 0.4 kcal/mol with the orthorhombic being slightly more stable.

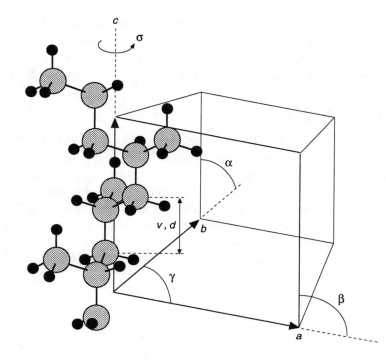

FIGURE 6.5.
Polymer helical coordinate system. [Reproduced with permission from Sorensen, R. A.; Leau, W. B.; Kesner, L.; Boyd, R. H. (1988), *Macromolecules* **21**, 200. Copyright © 1988 American Chemical Society.]

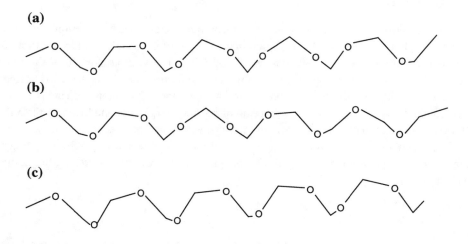

FIGURE 6.6.
Comparison of (a) an isolated strand of **6.4**, (b) a (elongated) strand from the orthorhombic crystal, and (c) a (compressed) strand from the hexagonal crystal.

Sorensen et al. (1988) found both crystalline forms of poly(oxymethylene) to be stabilized relative to an isolated chain by interchain electrostatic interactions (by 0.5 kcal/mol) as well as by attractive interchain van der Waals interactions. These two effects were not included in the Natta–Corradini procedure or in the procedure discussed in Section 6.2.2.

6.2.4 Poly(vinylidene fluoride): A General Structure Solution

While it would be possible to add other polymer chains and associated helical chain coordinates to the unit cell of the helical coordinate procedure, a more general approach to predicting polymer crystal structures would be to eliminate the helical restriction and simultaneously optimize the unit cell coordinates and the coordinates of each atom in the unit cell without symmetry restriction. This approach was reported by Karasawa et al. (1991) for polyethylene, where the experimental structure and packing energy were well reproduced. Karasawa and Goddard (1992) of the California Institute of Technology applied this method to a study of several crystalline phases of poly(vinylidene fluoride), **6.5**.

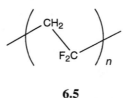

6.5

Application of a voltage across a sample of **6.5** causes a strain in the system and in reverse, the application of a stress induces a voltage. Such materials are called piezoelectric and have a number of interesting applications.

METHODS

For these studies, the Ewald summation technique, as discussed in Section 2.5.3, was used for the lattice nonbonded energy summation, rather than a finite real space summation, as used in the studies in Sections 6.2.2 and 6.2.3. The valence portion of the force field was obtained by fitting the matrices of the second derivatives of the ab initio Hartree–Fock energy with respect to coordinates for n-butane and 2,2-difluoropropane. Electrostatic interactions were represented in terms of partial charges obtained from a fit to the electrostatic potential of 1,1,1,3,3-pentafluorobutane. The van der Waals interactions were described with an exponential-6 potential with carbon parameters fit to graphite, hydrogen parameters fit to polyethylene, and fluorine parameters fit to CF_4 and poly(tetrafluoroethylene) crystals.

RESULTS

Experimentally, **6.5** has been found in four crystal forms (Takahashi and Tadokoro, 1980; Bachmann et al., 1980; Lovinger, 1981; Takahashi et al., 1983). The known forms, as well as several others calculated by Karasawa and Goddard (1992), are shown in Figure 6.7. The experimentally known forms are I_p, II_a, II_p, and III_{pu}. The II_a form is a statistical

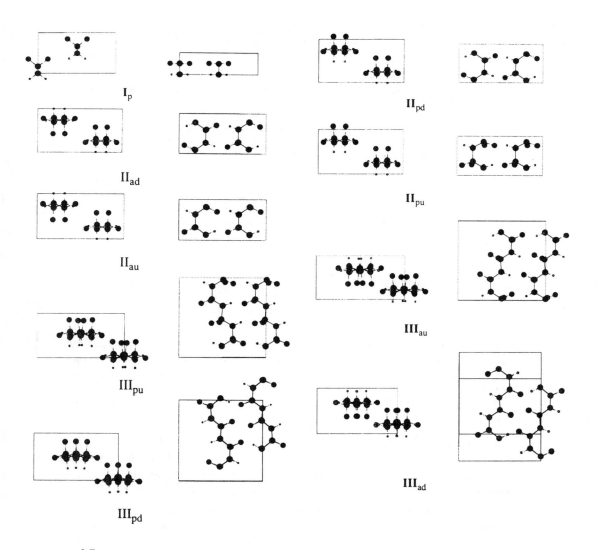

I_p

II_{pd}

II_{ad}

II_{pu}

II_{au}

III_{au}

III_{pu}

III_{ad}

III_{pd}

FIGURE 6.7.
Crystal packing of nine stable crystal forms of poly(vinylidene fluoride), **6.5**. The structural types are denoted I, II, and III. The subscripts a and p indicate antiparallel and parallel orientation of the polymer chains. The subscripts u and d signify up–up or up–down relative directions of adjacent chains. [Reproduced with permission from Karasawa, N.; Goddard, W. A. III (1992), *Macromolecules* **25**, 7268. Copyright © 1992 American Chemical Society.]

average of II_{au} and II_{ad}, and II_p is a statistical average of II_{pu} and II_{pd}. Piezoelectric properties are observed experimentally in two forms (I_p and II_p).

All nine of the structures shown in Figure 6.7, including the four experimentally observed structures of (**6.5**) poly(vinylidene fluoride), were calculated to have total (and free) energies within 2 kcal/mol of the lowest structure, indicating that they all should be experimentally observable. The calculated cell parameters were in good agreement with experiment. In addition, the energetic results, as well as presence or absence of polarity within the unit cell, were consistent with the processing conditions used to prepare the

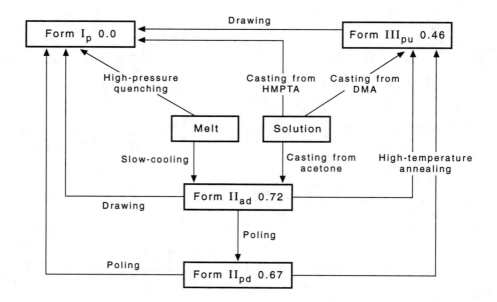

FIGURE 6.8.
Processing relationships between the observed forms of **6.5**. The energies are per CH_2CF_2 unit. Hexamethylphosphorictriamide = HMPTA; dimethylacetamide = DMA. [Reproduced with permission from Karasawa, N.; Goddard, W. A. III (1992), *Macromolecules* **25**, 7268. Copyright © 1992 American Chemical Society.]

experimentally observed materials (see Fig. 6.8). Cooling from the melt led to II_a ($E = 0.72$ kcal/mol). Drawing at room temperature led to I_p ($E = 0.0$). Poling (application of an external electric field) led to the polar form II_p ($E = 0.67$). High-temperature annealing from II_p led to III_p ($E = 0.46$). Poling from II_p or drawing from III_p led to I_p.

Karasawa and Goddard (1992) found nine stable crystal forms of **6.5**, all within a 2-kcal/mol range, two displaying piezoelectric behavior. Furthermore, they found each of the nine stable crystal forms of **6.5** contained two distinct polymer chains in the unit cell. It would have been possible to add helical coordinates for the second strand, but the resulting computational model would have had as many degrees of freedom as the general approach. In addition, the piezoelectric attribute of **6.5** is a long-range electrostatic property of the system certainly requiring the present Ewald summation approach rather than the local representation used in Section 6.2.3.

6.3. ESTIMATION OF ELASTIC MODULI

6.3.1 Background

The deformability of homogenous, isotropic, elastic materials can be described in terms of their elastic constants or moduli, which are physical constants of a material analogous to the force constant of a harmonic bond stretch potential. The elastic constants describe the harmonic restoring forces associated with the six lattice degrees of freedom (6 diago-

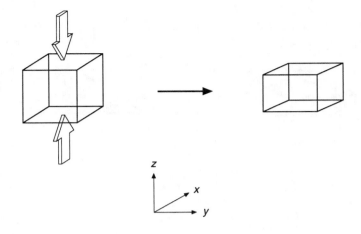

FIGURE 6.9.
Stress-induced structural distortion.

nal terms, and 30 cross terms). The related elastic moduli indicate the degree to which a polymeric material changes its size or shape in response to a specific applied force (and then relaxes back to the original size or shape). These moduli can be measured by applying tension to a sample of the material, by applying a shear force to the material, or by providing a uniform compression to the material. The magnitudes of the moduli of a material are related to both intramolecular and intermolecular attributes (stretch, bend, and torsion vibrational force constants, electrostatics, hydrogen bonding, volume, etc.) of the material, and thus are of intrinsic interest as well as being of practical importance.

If a moderate force is applied to both ends of a sample of material (call it the z axis) the material will shorten slightly along z, see Figure 6.9. For elastic materials, that is, materials that will bounce back upon relief of the force, and for small forces, Eq. (6.9) holds, where σ is the tensile stress (force per unit area), ε is the tensile strain (extension per unit length), and E is Young's modulus (or the tensile modulus).

$$\sigma = E \, \varepsilon \tag{6.9}$$

The parameter E is a constant or property specific to a particular material, presented in units of gigapascals [1 GPa $= 10^{10}$ dyn/cm^2 $= 10$ kbar $= 0.14393$ (kcal/mol)/Å^3]. Equation (6.6) suggests that there will be a linear relationship between the stress applied and the strain observed; this is experimentally observed for elastic materials over a wide range of applied stresses, though the range of applicability varies from material to material.

The compression in the z direction will be accompanied with a change in dimension in the directions perpendicular to z. For isotropic materials the perpendicular dimensions will generally undergo an inverse change (if z is decreased then x and y will increase). The relationship between the change in length to the change in width is Poisson's ratio, see Eq. (6.10).

$$v_p = \frac{dy/y_o}{dz/z_0} \tag{6.10}$$

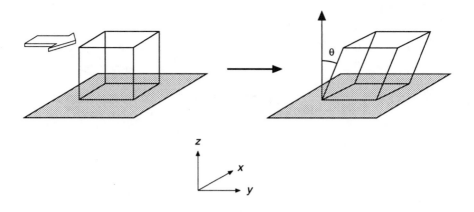

FIGURE 6.10.
Transverse stress-induced structural distortion.

If one side of a sample of material is firmly secured to a surface and the opposite side of the sample is subjected to a transverse force the sample will deform through an angle θ, see Figure 6.10. Equation (6.11) describes this deformation.

$$G = \frac{\sigma_s}{\varepsilon_s} = \frac{F}{A \tan \theta} \tag{6.11}$$

In Eq. (6.11), F is the transverse force, A is the area, G is the shear modulus, σ_s is the sheer stress applied, and ε_s is the sheer strain observed.

If a uniform pressure is applied to a sample of material the volume will decrease, see Figure 6.11. The relation that describes the uniform compression due to applied pressure is given by Eq. (6.12) where P is the applied pressure, V_o is the initial volume, ΔV is the change in volume, and B is the bulk modulus.

$$B = \frac{P}{\Delta V / V_0} \tag{6.12}$$

Since isotropic materials lack material asymmetry they only possess two independent elastic moduli and these constants can be interrelated by Eq. (6.13).

$$E = 3B(1 - 2\nu_p) = 2G(1 + \nu_p) \tag{6.13}$$

Using the energy minimization techniques discussed in Section 6.2, it would be easy to take a minimized structure, induce a strain by stretching or compressing one of the lattice coordinates, and calculate the intrinsic stress for a crystal of a polymeric material. This has been done for several polymeric systems. Unfortunately, mechanical properties of real world solid polymer samples are largely determined by the degree of crystallinity present in the samples and by the orientation of the polymer chains in the samples with respect to the direction of applied stress. Even though the intrinsic moduli discussed above are rarely achieved in real materials, using molecular mechanics to predict the modulus

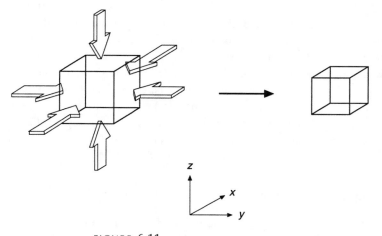

FIGURE 6.11.
Pressure-induced structural distortion.

of a perfectly crystalline sample is an aid in the preparation of high-modulus materials. In addition, understanding the molecular source of the rigidity or stiffness of a material can aid in the construction of new materials.

The intrinsic rigidity or stiffness of a polymeric material can be related to the energetics of elongation or compression along the polymer strand (the longitudinal tensile modulus) and to the energetics of elongation or compression perpendicular to the polymer strand (the lateral moduli). Chemically similar materials have significant differences in the longitudinal modulus. For example, polyethylene, **6.6**, which adopts the fully extended all-*trans* configuration, has a longitudinal Young's modulus of 320–360 GPa.

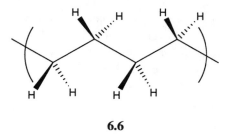

6.6

In contrast isotactic polypropylene, **6.7**, which adopts a *trans, gauche* conformation, has a longitudinal Young's modulus of only 40 GPa.

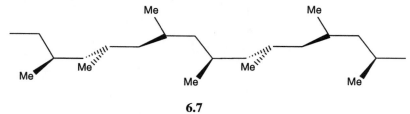

6.7

This difference in Young's modulus between **6.6** and **6.7** is due to a difference in vibrational modes that are accessed during elongation or compression. Because there are *gauche* conformations in isotactic polypropylene chains in a crystal, the stress, due to elongation or compression, can be relieved by changes in the low-energy C−C torsional modes. Since polyethylene is in the fully extended conformation in the crystal, when a sample of polyethylene is elongated the stress must be relieved by changes in the higher energy C−C−C bend and C−C stretch coordinates. As can be seen from **6.6**, if a polyethylene strand is stretched it cannot lengthen by changing C−C torsional angles, because the torsional angles are already at their maximal values. For **6.7** the strand can get longer simply by increasing C−C torsional angles from 60° to slightly larger values.

6.3.2 Poly(ethylene oxybenzoate) and Poly(ethylene terephthalate)

In addition to determining the crystal structure of the α form of poly(ethylene oxybenzoate), Kusanagi et al. (1977) explained the large difference in longitudinal elastic (Young's) modulus for α-poly(ethylene oxybenzoate), **6.1**, and the related poly(ethylene terephthalate), **6.2**. The Young's modulus for **6.1** is 6 GPa and for **6.2** is 106 GPa. They attributed the smaller modulus of α-poly(ethylene oxybenzoate) to the more coiled or compressed conformation of the polymer chains of α-poly(ethylene oxybenzoate) (see Fig. 6.12). This more compressed conformation is likely due to the preference of ether (C−O−C) linkages for *gauche* conformations compared to the *trans* conformations usually found for ester (O=C−O−C) linkages.

6.3.3 Polyethylene and Poly(oxymethylene)

By using a second derivative minimization procedure Sorensen et al. (1988) obtained the elastic moduli for polyethylene and poly(oxymethylene) (in addition to the structures and packing energies discussed in Section 6.2.3). By way of benchmark they reproduced the longitudinal Young's modulus of polyethylene (341 GPa). For the two forms of poly(oxymethylene) (one being compressed and the other elongated compared to single-strand calculation) they found comparable elastic moduli (64 GPa for the elongated chain and 81 GPa for the compressed chain) suggesting that there was not a significant stiffening of the crystal by either compression or elongation.

6.3.4 Syndiotactic Polystyrene

Sun et al. (1992) of the Michigan Molecular Institute reported a calculation of the longitudinal Young's modulus for the syndiotactic form of polystyrene, (Ph = Phenyl) **6.8**.

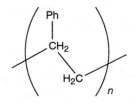

6.8

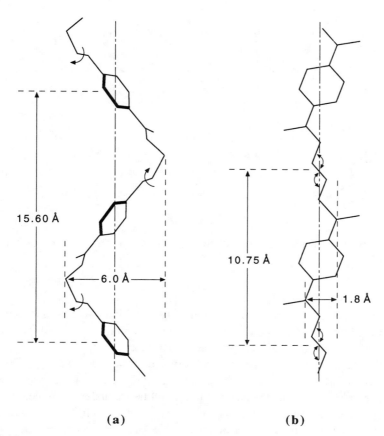

15.60 Å

6.0 Å

10.75 Å

1.8 Å

(a)   **(b)**

FIGURE 6.12.
Comparison of poly(ethylene oxybenzoate) (a) and poly(ethylene terephthalate) (b) chain structures in a crystalline environment. [Reproduced with permission from Kusanagi, H.; Tadokoro, H.; Chatani, Y.; Suchiro, K. (1977), *Macromolecules* **10**, 405. Copyright © 1977 American Chemical Society.]

Syndiotactic polystyrene was discovered in 1986. It has several useful properties including a higher melting temperature than the corresponding isotactic polystyrene ($>$ 270°C compared to 220°C), a more rapid rate of crystallization, and probably enhanced mechanical properties. In isotactic polystyrene the phenyl group substituents are all on the same side when the polymer is drawn in an extended form, Figure 6.13(a). For a syndiotactic sample of the material the substituents are on alternating sides, Figure 6.13(b). Several crystalline phases of syndiotactic polystyrene are observed, with the most stable possessing fully extended all-*trans* chains.

In order to estimate the ideal or limiting Young's modulus of syndiotactic polystyrene, Sun et al. (1992) applied a tensile stress along the chain direction, c, minimized the system under this stress, and calculated the strain (change in c cell dimension). The Dreiding force field (Mayo et al., 1990) was used within the polymer simulation program POLY-GRAF (available from Molecular Simulations Inc.). A unit cell consisting of a single

(a)

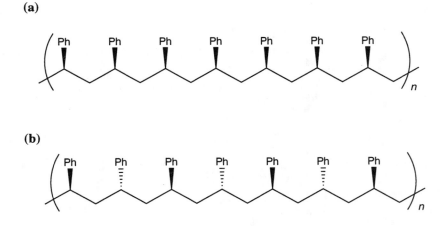

(b)

FIGURE 6.13.
Stereochemistry of stereoregular polystyrene chains. (a) Isotactic polystyrene and (b) syndiotactic polystyrene.

polymer strand was constructed, periodic boundary conditions were used to place the cell in the correct, infinite solid state environment, and a minimization of all the coordinates including the lattice degrees of freedom was carried out.

The calculated stress–strain curve of syndiotactic polystyrene (see Fig. 6.14) yielded a modulus of 83 GPa, in excellent agreement with the X-ray diffraction estimate of 86 GPa (obtained by determining the changes in the cell lattice under an applied stress). This modulus is more than an order of magnitude larger than 7 GPa, the largest modulus determined by bulk mechanical measurement. The large difference between the X-ray

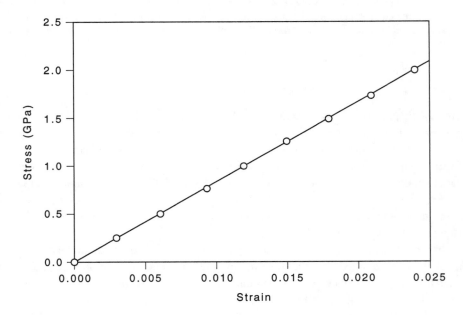

FIGURE 6.14.
Calculated stress–strain curve for syndiotactic polystyrene. Crystalline modulus = 83 GPa.
[Reproduced with permission from Sun, Z.; Morgan, R. J.; Lewis, D. N. (1992), *Polymer* **33**, 725.]

diffraction estimate of the modulus and that measured mechanically is due to stereochemical and conformational imperfections in the polymeric material. X-ray diffraction monitored the stress-induced distortions in crystalline portions of the sample, whereas application of a mechanical stress sampled amorphous regions as well as the crystalline parts of the material.

It is also interesting that this modulus is four times smaller than that of polyethylene, which also adopts a fully extended all-*trans* configuration. Sun et al. (1992) attributed this factor of 4 difference to the higher density of polyethylene, which lacks the pendent phenyl groups of polystyrene [the cross-sectional area of a polystyrene strand (~70 Å^2) is 3.9 times larger than a polyethylene strand (~18 Å^2)]. Because the individual strands are the source of "strength," polyethylene, with a smaller cross-sectional area, will have four times more strands per unit area, and hence has about four times more "strength" per unit area and will be four times less deformable.

6.4. ROTATIONAL ISOMERIC STATE ANALYSIS

6.4.1 Background

The size and shape of polymer chains form the basis for our understanding of the properties of polymeric materials. Because of the distribution of chain lengths in a polymeric material and the large number of energetically accessible conformations, we can only define polymer sizes and shapes in terms of average quantities. Two averages that tell us information about length and shape, respectively, are the average end-to-end root-mean-square distance, $<\bar{r}^2>^{1/2}$, and the average root-mean-square radius of gyration $<\bar{S}^2>^{1/2}$. The first tells us, on average, how far the ends of the polymer are from each other. The second provides a measure of how far polymer segments are from the center of mass of the strand. The brackets, $< >$, indicate averaging over chain lengths in the sample and the overbar indicates averaging over the conformations available to a given polymer chain length.

6.4.2 Polyethers

To understand the degree of extension or compression present in the structures of polymeric liquid crystals containing the sequence $-O(CH_2)_yO-$ ($y = 2-10$), Inomata et al. (1991) of the Tokyo Institute of Technology reported a rotational isomeric state analysis of the shapes of polyethers. The conformations of such polyether sequences are substantially different than the all-carbon analogs, with polyethers roughly half as extended as the all-carbon analogs. The explanation is thought to rest in the observation that dimethoxyalkanes $MeO(CH_2)_yOMe$ ($y = 1$ and 2) prefer a *gauche* conformation for C−C bonds adjacent to an oxygen atom, **6.9** (rather than the *trans* preference observed in alkanes, **6.10**).

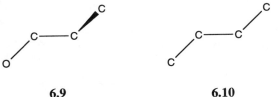

6.9   **6.10**

The degree of extension or compression in a polymer chain can be described by various models, the most complete of which is the rotational isomeric state model of Flory (reviewed in Flory, 1974). A brief description of the logic leading to the development of the rotational isomeric state model is presented in Box 6.1 to aid in understanding the analysis that Inomata et al. (1991) carried out.

BOX 6.1   The Rotational Isomeric State Model

The simplest description of the end-to-end distance of a polymer chain is obtained from the freely jointed chain model. This model assumes the bonds of a polymer chain are connected by free rotating joints. If we consider a three atom, two bond segment of polymer with bond distance d, the end-to-end distance r is obtained from the law of cosines as given by Eq. (6.14), see **6.11** for the geometry.

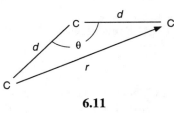

6.11

$$r^2 = 2d^2 - 2d^2 \cos \theta \tag{6.14}$$

When a long polymer segment is considered in the free jointed chain model, the angle θ will take on all values, so that when summed over all segments the $\cos \theta$ term will sum to zero. The resulting end-to-end distance, given by Eq. (6.15), is $\sqrt{n}$ shorter than a fully extended chain.

$$\langle r^2 \rangle = nd^2 \tag{6.15}$$

The ratio of $\langle r^2 \rangle$ to nd^2 provides a useful reference point and is called the characteristic ratio, C_n. As given in Eq. (6.15) the freely jointed chain has $C_n = 1$; for reference, polyethylene has an experimental C_n of 6.7, and the experimental C_n for poly(oxytrimethylene) is 3.9. The freely jointed chain model ignores both short- and long-range molecular interactions, resulting in the actual polymer being 4–7 times longer than predicted. For example, $\cos \theta$ will not take on all possible values because there are specific natural angles about any center in a molecule. In addition, the long-range nonbonded interactions between polymer segments that can be separated by many units in the polymer chain are not accounted for in this model. However, the freely jointed chain model provides a reference point for improved, more physical models. For example, the presence of natural bond angles can be accounted for by considering a specific angle, Eq. (6.16).

$$\langle r^2 \rangle = nd^2 \frac{(1 - \cos \theta)}{(1 + \cos \theta)} \tag{6.16}$$

For the case of a chain consisting of tetrahedral backbone atoms $\cos \theta = -\frac{1}{3}$ and Eq. (6.16) reduces to Eq. (6.17) suggesting the chain will be twice as long as the freely jointed chain model would predict.

$$r^2 = 2nd^2 \tag{6.17}$$

Addition of the angle term raises C_n to 2, which is still more than a factor of 3 smaller than observed for polyethylene.

The presence of finite torsional potential barriers for simple polymer chains can be accounted for by considering an average torsional term $\langle \cos \phi \rangle$, Eq. (6.18).

$$\langle r^2 \rangle = nd^2 \frac{(1 - \cos \theta)(1 - \langle \cos \phi \rangle)}{(1 + \cos \theta)(1 + \langle \cos \phi \rangle)} \tag{6.18}$$

The average torsional term $\langle \cos \phi \rangle$ can be developed by assuming that polymer chains predominantly reside near the minima of the backbone torsional potentials. For a chain of sp^3 centers, be it polyethylene or polymethylene oxide, the backbone torsional potentials are threefold periodic with minima at 180° and $\pm 60°$. In this case the average torsional term $\langle \cos \phi \rangle$ can be obtained from Eq. (6.19), where z is the rotational partition function and u_n is the statistical weight for the nth conformational state.

$$\langle \cos \phi \rangle = \frac{1}{z} \sum_n u_n \cos \phi_n \tag{6.19}$$

If, for our chain of sp^3 centers, we take the statistical weights u_n for each torsional state to be the same, then the average is given by Eq. (6.20) and Eq. (6.18) reduces to Eq. (6.16).

$$\langle \cos \phi \rangle = \cos(180°) + \cos(60°) + \cos(-60°) = 0 \tag{6.20}$$

A more realistic set of statistical weights for an sp^3 chain can be obtained by considering that the *gauche* ($\phi = \pm 60°$) conformation is higher in energy than the *trans* ($\phi = 180°$) (for butane, the *gauche* conformation is higher by 0.7 kcal/mol). Using 0.5 kcal/mol for the energy difference raises C_n to 3.4 (Flory, 1988 p. 59) and yields an end-to-end distance one-half the size observed experimentally for polyethylene. The factor of 2 error remaining in the model is thought to be due to the "pentane effect." If we consider rotation about the central two C—C bonds of pentane four unique conformations result, the *trans–trans* conformation (*tt*) Figure 6.15(a), the *gauche*$^+$*–trans* conformation (g^+t, and equivalent g^-t, tg^+, tg^-), Figure 6.15(b), the *gauche*$^+$*–gauche*$^+$ conformation (g^+g^+, and equivalent g^-g^-), Figure 6.15(c), and the *gauche*$^+$*–gauche*$^-$ conformation (g^+g^-, and equivalent g^-g^+), Figure 6.15(d). As should be evident from Figure 6.15(d) there will be

(continued)

BOX 6.1   The Rotational Isomeric State Model *(continued)*

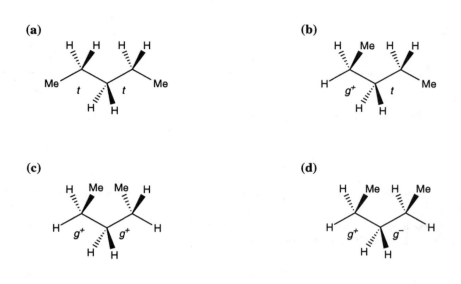

FIGURE 6.15.
Conformations of pentane. (a) The *trans–trans* conformation (*tt*), (b) the *gauche⁺–trans*
conformation, (c) the *gauche⁺–gauche⁺* conformation, and (d) the *gauche⁺–gauche⁻*
conformation.

substantial interaction between the terminal methyl groups in the g^+g^- conformation and this conformation (and the equivalent g^-g^+) will be substantially higher in energy than the *tt*, $tg^\pm$, $g^\pm t$, or
$g^\pm g^\pm$ conformations. This set of conformational energy differences, and hence statistical weights,
can be compactly represented in matrix notation, see Eq. (6.21).

$$
\begin{array}{c c}
 & \begin{array}{c c c} t & g^+ & g^- \end{array} \\
\begin{array}{c} t \\ g^+ \\ g^- \end{array} &
\left[\begin{array}{c c c}
1 & \sigma & \sigma \\
\sigma & \psi & 0 \\
\sigma & 0 & \psi
\end{array} \right]
\end{array}
\tag{6.21}
$$

The statistical weights are taken with respect to the lowest energy conformation, the *tt* conformation; the g^+g^- and g^-g^+ conformations are assumed to be high enough energetically so as not to
contribute. When this nearest-neighbor set of statistical weights is used, reasonable estimates for
end-to-end distances are obtained [the experimental C_n of 6.7 for polyethylene can be obtained
using reasonable estimates for the statistical weights (Flory, 1988, p. 147)]. Use of conformational
states and statistical weight matrices is known as a rotational isomeric state model. Operationally,
the weight matrices are multiplied together to obtain the stastics for a long change; an alternate
form to Eq. (6.21) with 1's in the first column is used to generate the rotational isomeric state
models. Estimates of the statistical weight matrix elements can be obtained either from an analysis
of experimental conformational populations or from molecular mechanics on small model chains.

METHODS

The observed characteristic ratios for polyethers range from 3.9 to 6.2, suggesting a large variation in extension ranging from slightly less than polyethylene to nearly one-half as extended as polyethylene. To develop the statistical weight matrices for a rotational isomeric state model Inomata et al. (1991) needed estimates of the conformational preferences of the OC−CC linkage. The vicinal coupling constants were measured for a set of α, ω-dimethyoxyalkanes by NMR. Vicinal coupling constants are three bond nuclear spin couplings between protons as illustrated in **6.12**.

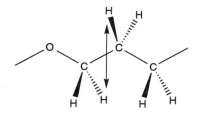

6.12

Molecular mechanics was used to find the local minima on the torsional potential surface and the Karplus relation, discussed in Section 2.3, was used to estimate the fractional contributions of the low-energy conformations to the equilibrium distribution.

RESULTS

The statistical weight matrices obtained from molecular mechanics and NMR spectroscopy were used to estimate C_n for three polyethers. The results of the rotational isomeric state analysis were in reasonable accord with experiment (see Table 6.1). Comparison of poly[oxy(2,2-dimethyltrimethylene)] **6.13**, with poly(oxytrimethylene) **6.14**, where the C_n values are quite similar, suggested to Inomata et al. (1991) that methyl substitution on the carbon framework had little effect on the C−C bond rotational characteristics.

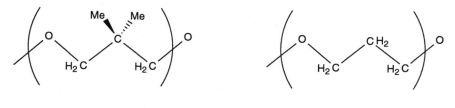

6.13   **6.14**

TABLE 6.1. Characteristic Ratios (C_n) for a Series of Polyethers

| | C_n | |
Polymer	Calc	Obs
Poly(oxytrimethylene), **6.14**	4.6	3.9
Poly[oxy(2,2-dimethyltrimethylene)], **6.13**	3.9	4.3
Poly(oxytetramethylene), **6.16**	5.9	5.4–6.2

The compressed nature of the polyethers **6.13** and **6.14** compared to polyethylene was ascribed to a weak attractive electrostatic interaction between the negatively charged oxygen and slightly positively charged γ-CH_2 group, see **16.5**.

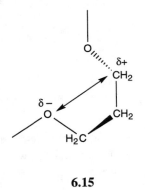

6.15

The charge on the γ-CH_2 group was due to the presence of an adjacent electronegative oxygen. The attractive 1,4 electrostatic interaction caused the *gauche* conformation to be preferred over the *trans* conformation by about 0.5 kcal/mol. This resulted in a more compressed structure.

For poly(oxytetramethylene), **6.16**, the charge on the γ-CH_2 group was reduced because it lacked an adjacent oxygen substitutent. Hence, the polymer was observed to adopt a more extended structure.

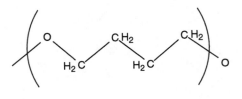

6.16

6.5. POLYMER SOLUTIONS AND BLENDS

6.5.1 Background

Knowledge of the mutual solubility or miscibility of binary mixtures is useful in choosing plasticizers, in understanding the degree of polymer swelling, and in developing a phase diagram for a polymer blend. The thermodynamic properties of polymer solutions and blends can be assessed by first considering the thermodynamics of normal solutions and then extending this analysis to polymer solutions and finally to polymer blends. The most noteworthy of such analyses is the Flory–Huggins theory (reviewed in Hill, 1960).

6.5.2 Molecular Mechanics Based Flory–Huggins

In 1992 Fan, Olafson, and Blanco of Molecular Simulations Inc., and Hsu of the University of Massachusetts, Amherst, reported a molecular mechanics procedure for obtaining the parameters used in a Flory–Huggins analysis and tested the procedure for an n-hexane–nitrobenzene mixture, a diisobutyl ketone–polyisobutylene solvent–polymer mixture, and a polyisoprene–polystyrene polymer blend (Fan, 1992). Flory–Huggins theory is reviewed in Box 6.2.

Fan et al. (1992) reported a procedure for calculating the differential energy of solvation, $\Delta\epsilon_{12}$, and the average coordination number z from molecular mechanics.

BOX 6.2   Flory–Huggins Theory

The thermodynamic properties of normal solutions are best described with respect to the thermodynamics of ideal solutions. For an ideal solution, the molecules of solute and solvent are assumed to be of comparable size. In addition, the intermolecular interactions between all particles are assumed to be the same ($\Delta H^M = 0$). The remaining component of the free energy of mixing, ΔG^M, is the entropic term, ΔS^M. This entropy of mixing term can be calculated from the Boltzmann equation, Eq. (6.22), where W is the number of distinguishable arrangements in the system.

$$S = k \ln W \tag{6.22}$$

These arrangements are called configurations, not the configurations of stereochemistry, but an assembly of distinct placements of molecules. If we adopt a lattice model with N_1 molecules of solute and N_2 molecules of solvent there are N_0 ($= N_1 + N_2$) sites in the system. There are $N_0!$ ways for the solute and solvent molecules to be arranged on the lattice. This number of arrangements has to be modified by the fact that interchange of solute molecules does not provide distinguishable configurations. The same is true for solvent molecules; thus the number of distinguishable configurations is given by Eq. (6.23).

$$W = \frac{N_0!}{N_1! N_2!} \tag{6.23}$$

The resulting configurational entropy S_c is given by Eq. (6.24).

$$S_c = k \ln\frac{N_0!}{N_1! N_2!} \tag{6.24}$$

Because of the large values of N associated with macroscopic samples, Stirling's approximation can be applied ($\ln N! \leqslant N \ln N - N$). Equation (6.24) becomes Eq. (6.25).

$$S_c = k[(N_0\ln N_0 - N_0) - (N_1\ln N_1 - N_1) - (N_2\ln N_2 - N_2)] \tag{6.25}$$

Substituting $N_1 + N_2$ for N_0 yields Eq. (6.26), which can be factored to give Eq. (6.27).

$$S_c = k[(N_1 + N_2)\ln N_0 - N_1\ln N_1 - N_2\ln N_2] \tag{6.26}$$

$$S_c = -k\left[N_1\ln\frac{N_1}{N_0} + N_2\ln\frac{N_2}{N_0}\right] \tag{6.27}$$

The mole fractions x_1 and x_2 can be substituted for the ratios $\dfrac{N_1}{N_0}$ and $\dfrac{N_2}{N_0}$, to give the final expression for the configurational entropy, Eq. (6.28).

$$S_c = -k[N_1\ln x_1 + N_2\ln x_2] \tag{6.28}$$

Since for a pure substance $x = 1$ ($x_1 = 1$ and $x_2 = 0$ or $x_1 = 0$ and $x_2 = 1$), the configurational entropy of mixing ($\Delta S^M = S_c - S_1 - S_2$) is simply the expression for S_c, Eq. (6.28). Furthermore, for an ideal solution we have set $\Delta H^M = 0$, so the free energy for mixing is given by Eq. (6.29).

$$\Delta G^M = -T\Delta S^M = Tk[N_1\ln x_1 + N_2\ln x_2] \tag{6.29}$$

By construction, x_1 and x_2 are less than 1, so $\Delta G^M < 0$ and the process of mixing will be spontaneous. Deviations of real solutions from ideal behavior can occur because $\Delta H^M \neq 0$ (regular solution behavior if ΔS^M is ideal, irregular solution behavior if ΔS^M is not ideal), or because of a nonconfigurational contribution to ΔS^M (athermal solutions).

A similar lattice model was used by Flory and Huggins independently to develop an equation for the entropy of mixing, ΔS^M, for ideal polymer solutions (Huggins, 1941, 1942; Flory, 1941, 1942). They began by randomly placing N_1 polymer molecules on a lattice so as to represent an amorphous arrangement of the polymer. Each polymer molecule was taken as a set of r segments each with the same volume as a solvent molecule. The resulting entropy of mixing equation, Eq. (6.30), is quite similar to the ideal solution equation presented in Eq. (6.27), where now $N_0 = (rN_1 + N_2)$ with N_1 polymer molecules of r segments each and N_2 solvent molecules.

$$\Delta S^M = -k\left[N_1\ln\frac{rN_1}{N_0} + N_2\ln\frac{N_2}{N_0}\right] \tag{6.30}$$

Instead of expressing the entropy of mixing in terms of mole fractions (x_i) as in Eq. (6.28), it is preferable, due to the extreme differences in size between solute (polymer) and solvent, to use the volume fraction ϕ_i since the number of sites occupied by the solvent and polymer are proportional to their volumes, see Eq. (6.31).

$$\Delta S^M = -k[N_1\ln\phi_1 + N_2\ln\phi_2] \tag{6.31}$$

Equation (6.31) is the equation for an *athermal* polymer solution because ΔS^M is not the same as Eq. (6.28) for an ideal solution. The difference can easily be seen by expressing the ideal solution entropy of mixing in terms of volume fractions $\phi_i = \dfrac{N_iV_0}{V}$, where V_0 is the volume for each solute or solvent molecule and V is the total volume, see Eq. (6.32).

$$\Delta S^M = -\frac{kV}{V_0}[\phi_1\ln\phi_1 + \phi_2\ln\phi_2] \tag{6.32}$$

For the case of a polymer solution, the volume of polymer V_1 is dependent on the number of segments as well as the number of polymer molecules, as given by Eq. (6.33).

$$V_1 = rN_1V_0 \tag{6.33}$$

(*continued*)

Box 6.2   Flory-Huggins Theory (continued)

Substitution for N_1 (the polymer) and N_2 (the solvent) in Eq. (6.31) yields Eq. (6.34).

$$\Delta S^M = -\frac{kV}{V_0}\left[\frac{\phi_1}{r}\ln\phi_1 + \phi_2\ln\phi_2\right] \tag{6.34}$$

As r, the degree of polymerization increases, the first term in Eq. (6.34) becomes smaller and ΔS^M deviates from ideal solution behavior by becoming smaller in magnitude. As for real, conventional solutions there are deviations from the above idealized behavior due to enthalpic effects and non-configurational entropic contributions; nonetheless this configurational model provides a reasonable reference point to which deviations of real polymeric materials can be compared.

The simplest model for ΔH^M can be developed by assuming that the energy change upon mixing is due to the breaking and making of specific solvent–solvent, solute–solute, and solvent–solute interactions or "bonds." If there is no volume change upon mixing, ΔH^M can be approximated by Eq. (6.35), where q is the number of new "bonds" formed in solution, ε_{11} the energy of polymer–polymer interactions, ε_{22} the energy of solvent–solvent interactions, and ε_{12} the energy associated with the new polymer–solvent interactions.

$$\Delta H^M = q\left[\varepsilon_{12} - \frac{1}{2}(\varepsilon_{11} + \varepsilon_{22})\right] = q\Delta\varepsilon_{12} \tag{6.35}$$

The number of interactions, q, can be obtained by using the above lattice model. Each polymer molecule will be surrounded by $\phi_2 rz$ solvent molecules where, as above, ϕ_2 is the volume fraction for the solvent, r is the degree of polymerization, and z is the coordination number of the chosen lattice. The volume fraction for the solvent can be substituted for the volume fraction of the polymer, which leads to Eq. (6.36).

$$\Delta H^M = N_2\phi_1 z\Delta\varepsilon_{12} \tag{6.36}$$

A new, dimensionless, parameter χ can be defined in terms of z and $\Delta\varepsilon_{12}$, see Eq. (6.37).

$$\chi = \frac{z\Delta\varepsilon_{12}}{kT} \tag{6.37}$$

This leads to the final expression for ΔH^M in terms of the volume fraction of the polymer ϕ_1, the number of solvent molecules N_2, and χ, see Eq. (6.38).

$$\Delta H^M = kT\phi_1 N_2 \chi \tag{6.38}$$

The parameter χ is usually obtained by fitting observed thermodynamic data but it could also be obtained from molecular mechanics.

METHODS

The solvent–solvent and solute–solute self-interaction energies and the solute–solvent cross-pair energies were obtained using a procedure similar to that used in the chiral chromatography case study in Section 2.5.2, where the energy of interaction between the chiral support and the eluent was calculated. As with the chiral chromatography example, when the interaction between two molecules such as a solvent and solute is considered, there is not an obvious specific site for interaction. Here it is even more complex because the solvent will certainly sample the entire solute surface. To efficiently roll the solvent around the solute and roll the solute around the solvent, and thus obtain a meaningful average interaction energy, Fan et al. (1992) used a spherical polar coordinate system (as was also done in the chiral chromatography example). One of the two molecules, say the solute, is placed at the origin, the center of the second molecule is placed on the surface of a sphere at spherical polar angles of θ and ϕ and a distance r. The orientation of the second molecule with respect to the first is defined by three Euler angles α, β, and γ. Rather than using a fixed grid as in the chromatography example, the angles were generated randomly where the random sampling was designed to evenly sample coordinate space, see Eq. (6.39).

$$
\begin{aligned}
\alpha &= 2\pi \, (\mathrm{rand}_1) \\
\cos \beta &= 1 - 2(\mathrm{rand}_2) \\
\gamma &= 2\pi \, (\mathrm{rand}_3) \\
\phi &= 2\pi \, (\mathrm{rand}_4) \\
\cos \theta &= 1 - 2(\mathrm{rand}_5)
\end{aligned}
\tag{6.39}
$$

The distance r was chosen so that the van der Waals surfaces of the two molecules just touched. For this geometry the energy of the combined system was evaluated and was included in the generation of the average interaction energy based on a Metropolis Monte Carlo sampling algorithm [see (Wood, 1957) and Appendix D].

The average coordination number z was obtained by repeating the above process for a second (or third, etc.) solvent molecule to see how many could be placed around the solute. For each additional solvent molecule the previous solvent molecules were included in the van der Waals distance evaluation but the new configuration was included only if the new solvent actually touched the solute.

Flory–Huggins theory only considers the configurational component of the entropy of mixing. Since Fan et al. (1992) could obtain the temperature dependence of the enthalpy of mixing, ΔH^{M}, the derivative of ΔH^{M} with respect to temperature provided an additional entropic contribution.

RESULTS

A coordination number of 7.8 was found for the n-hexane–nitrobenzene solvent–solvent test case. The temperature dependence of interaction energy was found to follow Eq. (6.40).

$$
\Delta \varepsilon_{12} = -7.66 \times 10^{-3} + \frac{1.05 \times 10^4}{T^2}
\tag{6.40}
$$

This led to a critical temperature T_{cr} (above which the two liquids would be miscible for any composition) of 268 K. The experimental T_{cr} for this system is 293 K. In this temperature range Fan et al. found T_{cr} to be nearly linearly dependent on the average coordination number, z. A z of 11.5 would have resulted in agreement with experiment.

The diisobutyl ketone–polyisobutylene solvent–polymer example provides an illustration of a difficulty with Flory–Huggins theory. As discussed above, because the approach uses a lattice model, the solvent and polymer repeat unit are assumed to be of nearly equal size (to fit in the lattice). Fan et al. dealt with this problem by taking the polymer repeat unit as twice the actual repeat unit so that it would be roughly the same size as the diisobutyl ketone solvent (within 5%). When polymer connectivity (the fact that the polymer continues on both sides of the repeat unit) was considered, an average coordination number of 6.5 was obtained. The temperature dependence of the interaction energy was found to follow Eq. (6.41).

$$\Delta\varepsilon_{12} = 8.20 \times 10^{-3} + \frac{5.45}{T} \tag{6.41}$$

This resulted in critical temperatures, T_{cr}, for three different molecular weights of the polymer of 295 K for MW = 22,700, 315 K for MW = 285,000, and 320 K for MW = 6,000,000. The corresponding experiment critical temperatures are 292, 319, and 329 K.

For the polyisoprene–polystyrene polymer blend the coordination number was calculated to be 5. The temperature dependence of the interaction energy was found to follow Eq. (6.42).

$$\Delta\varepsilon_{12} = 6.49 \times 10^{-3} + \frac{1.66}{T} \tag{6.42}$$

The parameter T_{cr} for five polymer–polymer MW combinations are collected in Table 6.2. If the calculated coordination number was reduced to 4 from the calculated value of 5, reasonable agreement with experiment was obtained.

The Fan et al. (1992) procedure generated reasonable estimates of interaction energies $\Delta\varepsilon_{12}$, but, as with any Flory–Huggins model, the procedure was quite sensitive to the relative sizes and shapes of the solutes and solvents. This result is evidenced by the difficulty in obtaining coordination numbers. Reasonable estimates of T_{cr} were calculated by making a small modification in the calculated coordination number parameter, z.

TABLE 6.2. Experimental and Calculated Critical Temperatures for Polyisoprene (pip) Polystyrene (ps) Blends

MW		T_{cr}(K)		
pip	ps	Exp.	Calc ($z = 4$)	Calc ($z = 5$)
1000	1000	243	230	267
2000	2700	329	397	473
2700	2100	408	410	495
2700	2700	448	450	550

6.6. ESTIMATION OF GLASS TRANSITION TEMPERATURES

6.6.1 Background

At low temperatures polymeric materials, though not generally crystalline, are solids. At sufficiently high temperatures, noncross-linked polymeric materials behave like viscous liquids. Between these two temperature limits amorphous polymeric materials soften and become deformable at well-defined temperatures. The material specific temperature at which this occurs is called the glass transition temperature, T_g. The solid has not melted at this temperature (the melting temperature or T_m can be $100°$ above T_g) but because of local motion, the polymer strands can slip past each other. The occurrence of a glass transition is dependent on the solid sample having significant amorphous or glassy regions. This makes the atomistic modeling of such systems difficult since a glassy material does not have long-range order; periodic boundary conditions thus supply an artifical periodicity to the system. The size of the unit cell needed to simulate an ''amorphous'' material is still an open research question. In atomistic simulations of the glasses it is hoped that making the unit cell as large as can be afforded is sufficient to properly simulate the amorphous nature of the material. Furthermore, T_g is not a first-order thermodynamic property, as can be seen by comparing the curve of Figure 6.16(a) (for an amorphous or glassy polymer) with the curve of Figure 6.16(c) (for a crystalline sample). The crystalline sample has a rather sharp break in the property versus temperature curve at the melting temperature, T_m. The amorphous sample has a bit of a bend at T_g, but not a sharp break. A semicrystalline sample (curve in Fig. 6.16(b)) has a bend at T_g and a sharp break at T_m.

Knowledge of T_g would aid in the design of materials with a wider temperature gap between T_g and T_m. An increased gap between T_g and T_m would mean an increased range of processing conditions. The standard approach for estimating T_g is the van Krevelen group additivity model (van Krevelen, 1990). Group additivies are discussed in Box 6.3.

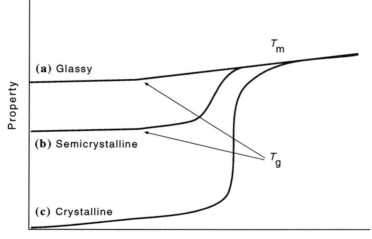

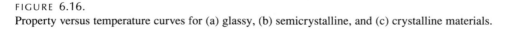

FIGURE 6.16.
Property versus temperature curves for (a) glassy, (b) semicrystalline, and (c) crystalline materials.

BOX 6.3   Group Additivities

In the van Krevelen group additivity model a property of a specific system P is taken as a sum of terms over a set of component subgroups with appropriate weighting terms, where P_i is the group contribution for group i, and s_i is the weight for group i, see Eq. (6.43).

$$P = \frac{\sum_i s_i P_i}{\sum_i s_i} \tag{6.43}$$

For the glass transition temperature T_g, van Krevelen and Hoftyzer in 1975 proposed (referenced in van Krevelen, 1990) that the T_g group additivity model should be taken as a simple sum of group terms (with corrections) weighted by the mass as described in Eq. (6.44), where M is the molar weight per structural unit, Y_{gi} is the molar glass transition function, and correction terms are needed in some cases.

$$T_g = \frac{\sum_i Y_{gi}}{M} + \text{Corrections} \tag{6.44}$$

For polypropylene there are two groups, CH_2 and $CH(Me)$. The CH_2 unit has a Y_g of 2700 K-g/mol and $CH(Me)$ has a Y_g of 8000 K-g/mol, giving a total of 10,700. For a polymer as simple as polypropylene there are no connection terms. The mass is $3 \times 12 + 6 \times 1 = 42$ so the calculated T_g is 255 K. This can be favorably compared with an experimental T_g range for atactic polypropylene of 238–299 K. Group additivity results are collected in Table 6.3 for a series of 30 polymers.

TABLE 6.3. Calculated and Experimental Glass Transition Temperatures for a Set of 30 Polymers

Polymer	T_g(K)			
	van Krevelen (1990)	Eq. (6.42)	Eq. (6.48)	Experimental van Krevelen (1990)
Poly(methylene oxide)	223	228	221	188–243
Poly(ethylene oxide)	214	218	222	206–246
Poly(trimethylene oxide)	209	213	211	195–228
Poly(tetramethylene oxide)	206	211	205	185–194
Polyethylene	193	198	177	143–250
Polypropylene	255	224	249	238–299
Polystyrene	200	345	334	353–380
Poly(vinyl fluoride)	298	246	398	253–314
Poly(vinylidene fluoride)	245	326	326	238–286
Poly(1,2-difluoroethylene)	344	340	329	323–371
Poly(vinyl chloride)	360	347	341	247–354
Poly(trifluorochloroethylene)	310	378	377	318–373
Poly(ethylene terephthalate)	341	328	351	346
Poly(4,4'-isopropylidene diphenylene carbonate)	413	407	381	414–423
Poly(isobutylene)	316	311	273	198–243
Poly(vinyl alcohol)	357	309	332	343–372
Poly(decamethylene terephthalate)	260	246	241	268–298
Poly(metaphenylene isophtalate)	400	398	438	411–428
Poly(hexamethylene adipamide)	321	242	277	318–330
Poly(1-butene)	238	222	218	228–249
Poly(1-pentene)	227	213	213	221–287
Poly(1-octene)	210	208	208	208–228
Poly(vinyl methyl ether)	252	215	242	242–260
Poly(vinyl ethyl ether)	237	206	234	231–254
Poly(vinyl hexyl ether)	210	194	217	196–223
Poly(methyl acrylate)	279	296	318	279–282
Poly(hexyl acrylate)	219	242	252	213–216
Poly(hexadecyl methacrylate)	290	315	259	288
Poly(methyl methacrylate)	378	385	354	266–399

6.6.2 Molecular Mechanics Based Group Additivity

As with any empirical method, a fundamental weakness of the group additivity scheme is the need for experimental data from which to obtain the ''group'' parameters. Since T_g is related to the flexibility of individual polymer chains (heights of torsional potential barriers) and the strengths of interstrand forces such as hydrogen bonding, molecular mechanics should be able to estimate T_g ''group'' parameters. Hopfinger et al. (1988) of the University of Illinois at Chicago, and Tripathy of the University of Lowell, published a molecular mechanics based method for obtaining group additivity parameters to use in the estimation of T_g.

METHODS

The initial model of T_g adopted by Hopfinger et al. (1988) was based on the assumption that, as with the rotational isomeric state model, intrastrand torsional motion is dominant in determining T_g, and interstrand interactions are small. The model consists of conformational entropy terms and mass moment terms both for polymer strand backbone torsions and for side-chain torsions. The parameter T_g can be expressed as in Eq. (6.45), where α, β, γ_i, δ_i, and ω are coefficients obtained from fitting the available experimental data; $b_i(\theta)$ and $s_{i,j}(\phi)$ are torsional entropy terms for the backbone and side-chain torsions, respectively; $m_i(\theta)$ and $w_{i,j}(\phi)$ are mass moments of the backbone and side-chain torsion angle units, respectively; and n_b and n_s are the number of backbone and side-chain torsion angles, respectively.

$$
T_g = \alpha\left(\frac{\sum_i b_i(\theta)}{n_b}\right) + \beta\left(\frac{\sum_i m_i(\theta)}{n_b}\right) + \sum_i \gamma_i\left(\frac{\sum_j s_{ij}(\phi)}{n_s}\right) + \sum_i \delta_i\left(\frac{\sum_j w_{ij}(\phi)}{n_s}\right) + \omega
$$

(6.45)

In this model the inverse mass dependence of the van Krevelen model is replaced by two mass moment terms, $m_i(\theta)$ and $w_{i,j}(\phi)$, and the molar glass transition functions, Y_{gi}, are replaced by the β and δ_i coefficients. Torsional entropy terms have also been added. Equation (6.45) is called a quantitative structure property relation (QSPR) and is analogous to the QSAR methodology discussed in Section 4.4.

The torsional entropy terms, $b_i(\theta)$ and $s_{i,j}(\phi)$, are obtained as individual 1-D rotational terms, each developed as a summation representation of the rotational probability integral, Eq. (6.46), where P_i, the probability of ''state'' i, is obtained from Eq. (6.47).

$$
b(\theta) = -R\sum_i P_i \ln P_i
$$

(6.46)

$$
P_i = \frac{\Delta\theta e^{\frac{-E_i(\theta)}{RT}}}{\sum_i \Delta\theta e^{\frac{-E_i(\theta)}{RT}}}
$$

(6.47)

The parameter $E_i(\theta)$ is the energy of the ith conformer state obtained by rotating about the torsion angle θ in increments of $\Delta\theta$.

The mass moments, $m_i(\theta)$ and $w_{i,j}(\phi)$, are taken as one-half of the masses of the torsional structural units. For example, the mass moment for vinylidene difluoride $+CF_2-CF_2+_n$ is given in Eq. (6.48).

$$m(\theta) = \frac{1}{2}\{M_C + 2M_F + M_C + 2M_H\} \tag{6.48}$$

For polypropylene, $b(\theta)$ for the $-CH_2CH(Me)-$ unit is 7.72 cal/deg-mol (3.86 cal/deg-mol for each of the two torsional angles). The parameter $M(\theta)$ is 42 amu (21 amu for each torsional unit, which comes from $\frac{1}{2}$ of 42). Dividing by the number of torsions (2 in this example) yields $b(\theta) = 3.86$ cal/deg-mol and $m(\theta) = 21$ amu. By using the best-fit parameters for Eq. (6.45) ($\alpha = -35.18$, $\beta = 1.55$, $\gamma = -18.26$, $\delta = 1.35$, and $\omega = 327.30$) T_g for polypropylene is given in Eq. (6.49), recall the experimental range is 238–299 K.

$$T_g = -35.18 \times 3.86 + 1.55 \times 21 + 327.30 = 224 \text{ K} \tag{6.49}$$

Shortly thereafter, Koehler and Hopfinger (1989) extended this methodology to include a model of interstrand interactions. Three average interaction energies $\langle E_D \rangle$, $\langle E_+ \rangle$, and $\langle E_- \rangle$, were added to the QSPR analysis to provide an initial estimate of interstrand interactions. These energies were obtained by placing one of three "atomic" probes in the plane perpendicular to the torsional bond axis, halfway between the end points of centers i and $i + 1$, see 6.17. In 6.17 the halfway point is labeled C_i.

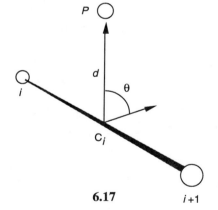

6.17

For each angle θ, out of a set of n (typically 12), the energy as a function of the perpendicular distance d was minimized and the resulting best energies averaged, Eq. (6.50), where p indicates the type of probe used ($d, +, -$).

$$\langle E_p \rangle = \frac{\sum_{j=1}^{n} E_p(\theta,d_{min})_j \exp[-E_p(\theta,d_{min})_j/RT]}{\sum_{j=1}^{n} \exp[-E_p(\theta,d_{min})_j/RT]} \tag{6.50}$$

TABLE 6.4. Group Entropy, Mass Moment, and Intermolecular Energy Terms for a Set of 12 Polymers

Polymer	S_B	M_B	E_D	E_+	E_-
$-CH_2-O-$	3.33	15.0	-0.86	-1.53	0.69
$-(-CH_2)_2-O-$	3.65	14.7	-1.04	-0.87	0.37
$-(-CH_2)_3-O-$	3.82	14.5	-1.13	-0.54	0.22
$-(-CH_2)_4-O-$	3.91	14.4	-1.18	-0.34	0.12
$-(-CH_2-)-$	4.30	14.0	-1.39	0.45	-0.26
$-CH_2-CH(Me)-$	1.93	14.0	-1.51	0.51	-0.37
$-CH_2-CH(\phi)-$	1.00	34.3	-3.26	-1.09	-0.59
$-CH_2-CH(F)-$	1.70	24.0	-1.52	-0.81	-0.38
$-CH_2-CF_2-$	1.48	32.5	-1.69	-1.27	-0.16
$-CH(F)-CH(F)-$	1.06	32.0	-1.36	-1.03	-0.50
$-CH_2-CH(Cl)-$	0.80	31.0	-1.93	-0.90	-0.46
$-CF_2-CF(Cl)-$	1.12	58.0	-2.18	-1.92	-1.08

For the van der Waals, or dispersion, probe (d) a united atom representation of a methyl group was used. For the positive probe ($+$) a hydrogen with unit positive charge was used, and for the negative probe ($-$) an oxygen with a unit negative charge was used. To facilitate the use of their molecular mechanics group additivity model and to describe all of the polymers with the same QSPR, Koehler and Hopfinger (1989) normalized mass moment and conformational entropy terms of Eqs. (6.45–6.48), calling the normalized terms M_B, S_B, and S_S. A subset of this normalized data is reported in Table 6.4. When this full model, including intermolecular data was fit, Eq. (6.51) resulted.

$$T_g = -25.3S_B - 7.66S_S + 1.40M_B - 0.14M_s - 0.77E_D -$$
$$31.6E_+ - 24.4E_- + 275.17 \quad (6.51)$$

The reported standard deviation was 14.9°. Given the small magnitude of the M_s and E_D coefficients (0.14 and 0.77), a T_g model was fit that ignored the effect of side-chain mass and average dispersion energy. This led to see Eq. (6.52).

$$T_g = -27.3S_B - 10.1S_S + 1.07M_B - 29.3E_+ - 15.1E_- + 288.83 \quad (6.52)$$

For this final model the standard deviation was 15.6°.

RESULTS

With a goal of developing a general group additivity model for T_g, Hopfinger et al. (1988) reasoned that intrastrand torsional motion would dominate T_g, and that molecular mechanics could be used to estimate torsional group parameters. Their initial study, Eq. (6.45), yielded a standard deviation of 17.9° for predicted T_g for a set of 30 polymers. The results, along with van Krevelen estimates and experimental data, are collected in Table

6.3. The largest reported error (76.5°) was for poly(hexamethylene adipamide), where interstrand hydrogen bonding is certainly important.

The errors found for materials with significant interstrand interactions led Koehler and Hopfinger (1989) to include estimates for these interactions in their group additivity model. For this model, Eq. (6.52), the standard deviation was 15.6° for a set of 35 polymers. The largest error again was for poly(hexamethylene adipamide), but the inclusion of intermolecular interactions dropped the error to 40.8° (see Table 6.3).

With this final model the T_g for polypropylene is predicted using the group parameters listed in Table 6.4. For polypropylene, the conformational entropy S_B for the $-CH_2-CH(Me)-$ unit is 3.06 cal/K-mol and M_B is 14.0 amu. There is no S_S term for propylene because the effect of the methyl was built into S_B. For polypropylene, $E_+ = 0.51$ and $E_- = -0.37$. The best-fit parameters from Eq. (6.50) can be used to obtain Eq. (6.53). Again, recall the experimental range for polypropylene is 238–299 K.

$$T_g = -27.3 \times 1.93 + 1.07 \times 14 - 29.3 \times 0.51 -$$
$$15.1 \times (-0.37) + 288.83 = 250 \text{ K} \quad (6.53)$$

Since the intermolecular interactions added to the model in going from Eq. (6.45) to (6.52) should also be phenomenologically responsible for polymer melting, Koehler and Hopfinger (1989) generated a QSPR for T_m for a set of 30 polymers (only the weighting coefficients were changed). For T_m the standard deviation was 31.6° and the maximum error was substantially larger, $-100.4°$ for poly(isobutylene). The difficulty in predicting T_m was ascribed to the presence of specific localized interactions in the crystalline environment not present in an amorphous polymeric material.

The Koehler and Hopfinger (1989) molecular mechanics based QSPR approach yielded estimates of T_g comparable to the strictly empirical van Krevelen approach. The Koehler and Hopfinger approach can be used for new polymers and to understand why a particular polymer has a low or high T_g. The molecular terms in the model are insufficient for estimating melting temperatures.

Homework

6.1. Use Eq. (6.4) to calculate the population at 200 K of the *trans* conformation of a 10 *mer* of polyethylene asuming the *trans* conformation is preferred over the gauche by 0.5 kcal/mol and assuming only a single defect structure per chain.

6.2. Use 3, 5, 7-trimethylnonane, **6.18**, as a model of a segment of isotactic polypropylene and carry out a conformational search to estimate the structure of a single-chain strand of isotactic polypropylene from the intermonomer and intramonomer dihedral angles of the central propylene unit.

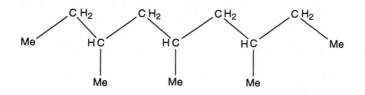

6.18

6.3. Use the crystal packing facility of your modeling software to take the single-chain strand of isotactic polypropylene from Problem 6.2 and generate a crystal fiber structure for isotactic polypropylene.

6.4. Minimize poly(methylene oxide) in a hexagonal crystalline environment using the isolated chain $O-C-O-C$ ϕ of 71.5° as a starting guess. Calculate the elastic moduli for this minimized structure. How do they compare to those reported by Sorensen et al. (1988)?

6.5. Use $MeOCH_2CH_2OMe$ as a model polyoxyalkane and apply your modeling software to determine the torsional profile for the $C-O-C-C$ unit. Is the *gauche* conformation preferred? If your modeling software has such a facility, generate the statistical weight matrix [see Eq. (6.21)] and obtain C_n for polyoxyethylene.

6.6. If your modeling software has such a facility, determine T_{cr} for a n-hexane–nitrobenzene mixture. How large is $\Delta\epsilon_{12}$? If you can use more than one set of partial charges or more than one force field, is $\Delta\epsilon_{12}$ sensitive to the choice of force field model?

6.7. Build and minimize 10-repeat unit strands of polyethylene ($C-C-C-C$ ϕ = 180°), and isotactic polypropylene, $[C-C-C(Me)-C$ ϕ = 180° and $C-C(Me)-C-C$ ϕ = $-60°$]. Measure the $C-C$ end-to-end distance of each. Add a constraint term to the energy expression to each polymer to lengthen the end-to-end distance by 1.0 Å and reminimize the structures. What happens to the $C-C$ distances, the $C-C-C$ angles, and the $C-C-C-C$ torsions for each material upon the addiition of the constraint term?

6.8. The premise of the molecular mechanics based QSPR of T_g was that those molecular effects important for the rotational isomeric state model were also responsible for changes in T_g. Use Eq. (6.52) and the data in Table 6.4 to estimate T_g for **6.13–6.15**. Which terms in Eq. (6.52) follow the trends for C_n established by the rotational isomeric state model?

References

Bachmann, M.; Gordon, W. L.; Weinhold, S.; Lando, J. B. (1980), The crystal structure of phase IV of poly(vinylidene fluoride), *J. Appl. Phys.* **51**, 5095.

Fan, C. F.; Olafson, B. D.; Blanco, M.; Hsu, S. L. (1992), Application of molecular dimulation to derive phase diagrams of binary mixtures, *Macromolecules* **25**, 3667.

Flory, P. J. (1941), Thermodynamics of high polymer solutions, *J. Chem.Phys.* **9**, 660.

Flory, P. J. (1942), Thermodynamics of high polymer solutions, *J. Chem. Phys.* **10**, 51.

Flory, P. J. (1974), Foundations of rotational isomeric state theory and general methods for generating configurational averages, *Macromolecules* **7**, 381.

Flory, P. J. (1988), *Statistical Mechanics of Chain Molecules*, Oxford University Press, New York.

Hill, T. L. (1960), *An Introduction to Statistical Thermodynamics*, Addison-Wesley, Reading MA.

Hopfinger, A. J.; Koehler, M. G.; Pearlstein, R. A.; Tripathy, S. K. (1988), Molecular modeling of polymers. IV. Estimation of glass transition temperatures, *J. Polymer Sci.: Part B: Polym. Phys.* **26**, 2007.

Huggins, M. L. (1941), Solutions of long chain compounds, *J. Chem. Phys.* **9**, 440.

Huggins, M. L. (1942), Theory of solutions of high polymers, *J. Am. Chem. Soc.* **64**, 1712.

Inomata, K.; Phataralaoha, N.; Abe, A. (1991), Conformational characteristics of α, ω-dimethoxyalkanes and related polymers: A combined use of molecular mechanics calculations and RIS simulations of NMR vicinal coupling constants, *Computational Polym. Sci.* **1**, 126.

Karasawa, N.; Dasgupta, S.; Goddard, W. A. III (1991), Mechanical properties and force field parameters for polyethylene crystal, *J. Phys. Chem.* **95**, 2260.

Karasawa, N.; Goddard, W. A. III (1992), Force fields, structures, and properties of poly(vinylidene fluoride) crystals, *Macromolecules* **25**, 7268.

Koehler, M. G.; Hopfinger, A. J. (1989), Molecular modelling of polymers: 5. Inclusion of intermolecular energetics in estimating glass and crystal-melt transition temperatures, *Polymer* **30**, 116.

Kusanagi, H.; Tadokoro, H.; Chatani, Y. (1975), A method of conformational and packing energy minimization: Application to X-ray structure analyses of crystalline polymers. *Rep. Prog. Polym. Phys. Jpn.* **18**, 193.

Kusanagi, H.; Tadokoro, H.; Chatani, Y.; Suehiro, K. (1977), Molecular and crystal structures of poly(ethylene oxybenzoate): α form, *Macromolecules* **10**, 405.

Lovinger, A. J. (1981), Unit cell of the γ phase of poly(vinylidene fluoride), *Macromolecules* **14**, 322.

Mayo, S. L.; Olafson, B. D.; Goddard III, W. A. (1990), DREIDING: A generic force field for molecular simulations, *J. Phys. Chem.* **94**, 8897.

Natta, G.; Corradini, P. (1959), Conformation of linear chains and their mode of packing in the crystal state, *J. Polym. Sci.* **39**, 29.

Sorensen, R. A.; Liau, W. B.; Kesner, L.; Boyd, R. H. (1988) Prediction of polymer crystal structures and properties. Polyethylene and poly(oxymethylene), *Macromolecules*, **21**, 200.

Sun, Z.; Morgan, R. J.; Lewis, D. N. (1992), Calculation of crystalline modulus of syndiotactic polystyrene using molecular modelling, *Polymer* **33**, 725.

Takahashi, Y.; Matsubara, Y.; Tadokoro, H. (1983), Crystal structure of form II of poly(vinylidene fluoride), *Macromolecules* **16**, 1588.

Takahashi, Y.; Tadokoro, H. (1980), Crystal structure of form III of poly(vinylidene fluoride), *Macromolecules* **13**, 1317.

Wood, W. W.; Parker, F. R. (1957), Monte Carlo equation of state of molecules interacting with the Lennard-Jones potential. A supercritical isotherm at about twice the critical temperature. *J. Chem. Phys.* **27**, 720.

van Krevelen, D. W. (1990) *Properties of Polymers: Their Correlation with Chemical Structure; Their Numerical Estimation and Prediction from Additive Group Contributions*, Elsevier, Amsterdam, The Netherlands.

Further Reading

Condensed State Simulations

Allen, M. P.; Tildesley, D. J. (1987), *Computer Simulation of Liquids*, Oxford University Press, Oxford, UK.

Bishop, M.; Clarke, J. H. R.; Rey, A.; Freire, J. J. (1991), Investigation of the end-to-end vector distribution function for linear polymers in different regimes, *J. Chem. Phys.* **95**, 4589.

Boyd, R. H., (1994) Prediction of polymer crystal structures and properties, *Adv. Polym. Sci.* **116**, 1.

Galiatsatos, V. (1995), Computational methods for modeling polymers: an introduction. In *Reviews in Computational Chemistry*, Vol. 6, Lipkowitz, K. B.; Boyd, D. B. Eds. VCH, New York. p. 149.

Hoover, W. G. (1991), *Computational Statistical Mechanics*, Elsevier Science, New York.

McKechnie, J. I.; Brown, D.; Clarke, J. H. R. (1992), Methods of generating dense relaxed amorphous polymer samples of use in dynamic simulations, *Macromolecules* **25**, 1562.

Rey, A.; Freire, J. J.; Bishop, M.; Clarke, J. H. R. (1992), Radius of gyration and viscosity of linear and star polymers in different regimes, *Macromolecules* **25**, 1311.

Yonezawa, F. (1993), Computer simulation methods in the study of noncrystalline materials, *Science* **260**, 635.

Polymers

Cowie, J. M. G. (1991), *Polymers: Chemistry and Physics of Modern Materials*, Chapman and Hall, New York.

Hopfinger, A. J. (1973), *Conformational Properties of Macromolecules*, Academic, New York.

Mattice, W. L.; Suter, U. W. (1994), *Conformational Theory of Large Molecules*, Wiley, New York.

Rosen, S. L (1993), *Fundamental Principles of Polymeric Materials*, Wiley, New York.

Supplemental Case Studies

Abraham, R. J.; Haworth, I. S. (1991), Molecular modelling of poly(aryl ether ketones): 2. chain packing in crystalline PEK and PEEK, *Polymer* **32**, 121.

Alemán, C.; Subirana, J. A.; Perez, J. I. (1992), A molecular mechanical study of the structure of poly (α-Aminoisobutyric Acid), *Biopolymers* **32**, 621.

Barino, L.; Scordamaglia, R. (1990), Conformational aspects of physical phenomena in polymeric materials, *Anal. Chim. Acta* **235**, 229.

Fan, C. F.; Hsu, S. L. (1991), Application of the molecular simulation technique to generate the structure of an aromatic polysulfone system, *Macromolecules* **24**, 6244.

Fan. C. F.; Hsu, S. L. (1992), Application of the molecular simulation technique to characterize the structure and properties of an aromatic polysulfone system. 2. Mechanical and thermal properties, *Macromolecules* **25**, 266.

Tashiro, K.; Tadokoro, H. (1981), Calculation of three-dimensional elastic constants of polymer crystals. 3. α and γ forms of nylon 6, *Macromolecules* **14**, 781.

CHAPTER 7

Inorganics

7.1. INTRODUCTION

Inorganic chemistry is a vast field encompassing the study of compounds of more than 100 elements in solution as well as in the solid state. Eight of the top 10 industrial chemicals, by annual tonnage, are inorganic. Inorganic materials are the basis of the semiconductor industry, high-temperature superconductors, and advanced ceramics. In addition, transition metal ions form the core of the catalysts used in the large-scale synthesis of most organic molecules and numerous polymeric materials. Silicates (SiO_2) with open lattice structures, such as zeolites, are used as the acid cracking catalysts for the production of more than 99% of the gasoline consumed worldwide.

Despite the importance of inorganic complexes in catalysis, molecular mechanics has not received the same general use in inorganic chemistry as in the other disciplines discussed in this text. Molecular modeling has a history that parallels the use of molecular mechanics in organic, biological, and polymer chemistry, but the general acceptance and use has been hindered by the variety of structures observed, the large number of elements, and hence atom types involved, and the paucity of experimental structures that can be used in parameterization. For example, there are only 4600 cobalt-containing structures in the Cambridge Structural Database, whereas there are more than 115,000 for carbon (Allen et al., 1991).

As with the molecules containing C, N, O, and H, molecular mechanics descriptions of inorganic molecules are motivated by bonding models. The primary bonding model used in organic chemistry is valence bond (Pauling, 1960). Here the lines drawn between atoms represent bonds, and hence stretch terms in a force field. The valence bond hybridizations ascribed to centers suggest particular angle bend values. For example, sp^3 hybridization implies bond angles of 109.5° and sp^2 hybridization implies bond angles of 120°. Numerous models to describe the structure and bonding of inorganic molecules have been devised. Valence bond (VB) and molecular orbital (MO) descriptions (Albright et al., 1985) have been developed, as well as nonorbital descriptions, such as the valence shell electron-pair repulsion (VSEPR) model (Gillespie and Hargittai, 1991). Each of these bonding models has played a role in the development of inorganic force fields. Valence bonds are still drawn between atoms, but VSEPR suggests that the angular preferences for metal centers is dictated strictly by nonbonded repulsions between the ligands. In contrast, MO theory suggests that for most transition metal complexes the preference for one coordination geometry over another is electronic in origin and should be mimicked by an angle bend term in the force field. The force field angle bend terms described in this chapter for transition metal complexes were motivated by either MO theory or VSEPR.

Inorganic chemistry can be broken down into the chemistry of the p block elements, where the p subshell is being filled with electrons as one progresses across the periodic table, and the chemistry of d and f block elements, where electrons are being added to d or f subshells as the periodic table is traversed.

7.1.1 The p Block Elements

The bonds between the p block, or main group, elements and their substituents can be characterized as having significantly more ionicity than the polar–covalent bonds in organic molecules. The ionicity in $Si-O$ and $Al-O$ bonds contribute to the acidity of hydroxylated metalloid–oxygen solids. Metalloid–oxygen solids, such as zeolites, are important Brønsted acid catalytic materials. The increased negative charge on oxygen in such materials amplifies the polarizablity of the oxide sites. This means the charge distribution at the oxide sites will shift as polar molecules approach. In the electrostatic model discussed in Section 3.2, the partial charges, q values, were taken as fixed and centered at the nuclei. To account for polarization, the partial charges must be able to move away from the nuclei. This polarization has been incorporated into a force field description through the use of a shell model. Both the shell model and a more conventional force field have been used to assess the acid sites in zeolites. These papers are discussed in Section 7.2.

7.1.2 The d Block Elements

The bonding of the d block or transition metal elements is more complex than the bonding in organic or p block compounds. In addition to the s and p valence orbitals of organic and main group elements, d block elements have valence d orbitals. These valence orbitals of higher angular momentum are more compact than the corresponding s and p orbitals on the metal center, permitting nonbonding d orbitals to be variably filled and radical centers with numerous unpaired electrons to be stable. For example, the oxidation states of osmium complexes range from -2 in $[Os(CO)_4]^{2-}$ to $+8$ in OsO_4, and $[Fe(NH_3)_6]^{3+}$ with five unpaired electrons can be isolated (Shriver et al., 1990). Molecular orbital theory is typically invoked to explain the stability and preferred geometries of such complexes. A MO description of the preference for an octahedral geometry in certain six-coordinate complexes and for a square planar geometry in a class of four-coordinate complexes is provided in Section 7.3.

There are several other complications that are part and parcel of the molecular mechanics description of d block or transition metal complexes. The first is that there are a number of coordination modes unique to organometallic compounds including π-allyl ligands, cyclopentadienyl ligands, and complexed alkenes, see Figure 7.1. These coordination modes require ingenuity in incorporating them into a force field. For some studies, specific, highly distorted angle bend terms have been included to attach π ligands to metal

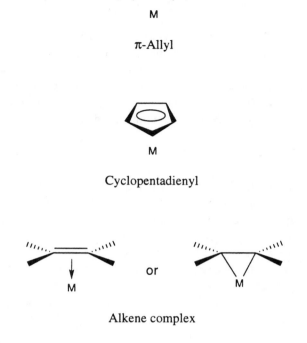

FIGURE 7.1.
The π-allyl, cyclopentadienyl, and alkine coordination modes.

centers, in other investigations the metal–ligand coordination environment has been held fixed. Alternatively, a third approach is to add fictitious or pseudoatoms to the force field to provide for the attachment of the π ligand to the metal center. The use of pseudoatoms appears to be prevailing and is overviewed as a part of the discussion of vinyl alkene polymerization in Section 7.5.

The second complication, that comes with the observation of coordination geometries such as octahedral, is the phenomenon of multiple natural angles. For an octahedral complex, a given ligand has bond angles of 90° with respect to ligands cis to it and 180° bond angles to ligands that are trans. The presence of these two minima has been dealt with in some force fields for metal complexes by simply ignoring all angle terms at metal centers. At times 90° minima have been chosen for angles near 90°, and the 180° minimum selected for angles near 180°. This approach eliminates the possibility of cis–trans interchange, which could lead to a more stable structure. Other force fields have incorporated new functional forms that have minima for both 90° and 180°. Another, somewhat restrictive approach, is to fix the positions of the atoms in the metal coordination environment and only examine the steric distortions somewhat removed from the metal coordination sphere. A review of angle functional forms for metal centers is provided during the course of discussing asymmetric hydrogenation in Section 7.6.

The bonding at metal centers is clearly more complex than the bonding between C, N, O, and H. The fact that more than sp^n hybrids are needed for a complete bonding picture leads to another obstacle to the development of metal ligand force fields: for three, four, five, and six coordination there is more than one observed structural preference. For three coordination, T-shaped structures as well as conventional trigonal planar structures are observed or invoked. For four coordination, square planar as well as tetrahedral structures have been observed. For five coordination, square pyramidal as well as trigonal bipyramidal structures are found. For six coordination, trigonal prismatic and octahedral structures need to be considered. This structural variety is central to the bioinorganic chemistry of Cu(II) and Zn(II). One solution to this problem was discussed in Section 3.3.2 for carbonic anhydrase. Alternative solutions are discussed in Section 7.7.

7.1.3 Transition Metal Applications

One of the first uses of molecular mechanics in inorganic chemistry was a report by Corey and Bailar (1959) on the stereochemistry of metal ethylenediamine complexes where the ligand, ethylenediamine(en), occupies two metal coordination sites, and hence is chelated. Metal chelate complexes have continued to be of interest, and a study on the classic system, tris(ethylenediamine)cobalt(III), is discussed in Section 7.4.

Molecular mechanics has been used to increase our understanding of the contributions that sterics make to stereoselectivity in catalytic reactions. A study of the stereoselectivity in propylene polymerization is discussed in Section 7.5. A combined molecular mechanics–NMR examination of the intermediates in asymmetric hydrogenation is reviewed in Section 7.6.

One of the more structurally challenging aspects of inorganic chemistry is the observation of "plastic" complexes such as those of four coordinate Cu(II). One might assume that four-coordinate metal complexes could be grouped into either the idealized tetrahedral coordination or the square planar coordination limits. However, a large number of the observed Cu(II) complexes are somewhere in between. In fact, square planar and distorted tetrahedral complexes of the same molecule $CuCl_4^{2-}$ have been observed in the same crystal lattice (Battaglia et al., 1982). Because four coordinate Cu(II) complexes adopt a variety of coordination geometries dependent on environment, they are often called "plastic." The plasticity of four-coordinate Cu(II) complexes is discussed in Section 7.7. Another class of "plastic" complexes are the hexamine complexes, such as **7.1**, M^{n+}(sar) (sar = sarcophagine) where the metal center is surrounded by an organic cage.

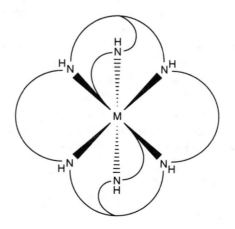

7.1

For this class of complex the degree of distortion from an idealized octahedron toward a trigonal prismatic structure varies dramatically with the metal, even for the same hexadentate ligand. Molecular mechanics studies of the preferred coordination geometries of Co(III) cage complexes are reviewed in Section 7.7.

7.2. ACID SITES IN ZEOLITES

7.2.1 Background

Zeolites are 3-D porous materials of idealized general formula $M_{x/n}^{n+}[(AlO_2)_x(SiO_2)_y]\cdot nH_2O$ (Cotton and Wilkinson, 1988). The solid consists of an open framework of SiO_4 tetrahedra. The tetrahedral silicon ion sites in the lattice are substituted by varying numbers of aluminum ions (depending on the material composition). The tetrahedra sites, denoted T (T = Si or Al) are connected by nearly linear T−O−T linkages, **7.2**.

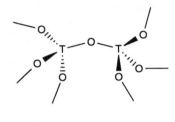

7.2

Since Si is formally +4 and Al +3, each aluminum substitution is accompanied by the addition of a cation such as Na^+ or H^+. The presence of large, hydrated ions such as tetraalkyl ammonium ions during synthesis and subsequent heating of the material results in large channels and cavities. The 3-D structures of zeolites are made up of a variety of building blocks.

The fundamental unit of one family of zeolites is the nearly spherical sodalite cage, see Figure 7.2(a). Since the T−O−T bond angles in zeolites are greater than 140°, model structures of zeolites often only display the T sites that are connected as though they were bonded, see Figure 7.2(b). These model structures do provide a reasonable representation of the cavities and channels in the zeolite structures without being overly congested. In the zeolite sodalite, eight such cages are stacked in a cubic packing pattern as shown in Figure 7.3(a). Zeolite A, used in water softeners, adopts a more open arrangement of sodalite cages, again with a cubic packing pattern, but with connecting oxygen bridges, is shown in Figure 7.3(b). The sodalite cages can also be stacked in a diamond arrangement with oxygen bridges as shown in Figure 7.3(c). This is the structure of faujasite or zeolite Y. Faujasite is the zeolite commonly used in petroleum processing. For faujasite, the diamond lattice arrangement of sodalite cages gives a network of large cavities, called super cages, connected to form a rather open diamond structure. A faujasite supercage is illustrated in Figure 7.3(d).

The ZSM-5 zeolite is representative of a different structural type. Rather than consisting of nearly spherical cavities, the porous structure of ZSM-5 is built from an array of nearly cylindrical channels connected by a perpendicular set of sinusoidal channels that

(a) **(b)**

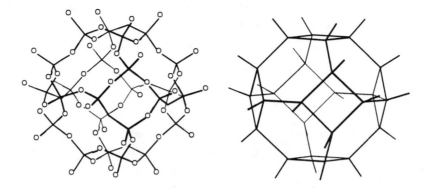

FIGURE 7.2.
Sodalite cage, (a) all atom representation and (b) T only representation.

(a) **(b)**

(c) **(d)**

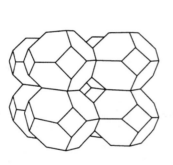

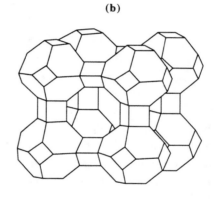
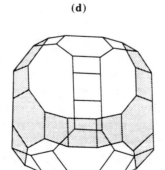

FIGURE 7.3.
Idealized cavity structures for a set of zeolite structures built from sodalite cages. Shown are (a) the T-only structure of the parent, sodalite; (b) the structure of zeolite A; (c) the structure of zeolite Y (faujasite); and (d) a supercage of faujasite.

weave through the array of elliptical channels. The parallel, elliptical channels are shown in Figure 7.4(a). The perpendicular set of sinusoidal channels are shown in Figure 7.4(b). The interconnected pattern created by these two sets of channels is shown in Figure 7.4(c).

The cavities or channels of zeolites are potentially acidic and have been exploited as molecular sieves, ion exchange materials, catalysts, and catalyst supports (for a general discussion see, Dyer, 1988).

Catalytic cracking, the conversion of long-chain hydrocarbons into useful C_1–C_6 fractions, is carried out by faujasites. Hydrocracking, the upgrading of heavy oils under hydrogen pressure, utilizes a bifunctional catalyst (hydrogenation activity from a noble metal and acidic functionality from a zeolite such as ZSM-5). Dewaxing, the removal of C_{18} and larger hydrocarbons by selective hydrocracking, can be accomplished by ZSM-5 without the need for a noble metal. The production of ethylbenzene from ethylene and benzene (a step in the synthesis of styrene for polystyrene formation) can be accomplished by the Mobil "Badger" process, which uses a ZSM-5 catalyst. The isomerization of o-xylene to p-xylene (a precursor to terephthalic acid, used as a monomer in polyester synthesis) can be carried out by ZSM-5. The greater diffusion rate of p-xylene out of ZSM-5 causes a shift in the steady-state distribution between the ortho and para isomers established inside the acidic material. One particularly intriguing application of zeolite catalysis is the use of ZSM-5 as a catalyst for the conversion of methanol to gasoline.

In each of the above processes the catalytic activity is ascribed to the presence of Brønsted acid (proton donor) sites in the interior of the solid material. It is generally recognized that the protons are present as hydroxyl groups that are positioned adjacent to Al^{3+} ions. When the proton is attached to the T$-$O$-$T linkage the T$-$O$-$T angle drops from about 140° to about 112°, **7.3**.

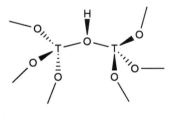

7.3

One of the T sites in the T$-$O$-$T linkage is always Al, the other is Si unless the percent Al approaches 50%. The cracking efficiency of the acid sites is thought to be influenced by the chemical composition (%Al) and the structure of the zeolite lattice. As the Al concentration increases, the cracking activity increases (Wielers et al., 1991). Two materials of similar composition, faujasite and ZSM-5, differ in cracking activity by two orders of magnitude (Wielers et al., 1991).

The cracking process is thought to occur through a number of pathways, the most common being the β-scission route (Gates et al., 1979). The β-scission pathway is thought to proceed through a cationic chain mechanism. The chain reaction is initiated by a num-

316

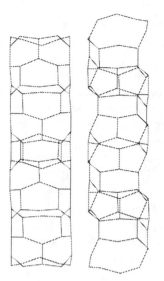

(a)

(b)

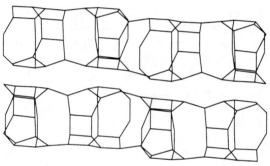

FIGURE 7.4.
Idealized, T-only, drawings of the interconnecting channels of ZSM-5. Nonchannel
atoms have been removed for clarity.
Shown are (a) the linear, elliptical main
channel structure, (b) the perpendicular
sinusoidal nearly circular channel structure,
and (c) a combination of the two interconnecting channel structures.

(c)

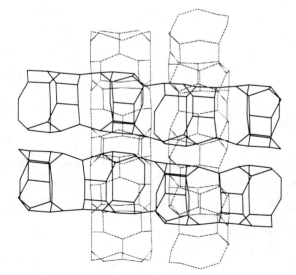

ber of mechanisms, one of which is protonation of an alkene present in the feedstock, Eq. (7.1).

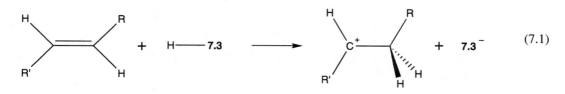

$$(7.1)$$

The cracking event is thought to be scission of a $C-C$ bond β to the carbonium ion center, Eq. (7.2), yielding an alkene and a new carbonium ion.

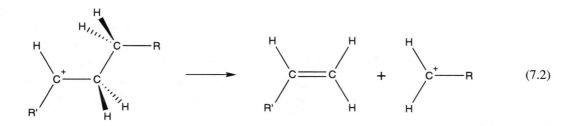

$$(7.2)$$

Chain propagation occurs when the carbonium ion carrier abstracts a hydride from a longer chain alkane generating a new carbonium ion center, Eq. (7.3), which in turn can undergo β scission, Eq. (7.2).

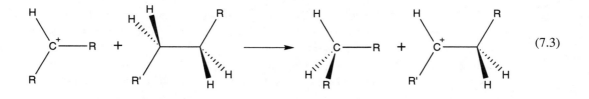

$$(7.3)$$

Chain termination can occur through a proton transfer from the carbonium ion to the conjugate base of a zeolitic acid site generating an alkene, Eq. (7.4).

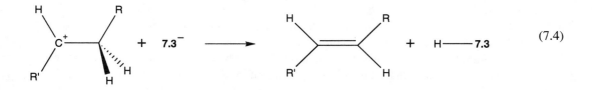

$$(7.4)$$

The acidity of the acid sites affects the rates of initiation and termination but not the rate of chain transfer.

The numbers of sites and types of sites differ from zeolite to zeolite. Because all of the T sites (Si or Al) in faujasite are equivalent, there are only four possible sites for the hydroxyl group (the four O atoms bound to the Al), labeled 1–4 in Figure 7.5. A descrip-

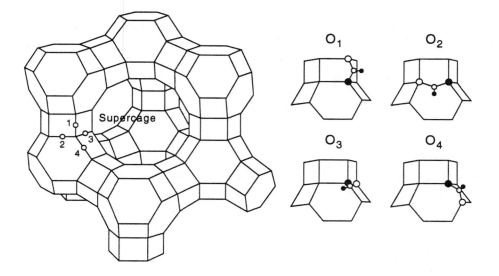

FIGURE 7.5.
Crystal structure of faujasite. The four unique oxygen sites of the lattice are labeled 1–4. The likely locations for a proton adjacent to Al in the lattice are shown in the cutout drawings. [Reproduced with permission from Krama, G. J.; van Santen, R. A. (1993), *J. Am. Chem. Soc.* **115**, 2887. Copyright © 1993 American Chemical Society.]

tion of the acid sites in ZSM-5 is more complex. High silicon ZSM-5 adopts different crystal structures at low (below 350 K) and high temperature. The low-temperature monoclinic form contains 24 T sites and 48 unique oxygen atoms yielding 96 possible H positions. The high-temperature orthorhombic form contains 12 unique T sites and 26 unique oxygen atoms yielding 48 possible H positions.

In order to further understand the structural sources for the dependence of acidity on chemical composition and lattice structure, Schröder et al. (1992a, 1992b), and Kramer and van Santen (1993) calculated the relative proton affinities for the possible O–H sites in faujasite and ZSM-5 using a shell model (Schröder et al., 1992a, 1992b) and a more conventional force field (Kramer and van Santen, 1993).

7.2.2 Shell Model Study of Faujasite and ZSM-5

Schröder and Sauer, of the Zentralinstitut für Physikalische Chemie, Berlin, Germany; Leslie of the Daresbury Laboratory, United Kingdom; and Catlow of the Davy-Faraday Research Laboratory, Royal Institution of Great Britain (Schröder et al., 1992a) reported a shell model force field study of the acid sites of faujasite. Also in 1992 the same authors along with Thomas of the Davy-Faraday Research Laboratory, Royal Institution of Great Britain (Schröder et al., 1992b) reported a shell model study of the acid sites of the low-temperature form of ZSM-5.

METHODS

Because the bonding between Si/Al and O is substantially ionic and because the Si−O−Si bond angles are nearly linear ($>140°$), a simplified force field was developed that permitted polarization of the oxygen charge distribution within the context of a formal charge representation of the Si^{4+}, Al^{3+}, and O^{2-} ions. Short-range ''bonding'' interactions were described with an exponential-6 potential, Eq. (7.5), where the A, B, and C_6 parameters were chosen to reproduce the structural and energetic properties of the ionic solids α-quartz (SiO_2) and Al_2O_3.

$$V_{bonding} = Ae^{-B\rho} - \frac{C_6}{\rho^6} \qquad (7.5)$$

A harmonic angle bending term for O−Si−O angles was also included. To permit polarization (off-center distortion) of the oxide ions, a shell model (discussed in Box 7.1) was used for the O^{2-} ions. Addition of a hydroxyl group, with largely covalent bonding, dictated the addition of a covalent term between the oxygen and hydrogen of the hydroxyl group to this ionic bonding model. Schröder et al. used a Morse function to describe the hydroxyl bond. Furthermore, the charge on H was set at $+0.4$ rather than $+1$, as dictated by formal charges. The charge on the oxygen bonded to the hydrogen was also reduced from its formal charge of -2 to -1.4.

RESULTS

To probe the structural dependence of acidity on chemical composition and on lattice structure Schröder et al. calculated the relative proton affinities for the possible oxide sites in faujasite and ZSM-5 using a shell model. For faujasite, the complete unit cell structure was minimized for each of the four possible OH locations. Small energy differences were reported between the sites, with the O3 site lowest in energy. The relative energies are collected in Column 2 of Table 7.1. The two sites of lowest energy calculated by the shell model are in agreement with the experimentally observed site occupancies. For ZSM-5, with 96 possible sites, a prescreening was carried out. For each of the 96 possible H

BOX 7.1   Shell Model

In a shell model of ionic bonding, ions are described in terms of nuclear cores and massless shells of electron density. The shells and cores are connected by harmonic potentials, the force constants of which are related to the electric polarizibilities of the free ions. The van der Waals and angle bend potentials, which describe electronic interactions, use the coordinates of the shells rather than of the cores. For these studies Shröder et al. used the shell model for the O^{2-} ions only. The charge on the O^{2-} core was set to $+0.86902$, which led to a charge on the shell of -2.86902. Long-range electrostatic interactions were calculated using the Ewald summation technique discussed in Section 2.5.3.

TABLE 7.1. Relative Energies (kcal/mol) for the Four Possible Hydroxyl Locations in Faujasite

Site	Shell Model	Conventional FF	Experimental (Czjzek et al, 1992) Fractional Occupation	Estimated ΔE (573 K)
O1H	1.3	8.3	0.436	0.0
O2H	4.7	17.1	0.233	0.71
O3H	0.0	0.0	0.330	0.32
O4H	5.7	21.2	0.	

locations lattice relaxation was permitted for ions up to 6 Å away from the site of protonation. For each of the 24 possible Al sites and the lowest energy H placement for each Al site, the radius of relaxation was expanded to 8 Å. Finally, the radius of lattice relaxation was expanded to 10 Å for the 9 lowest energy hydroxyl sites. The relative energies for the 9 lowest energy AlOH sites are collected in Column 2 of Table 7.2. Little energetic preference was found for proton positioning in ZSM-5; the 9 lowest energy sites were within 1.5 kcal/mol.

7.2.3 Force Field Study of Faujasite and ZSM-5

Kramer of the Koninklijke/Shell-Laboratorium Amsterdam and van Santen of the Koninklijke/Shell-Laboratorium Amsterdam and the Schuit Institute of Catalysis, Eindhoven University of Technology all of the Netherlands (1993) reported a more conventional force field study of the acid sites of both faujasite and ZSM-5.

METHODS

A simplified force field was also used for this study. Short-range "bonding" interactions were described with a van der Waals potential as above, Eq. (7.5). A harmonic angle

TABLE 7.2. Relative Energies for the Nine Lowest Energy Hydroxyl Locations in ZSM-5 for the Shell Model Force Field (Monoclinc Crystal)

Al Site	O Site	Energy (kcal/mol)
7	17	0.0
19	43	0.6
20	33	0.8
16	17	0.9
14	32	0.9
1	1	1.0
18	45	1.0
15	20	1.2
19	33	1.5

TABLE 7.3. Relative Energies for the Nine Lowest Energy Hydroxyl Locations in ZSM-5 for the Conventional Force Field (Orthorhombic Crystal)

Al Site	O Site	Energy (kcal/mol)
12	12	0.0
1	16	1.6
8	12	2.1
10	9	4.6
3	19	4.6
2	13	5.1
3	3	5.3
6	19	5.5
11	10	6.0

bending term for $Si-O_H-Al$ angles was also included (natural angle 88.62°); $O-Si-O$ and $Si-O-Si$ angle terms were not used, and the angular preferences were provided by nonbonded repulsions. The proton was not explicitly considered in the simulations; an O_H united atom representation was used. In a united atom representation the hydrogen atoms on the hydroxyl groups are ignored, and the mass and van der Waals radius of oxygen are expanded to approximately account for the hydrogen atoms in a spherically average way. Partial charges were included ($q_O = -1.2$, $q_{OH} = -0.2$, $q_{Si} = 2.4$, and $q_{Al} = 1.4$). For both faujasite and ZSM-5 full lattice minimizations were carried out.

RESULTS

Kramer and van Santen (1993) calculated the relative proton affinities for the possible oxide sites in faujasite and ZSM-5. For faujasite, the O3 site was predicted to be most stable, in agreement with the shell model, see Column 3 of Table 7.1. The energy ordering was also consistent with the shell model study. However, the energy differences calculated were substantially larger than with the shell model. For ZSM-5 there was little agreement between the shell model results and the Kramer and van Santen force field study (cf. Tables 7.2 and 7.3). In addition to differences between the two force fields, the shell model report was based on the low-temperature monoclinic crystal, whereas the Kramer and van Santen force field effort used the high-temperature orthorhombic crystal.

Kramer and van Santen (1993) sought a structural explanation for the substantial site differentiation obtained for ZSM-5. As presented in Figure 7.6, they found an inverse correlation between proton stability and average $T-O$ distance, a linear correlation between proton stability and average $O-O$ distance, and virtually no correlation between proton stability and $T-O-T$ angle. The inverse correlation between proton stability and average $T-O$ distance was ascribed to the fact that $T-O_H$ distances should be substantially longer than $T-O$ distances, and sites with longer $T-O$ distances were predisposed to lengthening. The linear correlation between $O-O$ distance and proton stability was thought to be due to the observation that O_H-T-O angles range from 90–100°, rather than the normal tetrahedral angle of 109.471°. Again, sites with short $O-O$ distances were thought to be better able to accommodate the O_H. The lack of a correlation between $T-O-T$ angle and proton stability was ascribed to the intrinsic flexibility, and hence energetic insensitivity of this angle.

Both the conventional force field and the shell model found proton site preferences for both faujasite and ZSM-5. The range of energies was larger for the force field model than for the shell model. The ranges of energies for faujasite for both models were also substantially larger than would be consistent with the experimentally observed population of the O1 site, see Columns 4 and 5 of Table 7.1. The estimated ΔE values of Column 5 were obtained assuming an equilibrium distribution was established at 300°C, the calcining temperature. The site populations were assumed to be frozen in upon cooling. Both studies reported substantial geometric reorganization upon protonation. Kramer and van Santen (1993) attributed the site preferences to a predispositon of certain oxide sites toward the structural distortions associated with protonation. Direct correlation of the results of these investigations to cracking efficiency in faujasite and ZSM-5 awaits further work.

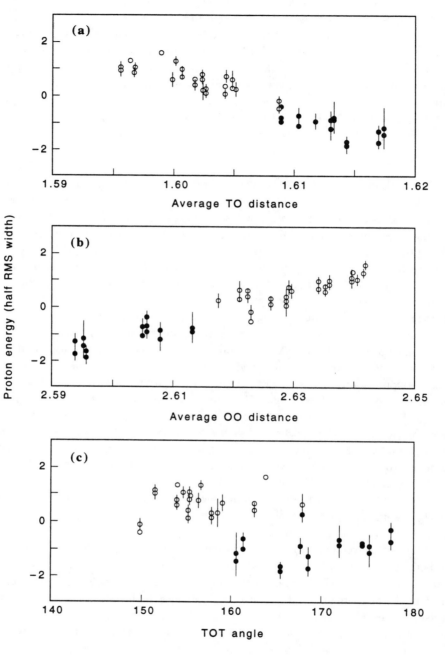

FIGURE 7.6.

Relation between proton stability and (a) average T−O distance, (b) average O−O distance, and (c) average T−O−T angle. Proton energies are in electronvolts, distances are in angstroms, and angles are in degrees. [Reproduced with permission from Krama, G. J.; van Santen, R. A. (1993), *J. Am. Chem. Soc.* **115**, 2887. Copyright © 1993 American Chemical Society.]

7.3. ELECTRONIC SOURCES OF OCTAHEDRAL AND SQUARE PLANAR COORDINATION

As in organic chemistry, the developers of molecular mechanics models for transition metal complexes have relied on bonding models for insight about functional forms. In deciding what type of angle bend term to use at six-coordinate transition metal centers, VSEPR could be relied on: It suggests octahedral coordination. If the nonbonded repulsions implicit in the VSEPR description are the entire source of the preference for an octahedral geometry, then simply including 1,3 nonbonded interactions for six-coordinate complexes would be correct. For four-coordinate complexes simply including 1,3 nonbonded interactions does not work. Four-coordinate complexes with eight valence d electrons, a d^8 configuration, such as Pt(II) are known to adopt a square planar geometry rather than the tetrahedral geometry dictated by 1,3 nonbonded interactions. There must be an electronic preference for this square planar geometry. Molecular orbital theory provides a convenient framework for describing this electronic effect. Furthermore, MO theory suggests an analogous electronic factor favors octahedral geometries for six-coordinate complexes with a d^6 electronic configuration.

7.3.1 Electronic Structure of Octahedral Coordination

The electronic preference of a metal center with six ligands and six nonbonding d electrons for octahedral coordination can be seen by considering the MO bonding description for a prototypical octahedral complex. The bonding in an octahedral complex can be separated into components by shape (or symmetry). The valence s, p, and d metal orbitals can be classified under octahedral symmetry and the set of 6 σ orbitals on the 6 ligands and the set of 12 π orbitals on the 6 ligands can be classified under octahedral symmetry. Only the metal and ligand orbitals of the same symmetry will interact; the degree of interaction being governed by the relative energetic stabilities of the orbitals. The symmetry classifications and relative stabilities of the orbitals can be used to construct what is called an orbital correlation diagram.

Construction of an ML_6 correlation diagram, considering σ interactions between the metal and a set of six ligands, begins with the assessment that the ligand σ orbitals are more stable (lower in energy) than the metal valence orbitals; this leads to the energy level diagram of Figure 7.7(a). The six ligand σ fragment orbitals are shown on the right; the set of five metal d orbitals are shown as the lowest energy metal orbitals, on the left. The metal s orbital is shown energetically above the d set and the set of three metal p orbitals listed highest in energy. For a fourth period or first-row transition metal, such as cobalt, the valence orbitals are $3d$ and $4s$ and $4p$. Labeling the metal valence orbitals by symmetry and combining the ligand orbitals into the symmetry combinations given in Figure 7.8(a) yields Figure 7.7(b). If we allow the metal and ligand fragment orbitals of the same shapes [cf. Fig. 7.8(a) for the ligand orbitals to Fig. 7.8(b) for the metal orbitals] to interact we obtain the full σ orbital correlation diagram in Figure 7.7(c). Often the diagram is simplified to the noninteracting or nonbonding metal t_{2g} d orbital set and the antibonding e_g d orbital set as shown in Figure 7.7(d). The precise placement of electrons in these nonbonding and antibonding d orbitals is in large measure responsible for the coordination geometries observed for six-coordinate complexes.

324

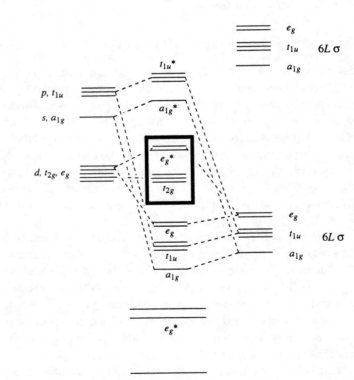

FIGURE 7.7.
Buildup of octahedral MO correlation diagram: (a) metal (on left) and ligand fragment orbitals, (b) symmetry combinations of metal and ligand fragment orbitals, (c) orbital correlation diagram with frontier orbitals highlighted and (d) octahedral frontier orbital diagram.

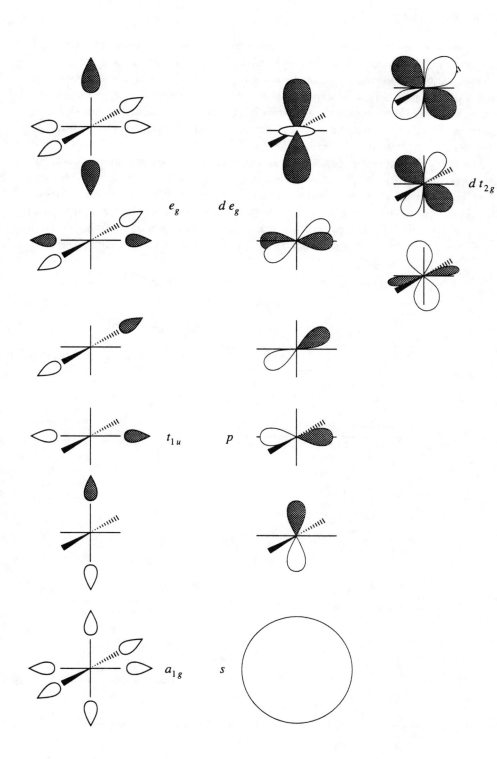

e_g $d\,e_g$

$d\,t_{2g}$

t_{1u} p

a_{1g} s

FIGURE 7.8.
Octahedral symmetry combinations of metal and fragment orbitals (ligand fragment orbitals on the left).

If we consider an octahedral Co(III) complex, there are six valence electrons to be placed in the five frontier orbitals of Figure 7.7(d). If all six electrons are placed in the t_{2g} set, then antibonding orbitals will not be occupied and metal–ligand antibonding interactions will be avoided. This combination of 12 electrons in ligand-based bonding orbitals and 6 electrons in three nonbonding d orbitals is certainly a stable configuration. Since 2 electrons are placed in each of the three t_{2g} orbitals, molecules with this electronic configuration are called low-spin d^6 complexes.

To see why octahedral coordination is preferred over other possible six-coordinate structures for Co(III) complexes, we will consider two modes of distortion where the M−L bond distances are kept constant. The first is elongation of the complex along the direction of one of the triangular faces of the octahedron—this is called a trigonal distortion. This distortion is accompanied by a change in the placement of the frontier orbitals, see Figure 7.9(a). Two of the three nonbonding orbitals become antibonding. These new

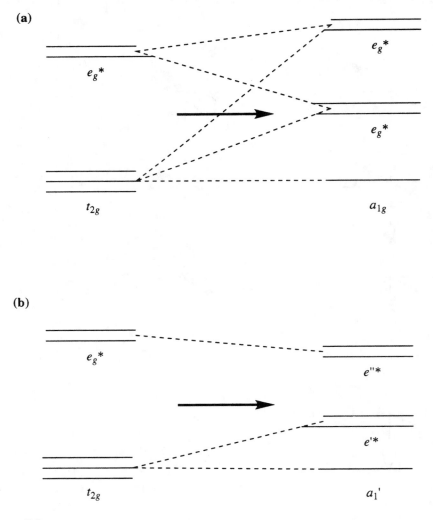

FIGURE 7.9.
Orbital correlations for distortion for (a) trigonal distortion and (b) trigonal prismatic distortion.

antibonding interactions raise the energy of the complex, and hence an octahedral geometry is favored over a trigonally distorted complex for molecules with six electrons in the frontier orbitals.

A second mode of distortion is to rotate one of the octahedral triangular faces with respect to the other. This distortion leads to trigonal prismatic coordination. This distortion is also accompanied by a change in the placement of the frontier orbitals, see Figure 7.9(b). Again, two of the three nonbonding obitals become antibonding. Again, these new antibonding interactions raise the energy of the complex, and hence an octahedral geometry is also favored over a trigonal prismatic geometry for complexes with six electrons in the frontier orbitals.

Force fields for low-spin d^6 complexes need to account for this electronic preference for octahedral coordination.

7.3.2 Electronic Structure of Square Planar Coordination

The preference of a metal center with four ligands and a d^8 configuration can, again, be seen by considering the MO description for a prototypical system. The metal valence s, p, and d orbitals can be classified under D_{4h} symmetry, see Figure 7.10(a). The four σ orbitals on the four ligands can also be combined and classified into combinations appropriate for D_{4h} symmetry, see Figures 7.10(b) and 7.11. The resulting correlation diagram, obtained by correlating metal and ligand orbitals of the same symmetry [see Fig. 7.10(c)] indicates that there is a single significantly antibonding d orbital, $d_{x^2-y^2}$. Eight electrons can be placed in dominantly nonbonding orbitals leaving the fifth (antibonding) d orbital empty. This can be compared with the case for tetrahedral coordination, developed in Figure 7.12 by using the fragment orbitals of Figure 7.13.

For a tetrahedral geometry, four of the eight electrons in the frontier d orbitals must be placed in the antibonding t_2 set. For a square planar geometry the antibonding d orbital can remain empty. Distortion from a square planar geometry towards a tetrahedral coordination geometry causes the highest nonbonding d orbital to become antibonding. The preference for a square planar geometry for Rh(I) and other d^8 complexes is thus electronic in origin. Sterics, that is 1,3 nonbonded repulsions, are not sufficient to maintain a square planar geometry because 1,3 nonbonded repulsions would lead to a tetrahedral geometry. Including only cis (90°) angle terms is not enough, because a highly unusual geometry with three cis ligands could result, see **7.4**.

7.4

The electronic origin of square planar structures must be mimicked by a proper angle term.

328

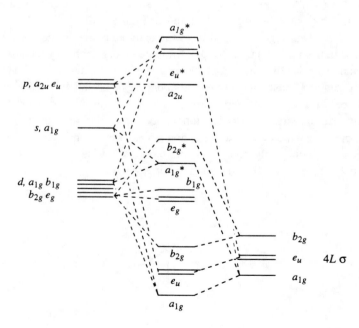

(a)

p

s

d

$4L\,\sigma$

(b)

$p,\,a_{2u}\,e_u$

$s,\,a_{1g}$

$d,\,a_{1g}\,b_{1g}$
$b_{2g}\,e_g$

b_{2g}

e_u $4L\,\sigma$

a_{1g}

(c)

$a_{1g}{}^{*}$

$e_u{}^{*}$

a_{2u}

$p,\,a_{2u}\,e_u$

$s,\,a_{1g}$

$b_{2g}{}^{*}$

$a_{1g}{}^{*}$

$d,\,a_{1g}\,b_{1g}$
$b_{2g}\,e_g$

b_{1g}

e_g

b_{2g}

b_{2g}

e_u $4L\,\sigma$

a_{1g}

e_u

a_{1g}

FIGURE 7.10.
Buildup of square planar MO correlation diagram: (a) metal (on left) and ligand fragment orbitals, (b) symmetry combinations of metal and ligand fragment orbitals, and (c) final orbital correlation diagram.

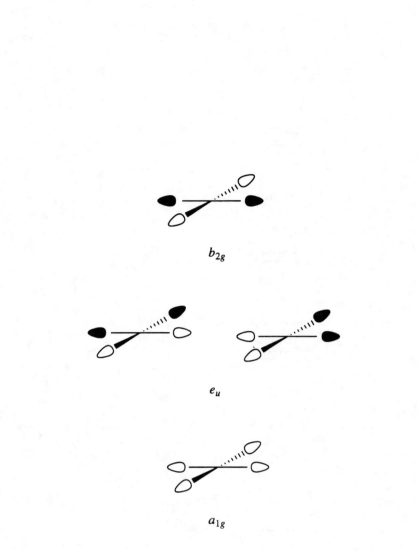

b_{2g}

e_u

a_{1g}

FIGURE 7.11.
Square planar symmetry combinations of fragment orbitals.

330

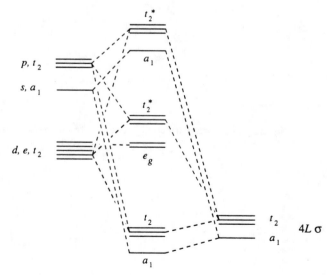

FIGURE 7.12.
Buildup of tetrahedral MO correlation diagram: (a) metal (on left) and ligand fragment orbitals, (b) symmetry combinations of metal and ligand fragment orbitals, and (c) final orbital correlation diagram.

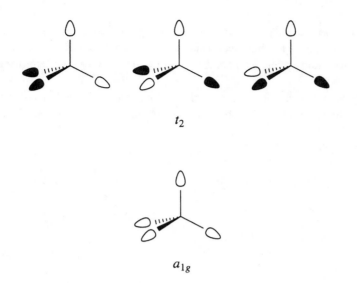

t_2

a_{1g}

FIGURE 7.13.
Tetrahedral symmetry combinations of fragment orbitals.

7.4. CONFORMATIONAL ANALYSIS

7.4.1 Background

Shortly after the presence of two low-energy conformations for decalin was established (Hückel, 1925), the conformational characteristics of five-membered chelate rings of inorganic chemistry were determined. Rosenblatt and Schleede (1933) suggested that the platinum–ethylenediamine chelate ring was not planar, see **7.5**.

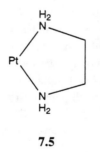

7.5

Theilacker (1937) and Kobayashi (1943) proposed a skew form for the ethylenediamine(en) chelate ring, see **7.6a**.

7.6a

By the 1950s, the presently accepted skew or *gauche* or half-chair conformation for ethyl-enediamine chelates, **7.6a**, had been established by infrared (IR), X-ray diffraction, and the presence of more optical isomers than possible for a flat ring system (for a review see, Saito, 1985).

Assuming a flat chelate ring system, tris(bidentate) octahedral complexes such as tris(ethylenediamine)cobalt(III) possess two optical isomers, designated Λ, and Δ, see **7.7Λ** and **7.7Δ**.

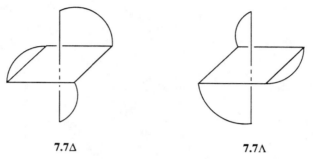

7.7Δ   7.7Λ

Relaxation of the assumption that the chelate rings are flat leads to puckered chelate rings, as in **7.6a** and **7.6b**.

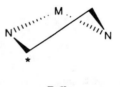

7.6b

The starred carbon of the chelate in **7.6b** can be either down or up, which leads to two conformations, δ and λ for each of the three chelate rings of the tris(bidentate) system. When ring puckering is added to the Λ and Δ optical isomer description, eight isomers result (a set of four diastereomers for Λ and a set of four diastereomers for Δ). The standard notation for the set of eight isomers is given in Table 7.4. A chirality invariant notation proposed by Corey and Bailer (1959) is listed as the last row of Table 7.4. The designation *lel* is short for paral*lel* and *ob* is short for *ob*lique. The shorthand notation refers to whether the C−C bond of the chelate ring is parallel to (or not parallel to) a threefold axis of the octahedral system. Cobalt(III) complexes tend to form the lel_3 isomer in the solid state, though all four isomers have been structurally characterized.

TABLE 7.4. Standard Notation for Tris(bidentate) Metal Chelate Complexes

Diastereomer	1	2	3	4
Λ Enantiomer	$\Lambda(\delta\delta\delta)$	$\Lambda(\delta\delta\lambda)$	$\Lambda(\delta\lambda\lambda)$	$\Lambda(\lambda\lambda\lambda)$
Δ Enantiomer	$\Delta(\lambda\lambda\lambda)$	$\Delta(\lambda\lambda\delta)$	$\Delta(\lambda\delta\delta)$	$\Delta(\delta\delta\delta)$
Corey–Bailer notation	lel_3	lel_2ob	$lelob_2$	ob_3

7.4.2 Cobalt(III) Chelate Complexes

Just as molecular mechanics was employed to help understand the conformational energy preferences of organic ring systems, molecular mechanics has been used to probe the conformational preferences of a number of chelate ring systems. Here we discuss the work of Niketic and Rasmussen (1978) and Hald and Rasmussen (1978) at the Technical University of Denmark on the octahedral tris(ethylenediamine)cobalt(III) cation.

METHODS

Niketic and Rasmussen (1978) and Hald and Rasmussen (1978) used the CFF force field for these studies. Here CFF is a conventional harmonic force field consisting of harmonic stretch and bend terms, threefold torsional potentials, and an exponential-6 non-bond potential. Since octahedral coordination is preferred for Co(III) complexes the problem was dealing with cis (90°) and trans (180°) angles at the metal center. Intrachelate angles (90°) were included, but they chose to ignore the trans angles at the metal center and to also ignore all interchelate angles (cis as well as trans). The metal–ligand van der Waals interactions (M ⃛ H and M ⃛ C) were either ignored or substituted by carbon parameters (C ⃛ H and C ⃛ C). Electrostatic interactions were ignored even though the metal center was formally $+3$. Initial structures were generated using standard bond distances and angles. Torsion angles were used to establish the proper chirality for each of the unique enantiomers. Following the initial structural study, Hald and Rasmussen (1978) used statistical thermodynamics (see Appendix B) to provide an estimate of the vibrational contribution to ΔG in order to compare with experimental isomer populations.

RESULTS

As an initial test of the force field, the molecular mechanics derived bond distances and angles of the experimentally observed lel_3 isomer of tris(ethylenediamine)cobalt(III), Co(en)_3, **7.8**, were compared with the corresponding X-ray diffraction data.

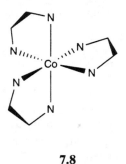

7.8

Since a number of Co(en)_3 structures had been reported that only differed in the counter-ion, Hald and Rasmussen (1978) compared the calculated results to an averaged set of experimental data. The results are collected in Table 7.5. The discrepancies between the calculated and observed Co–N (0.045 Å) and C–C (0.032 Å) distances were due to the force field having been fit to the early X-ray structures of Co(en)_3, where Co–N and C–C distances of 2.00 and 1.54 Å, respectively, were reported.

TABLE 7.5. Comparison of Average Bond Distances and Angles for CO(en)₃, **7.8**

Distance/Angle	Calculated	Average X-Ray
Co—N	2.018	1.973
N—C	1.475	1.486
C—C	1.541	1.509
N—Co—N	86.6	85.3
Co—N—C	106.7	109.1
N—C—C	107.7	107.4

The individual terms in the force field were used to assess the source of conformational differentiation for the four isomers of Co(en)₃. This data is collected in Table 7.6. There was little differentiation due to stretch and bend strain. The major differentiation was due to torsional strain (ranging from 0.3 to 0.6 kcal/mol) and nonbonded repulsions (ranging from 0.4 to 0.7 kcal/mol). The computed energy differences in Tables 7.5 and 7.6 corresponded to measurements carried out below 0 K (which is physically not possible). The effects due to differential zero-point vibrational motion and the temperature dependence of the population of vibrational levels needed to be included before a direct comparison with experiment at finite temperature would be valid (see Appendix B for a more detailed discussion). The vibrationally corrected results are collected in Tables 7.7 and 7.8 (Hald, 1978). The calculated conformational profile was in essential agreement with experiment, even when the small energetic differences due to the temperature dependent population of vibrational levels was considered.

In summary, just as with organic and bioorganic molecules, molecular mechanics can reproduce known molecular structures and conformational and configurational energy differences of transition metal chelate complexes. Furthermore, as with organic molecules, the molecular mechanics energy expression can be analyzed to understand the sources of conformational and configurational preferences.

7.5. ZIEGLER–NATTA ALKENE POLYMERIZATION

7.5.1. Background

Large scale, industrial processes for the preparation of commodity chemicals typically use heterogeneous catalysts. Homogeneous, or solution phase, transition metal containing catalysts are of interest because of the increased selectivity that can be found under the milder operating conditions accessible in a homogeneous solution: room temperature to 100°C, as compared to 100–300°C used for heterogeneous catalysts (Parshall and Itel, 1992). Homogeneous catalysts have proven particularly effective at controlling the stereochemical outcome of chemical transformations (Morrison, 1985). One example of stereochemically controlled catalysis is the use of transition metal complexes or metal halide solids to carry out the polymerization of vinyl alkenes.

TABLE 7.6. Individual Energy Terms for the Four Diastereomers of Co(en)$_3$, **7.8**

Term	ob_3	$lelob_2$	lel_2ob	lel_3
E_r	0.32	0.33	0.30	0.27
E_θ	1.93	1.97	2.00	2.11
E_ϕ	4.65	4.57	4.35	4.04
E_{nb}	−4.05	−4.02	−4.30	−4.70
Total E	2.85	2.85	2.35	1.72
ΔE	1.13	1.13	0.63	0.0

TABLE 7.7. Individual Energy Terms for the Four Diastereomers of Co(en)$_3$, **7.8**

Case	ob_3	$lelob_2$	lel_2ob	lel_3	lel_3:ob Ratio Calc	lel_3:ob Ratio Exp
$\Delta G_{298\,K}$	1.39	0.67	0.12	0.0	0.74:0.26	0.73:0.27[a]
$\Delta G_{313\,K}$	1.40	0.65	0.098	0.0	0.73:0.27	0.74:0.26[b]

[a] Harnung et al. (1974).
[b] Sudmeyer et al. (1972).

TABLE 7.8. Equilibrium Distributions for the Four Diastereomers of Co(en)$_3$, **7.8**

Case	ob_3	$lelob_2$	lel_2ob	lel_3
$\Delta G_{298\,K}$ Calc	0.04	0.15	0.36	0.45
$\Delta G_{298\,K}$ Exp[a]	0.03	0.16	0.40	0.40
$\Delta G_{400\,K}$ Calc	0.06	0.19	0.39	0.36
$\Delta G_{373\,K}$ Exp[a]	0.04	0.20	0.41	0.35

[a] From Lepard (1966).

When a string of vinyl alkenes are connected to form a polymer, the vinyl substituents, X, can be placed all in the same relative orientation, **7.9**. This arrangement of substituents is called *isotactic*.

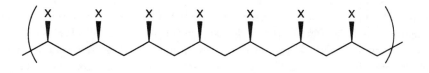

7.9

Another regular arrangement of vinyl substituents is for them to be placed in alternating orientations, **7.10**. This arrangement of substituents is called *syndiotactic*.

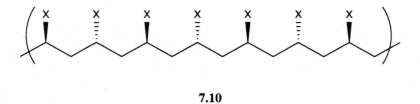

7.10

A third possibility is for the vinyl substituents to be placed randomly with respect to each other, **7.11**.

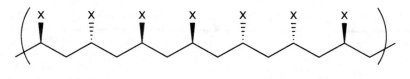

7.11

This arrangement of substituents is called *atactic*. This example of stereochemical control is unusual in that the product, in this case the polymer, is not chiral but the local environment is configurationally defined.

Controlling the stereochemical outcome of vinyl alkene polymerization reactions is of interest because the physical properties of polymeric materials are dependent on the stereochemistry of the polymerization reaction. As discussed in Chapter 6, the crystallinity of a polymer is dependent on the number of conformational defects along the chain. These conformational defects prevent an orderly packing necessary for crystal formation. Configurational defects also decrease the crystallinity of a sample by preventing long-range order.

The reaction sequence for stereoregular polymerization follows the general polymerization mechanism due to Cossee (1964), and Arlman and Cossee (1964) see Eq. (7.9). Polymer chain extension starts by complexation of an alkene to a vacant coordination site at a transition metal center, see Eq. (7.6). The alkene then inserts to the metal–polymer bond creating a new vacant coordination site.

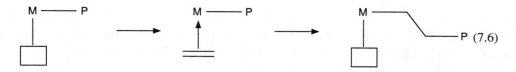

It is generally acknowledged that the steric environment around the metal contributes to stereocontrol. This control by the active site is referred to as enantiomorphic site control. The precise source for steoreocontrol was suggested by Pino et al. (1987, 1988). They said that the stereochemisty was set during the initial alkene complexation, which occurred with the vinyl substituent being placed in the least sterically congested place in the active site. Since alkenes are flat with four substituent positions, the vinyl substituent can be thought of as being placed in one of the four quadrants of an *x–y* coordinate system.

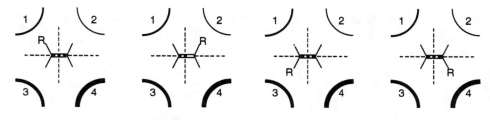

FIGURE 7.14.
Schematic representation of the four quadrants of an x–y coordinate system that could differentially interact with a vinyl alkene.

This is done schematically in Figure 7.14 where four varying congested quadrants are represented by lines of varying thickness. In this schematic drawing, the least sterically hindered quadrant is quadrant 2, so the most favorable alkene positioning places the R group in quadrant 2, which corresponds to the second diagram in Figure 7.14. In order to grow a stereoregular isotatic polymer, each complexation event must use the same enantiotopic alkene face. This condition can be achieved by constructing a single rigid sterically defined active site or a number of stereochemically identical active sites.

Both heterogeneous and homogeneous catalysts are known for the isostatic polymerization of propylene (X = Me in **7.9**). The homogeneous catalysts currently used for the isotactic polymerization of propylene are primarily of the general form Cp_2ML_2, (L = σ donor ligand) **7.12**, in which the cyclopentadienyl (Cp) ligands are stereoselectively substituted (Kaminsky et al., 1985).

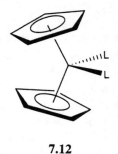

7.12

In order to create a stereochemically defined complex, the two cyclopentadienyl ligands could be connected together to provide a well-defined framework and then substituents placed on opposite sides of the two Cp rings to create a stereochemically defined dissymmetric but not chiral C_2 symmetric complex, **7.13**. These substituted complexes are referred to as *ansa*-metallocenes.

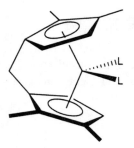

7.13

As indicated in **7.14a** and **7.14b**, if a C_2 symmetric active site is created, then the same alkene face will be presented to the metal center (M) independent of which side it approaches from.

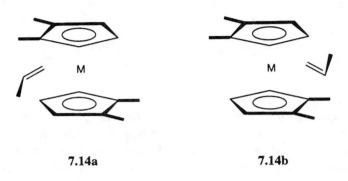

7.14a   **7.14b**

Let us follow propylene through a stereocontrolled insertion sequence using a C_2 symmetric ligand. We will indicate the steric encumbrances of the complexes in **7.14a** and **7.14b** with boldface lines, see Figure 7.15. The reaction sequence begins by propylene complexation, the methyl group being placed away from the steric encumbrance (boldface line). Stereocontrolled complexation is followed by the insertion of propylene into the metal–polymer bond. The next propylene complexation and insertion occurs, again placing the methyl group in the least sterically congested quadrant. Since the active site in Figure 7.15 is C_2 symmetric, the steric encumbrance on the back of the active site is at the top. Because of this the methyl group will be pointed down. Complexation and reaction occur using the same enantiotopic alkene face as the previous insertion. Use of the same alkene face is required for isotactic chain growth. Syndiotactic defects in the polymer occur when the propylene approaches from and reacts using the opposite enantiotopic face, see Figure 7.15. The defect complexation event occurs with the methyl group not being placed in the least sterically congested quadrant. Stereocontrol rests with differential steric interactions between the active site and the approaching alkene.

Ligand design for controlling the stereochemical outcome of vinyl alkene polymerization can thus be focused on providing a steric environment that will cause preferential reaction with a single alkene face for isotactic chain growth, or by providing a steric environment that will cause preferential reaction with alternating alkene faces for syndiotactic chain growth. This concept has guided numerous experimental efforts.

7.5.2 Stereoregular Polypropylene Catalysis

Hart and Rappé (1993) at Colorado State University, as a follow-up to the work of Castonguay and Rappé (1992), reported a molecular mechanics investigation of a systematic variation in substituents for the *rac*-[1,2-ethylenebis(η^5-indenyl)]Zr based propylene catalyst (see **7.15**).

Isotactic Chain Extension

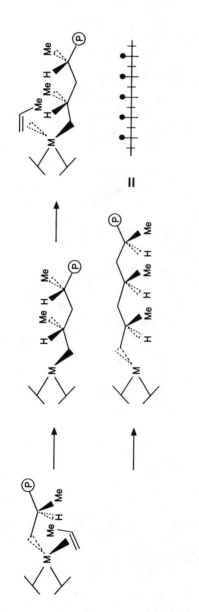

Syndiotactic Chain Defect

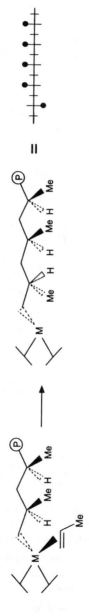

FIGURE 7.15.
Reaction sequence for isotactic propylene polymerization and generation of a syndiostatic stereodefect in an isotactic polymer chain. Bulky substituents are indicated by boldface lines and *P* indicates the polymer chain.

339

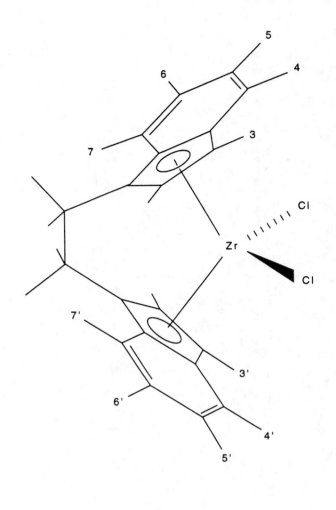

7.15

The study was undertaken to see if molecular mechanics could reproduce the experimentally observed stereoselectivities for a series of substituted catalysts and if molecular mechanics could suggest new catalysts with enhanced stereoselectivity.

The bonding between the metal center and the Cp rings in **7.12** and the indenyl rings of **7.15** is difficult to describe using simple covalent bonding ideas. Due to the size of transition metal ions and the shapes of transition metal valence *d* orbitals, transition metals are capable of forming delocalized bonds to organic ligands such as an alkene or a polyalkene such as cyclopentadiene. Delocaled ligands can be attached to transition metals through the construct of pseudoatoms. Pseudoatoms are discussed in Box 7.2.

METHODS

Ab initio calculations on a model system, $Cl_2ZrCH_3^+ + C_2H_4$ were used to determine the geometric position of an activated complex, which was taken as being along the reaction coordinate approximately halfway between the ground and transition states (Castonguay and Rappé, 1992). The degree of isotacticity was defined by the energy difference (ΔIS) between the activated complex leading to an isotactic insertion and the activated complex leading to a syndiotactic defect. The growing polymer chain in each activated complex was approximated by a 2,4,6-trimethyl heptyl group.

In addition to benchmark calculations on the known substituted systems [3,3'-dimethyl (Ewen et al., 1991; Lee et al., 1992), and 5,5',6,6'-, and 4,4',7,7'-, tetramethyl (Lee et al., 1992)], the nonbond contacts were examined in the catalytic pocket to choose the hydrogen atoms at the 3,3', 4,4', 5,5' positions of the indenyl rings for substitution in an attempt to enhance the isotactic selectivity of the catalytic polymerization of propylene. The 4,4',5,5'-tetramethyl derivative was also studied to see if the steric effects were additive.

The Dreiding force field (Mayo et al., 1990), augmented (Castonguay and Rappé, 1992) with an atom type for tetrahedral zirconium and pseudoatoms for a cyclopentadienyl centroid (Cp) and an alkene centroid (Ci), was used. The pseudoatoms used in this study were not projected out of the energy and derivative calculations because the version of the molecular mechanics program used, (Biograf, available from Molecular Simulations, Inc.) did not contain the pseudoatom technology discussed in Box 7.2.

The structures were minimized with partial charges obtained by the QEq charge equilibration scheme (Rappé and Goddard, 1991). Each minimized structure was then subjected to 10 cycles of annealed dynamics from 0 to 600 K with a symmetric energy ramp of $1°/fs$ ($10^{-15}s$). The lowest energy conformations obtained from the annealing process were minimized and their charges equilibrated. Of these conformations, the one with the lowest energy was used as input for annealed dynamics as above, and the whole process was repeated until no conformations with a lower total energy were found. All atomic and pseudoatomic positions were varied during the calculations.

The calculated energy difference, ΔIS, was related to the experimentally determined isotacticity in the following way. The difference in activation energies ($\Delta\Delta G^{\ddagger}$) for two competing processes was related to their rate constants k_1 and k_2 by Eq. (7.7).

$$\Delta\Delta G^{\ddagger} = -RT \ln(k_1/k_2) \tag{7.7}$$

The two approaches of propylene leading to either an isotactic or a syndiotactic insertion were the two competing processes. The percent occurrence of a series of five isotactic insertions (%mmmm), identifiable in the ^{13}C NMR spectrum of the polymer, was related to one isotactic insertion by Eq. (7.8).

$$1 \text{ event} = (\%mmmm/100)^{0.2} \tag{7.8}$$

On the assumption that $T\Delta\Delta S^{\ddagger}$ was small compared to $\Delta\Delta H^{\ddagger}$ for these systems, ΔIS was compared directly with $\Delta\Delta G^{\ddagger}$.

BOX 7.2   Pseudoatoms

To see the difficulty associated with π ligands and how pseudoatoms are used we will discuss the bonding of alkene complexes rather than the bonding in the more complex, but analogous, Cp complexes.

The bonding between a transition metal and an alkene, according to the Dewar–Chatt–Duncanson model (Dewar, 1951; Chatt and Duncanson, 1953) consists of a forward donation of electron density from the alkene π orbital into an empty σ orbital on the metal, see Figure 7.16(a), and back-donation (back-bonding) from a filled metal $d\pi$ orbital into the empty π* orbital of the alkene, see Figure 7.16(b). This interaction between the metal and the alkene can be strong, weak,

(a)   **(b)**

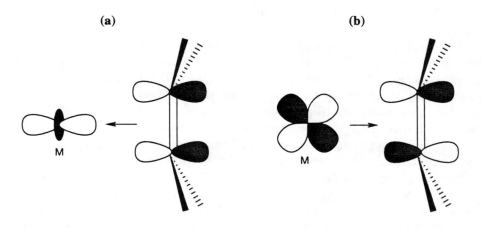

FIGURE 7.16.
Alkene bonding interactions of the Dewar–Chatt–Duncanson model. (a) Forward coordination and (b) back-donation.

or somewhere in between. This leads to two "limiting" valence bond structures for metal–alkene bonding interactions: A simple π complex **7.16a** for a weak metal–ligand interaction, and a metalla-cyclopropane **7.16b** for a strong metal–ligand interaction.

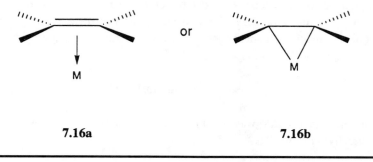

7.16a   **7.16b**

In experimental structures near the metallacyclopropane limit, the alkene substituents are bent back, consistent with sp^3 hybridization (Bender et al., 1992). The metallacyclopropane, **7.16b**, can be modeled in molecular mechanics as drawn, with two σ bonds between the metal and the alkene.

The π complex, **7.16a**, is more challenging to model. If σ bonds are used to bond the alkene to the metal, the alkene substituents will bend back and a metallacyclopropane will result, but this is not consistent with the observed geometry. Dramatically distorted M−C−X bond angles could be used to force the alkene to near planarity. This approach is not very general, as it requires extensive reparameterization for each system studied. Furthermore, it is difficult to imagine calculating proper vibrational frequencies with this approach. Incorrect vibrational frequencies would prevent one from carrying out the ΔG analysis used in the Co(III) study of Section 7.4, or possibly calculating kinetic isotope effects. An alternative approach is to attach the alkene to the metal through a pseudoatom—a fictious atom with neither mass, charge, nor van der Waals parameters. This result is best accomplished by expressing the coordinates of the pseudoatom in terms of the coordinates of the alkene, or other π-bound ligand, Eqs. (7.9)–(7.11), where n is the number of atoms participating in the π bonding (2 for alkenes, 3 for π allyls, and 5 for cyclopentadienyl ligands) (Doman et al., 1992).

$$x_\pi = \frac{1}{n} \sum_{i=1}^{n} C_i x_i \tag{7.9}$$

$$y_\pi = \frac{1}{n} \sum_{i=1}^{n} C_i y_i \tag{7.10}$$

$$z_\pi = \frac{1}{n} \sum_{i=1}^{n} C_i z_i \tag{7.11}$$

C_i, typically 1, can be used to shift the pseudoatom towards one of the π atoms as appropriate for asymmetrically coordinated alkenes. The x_π, y_π, and z_π coordinates can be used in conventional stretch, bend, and so on interactions without introducing artificial angles or force constants. As formulated, the forces of the pseudoatom are projected back onto the coordinates of the n π bonding ligands; the pseudoatom coordinates are not included as a part of energy minimization nor during molecular dynamics (Doman et al., 1992).

RESULTS

Hart and Rappé (1993) studied a family of substituted *rac*-[1,2-ethylenebis(η⁵-inde-nyl)]Zr based catalysts in order to see if molecular mechanics could reproduce observed stereoselectivities and predict energetic differentiation for a series of systematic substitutions. By using an activated complex model, based on an ab initio electronic structure study of $Cl_2ZrMe(C_2H_4)^+$, the geometry was determined for the activated complex for propylene insertion leading to isotatic chain growth and for the activated complex leading to a syndiotactic defect for a series of substituted indenyl systems (see Fig. 7.17). The

Isotactic Activated Complex

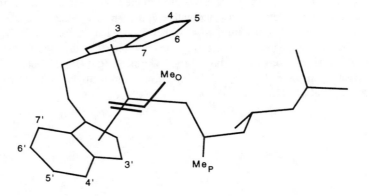

Syndiotactic Activated Complex

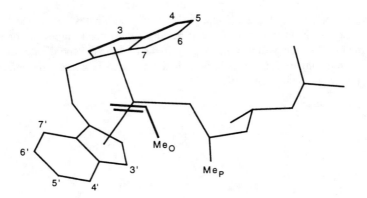

FIGURE 7.17.
Activated complexes for normal isotactic insertion, and for insertion leading to a syndiotactic defect. All hydrogen atoms have been removed for clarity, and the indenyl positions are marked. Me_O = methyl group of propylene and Me_P = 2-methyl group of the growing polymer chain.

TABLE 7.9. Calculated ΔΔG‡ Values from Experiment for a Series of C₂H₄-Bis(indenyl)Zr Catalysts, **7.15**

Catalyst	% mmmm[a]	% (1 Event)	ΔΔG‡ (kcal/mol)[b]
7.15[c]	89.4 (89)	97.8	2.4 (2.4)
7.15[d]	83 (89)	96	2.0 (2.4)
3,3'-Dimethyl[d]	18 (18)	71	0.6 (0.6)
4,4',7,7'-Tetramethyl[c]	92.2 (93)	98.4	2.6 (2.7)
5,5',6,6'-Tetramethyl[c]	79.4 (92)	95.5	1.9 (2.6)

[a] Estimated values from ΔIS values (Hart and Rappé, 1993) in parentheses.
[b] ΔIS values (Hart and Rappé, 1993) in parentheses.
[c] From Lee at al., 1992, $T = 40°C$.
[d] From Ewen et al., 1991, $T = 50°C$.

lowest energy structure for each substituted activated complex obtained from the conformational search placed the growing polymer chain (i.e., the 2,4,6-tetramethylheptyl group), the bridging ethylene group, and the indenyl rings in similar relative positions. For derivatives with alkyl substitution in the 5 and 5′ positions, the growing polymer chain was in a slightly different position than for the parent complex. A substantial twisting of the ethylene bridge was found for the 4,4′,7,7′-tetramethyl derivative. The differences between molecular mechanics energies for isotactic insertion and syndiotactic defect insertion, ΔIS, were compared to experimental differentiations obtained by ¹³C NMR. The energetic results for the experimentally known catalysts are gathered in Table 7.9, with the calculated ΔIS values in parentheses. The complete set of ΔIS results are gathered in Table 7.10.

TABLE 7.10. Calculated ΔIS Results for a Series of C²H₄-Bis(indenyl)Zr Catalysts

Catalyst	ΔIS (kcal/mol)
7.15	2.4
3,3'-Dimethyl	0.6
4,4'-Dimethyl	4.2
4,4'-Diethyl	2.4
4,4'-Diisopropyl	4.1
4,4'-Di(tert-butyl)	2.2
5,5'-Dimethyl	2.9
5,5'-Diethyl	2.7
5,5'-Diisopropyl	2.5
5,5'-Di(tert-butyl)	3.1
4,4',7,7',-Tetramethyl	2.7
5,5',6,6'-Tetramethyl	2.6
4,4',5,5'-Tetramethyl	3.3

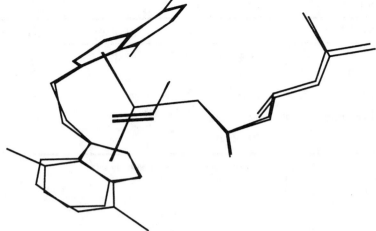

FIGURE 7.18.
Superimposed activated complexes leading to a normal isotactic insertion for the parent catalyst and the 4,4',7,7'-tetramethyl derivative. All hydrogen atoms have been removed for clarity.

The unsubstituted parent catalyst, C_2H_4-bis(indenyl)ZrCl$_2$, **7.15**, was experimentally observed to yield isotactic polypropylene (Ewen et al., 1991; Lee et al., 1992). With 2,4,6-tetramethylheptyl as a model for the growing polymer chain, the catalyst was predicted to produce isotactic polypropylene, with a ΔIS slightly larger than 2 kcal/mol, in agreement with experiment. The two activated complexes of the 3,3'-dimethyl derivative were calculated to be less than 1 kcal/mol apart, suggesting this catalyst would produce polypropylene with a large number of syndiotactic defects. This result was in agreement with experiment, where this catalyst was found to produce polypropylene with a mmmm of only 18% (Ewen et al., 1991).

The polymer produced by the 4,4',7,7'-tetramethyl derivative was experimentally observed to be slightly more stereoregular and the polymer produced by the 5,5',6,6'-tetramethyl derivative slightly less stereoregular than the unsubstituted parent catalyst (Lee et al., 1992). Hart and Rappé (1992) predicted both of these catalysts to produce polypropylene with a higher degree of stereoregularity than that produced by the parent. They rationalized that since the Dreiding force field (Mayo et al., 1990) produced heats of sublimation for hydrocarbons in error by up to 0.4 kcal/mol it would be reasonable to assume that errors of this magnitude would be present in stereoselectivity estimates making the 5,5'-, 5,5',6,6'-, and 4,4',7,7'-substituted catalysts all of comparable stereoselectivity to the parent catalyst. An interesting observation from the calculated molecular structures was that for the 4,4',7,7'-tetramethyl catalyst the methyl groups in the 7 and 7' positions interacted with the ethylene backbone, causing the molecule to twist significantly to relieve this strain (Fig. 7.18). This strain was relieved when the 7- and 7'-methyl groups were removed and the predicted stereoselectivity was increased. The 4,4'-dimethyl derivative was calculated to have a ΔIS of more than 4 kcal/mol. One would predict that by increasing the size of the alkyl substituents in the 4,4' positions, ΔIS would also increase. This is not what

Hart and Rappé (1992) found. For the 4,4'-diethyl and 4,4'-di-*tert*-butyl derivatives, ΔIS was approximately 2 kcal/mol, the same as for the unsubstituted parent, while ΔIS for the 4,4'-isopropyl derivative was about 4 kcal/mol.

These results were rationalized by a comparison of the nonbond interactions in the substituted and unsubstituted activated complexes leading to isotactic or syndiotactic insertion. For the unsubstituted case, the three-way interaction or double stereodifferentiation (Erker et al., 1991) of the methyl group of the coordinated propylene (Me_O) with the indenyl group and the first methyl side group of the growing polymer chain (Me_P) for the syndiotactic activated complex was enough to favor the isotactic activated complex by 2 kcal/mol. Since the growing polymer chain, the bridging ethylene group, and the indenyl rings were in the same relative positions for all of the 4,4'-dialkyl activated complexes, Hart and Rappé (1992) focused on the only difference between the derivatives: the interaction of the alkyl groups with the indenyl rings and the propylene.

In the 4,4'-dimethyl derivative, a methyl at the 4' position increased the unfavorable steric interactions for the syndiotactic activated complex. The methyl substituent at the 4 position did not interact with Me_P in the isotactic activated complex. For the 4,4'-diethyl substituted isotactic activated complex, the lowest energy conformation of the 4-position ethyl group placed a hydrogen pointed directly at Me_O, because the ethyl methyl group preferred to be perpendicular to the aromatic ring. This created a new repulsive interaction for the ethyl substituted isotactic activated complex that was not present for the methyl substituted isotactic activated complex, causing ΔIS to decrease relative to the 4,4'-dimethyl derivative (cf. **7.17** with **7.18**).

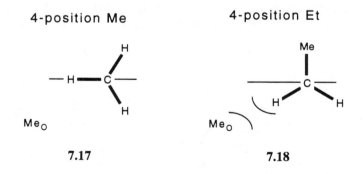

4-position Me   4-position Et

7.17   **7.18**

For the 4,4'-dimethyl and 4,4'-diethyl syndiotactic activated complexes, the three-way interaction of the 4'-position methyl (or ethyl, Et) group with Me_O and Me_P was schematically represented as in **7.19** and **7.20**.

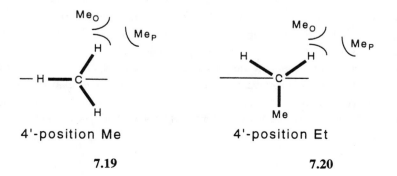

4'-position Me   4'-position Et

7.19   **7.20**

In the lowest energy conformations of each activated complex, determined in large part by alkyl–indenyl interactions, a hydrogen atom pointed into the space between Me_O and Me_P, so no significant new repulsive interactions were introduced in going from methyl to ethyl. The reduction in isotacticity of the ethyl-substituted catalyst relative to the methyl catalyst was suggested to be due to the differential introduction of repulsive steric interactions in the isotactic activated complex, **7.18** above.

The predicted energetics for the isopropyl substituted catalyst were explained by similar arguments. For the isotactic activated complex, **7.21**, the isopropyl (*i*-Pr) groups were found to be staggered relative to the indenyl rings.

4-Position *i*-Pr

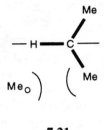

7.21

The isopropyl methyl and Me_O were close enough for a repulsive interaction to exist. For the syndiotactic activated complex, **7.22**, the 4'-isopropyl group rotated ($\sim 10°$) to avoid the steric interaction with Me_O in spite of the unfavorable Me–indenyl interaction.

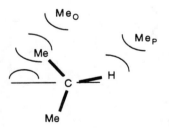

4-Position *i*-Pr

7.22

However, the Me_O–H–Me_P repulsive interaction was reduced relative to the methyl substituted catalyst. These competing factors resulted in no net change in the isotacticity for the isopropyl substituted catalyst relative to the methyl substituted one.

For *tert*-butyl substitution at the 4 position, the methyl groups of the 4-position *tert*-butyl group in the isotactic activated complex were in the same relative positions as the methyl hydrogen atoms in the 4,4'-dimethyl derivative. The methyl in the plane of the indenyl and the one below the plane both interact strongly with Me_O in a repulsive manner.

For the syndiotactic activated complex, the 4'-position *tert*-butyl group adopted a conformation in which one methyl group pointed between Me_O and Me_P. In going from the methyl to the *tert*-butyl substituted catalyst, the increase in the magnitude of the repulsive interactions was greater for the isotactic activated complex than for the syndiotactic activated complex, and the net effect was that the isotacticity for this catalyst was reduced to the level of the unsubstituted and ethyl substituted catalysts.

The 4,4',5,5'-tetramethyl derivative was also studied to determine if the effect of methyl substitution was additive. Since ΔIS was found to be only 3.3 kcal/mol the effects of methyl substitution were not additive. The adjacent methyl groups interacted with each other so they were no longer able to adopt their most favorable positions, and, as was the case for the 5,5'-dialkyl derivatives, the growing polymer chain was in a slightly different position for this catalyst compared to the 4,4'-dialkyl derivatives.

In summary, Hart and Rappé (1992) reproduced the observed propylene polymerization stereoselectivities for a set of four indenyl *ansa*-metallocene catalysts. They also provided visual support for the predicted, but unanticipated, trend in selectivity for a series of new catalysts. The agreement with experiment, and the visual observation of explanations for the differences is gratifying, but it should be remembered that the activated complex studied was not the transition state. Why was there such good agreement with experiement? There was likely an exquisite cancellation of errors. The ''early'' transition state was likely compensated for by a too stiff 6–12 van der Waals potential (Hart and Rappé, 1992).

7.6. ASYMMETRIC HYDROGENATION

7.6.1 Background

As discussed in Section 2.5.2 and Appendix A, the presence of two chiral centers in a molecule or complex leads to diastereomers, which are energetically different. If a metal complex contains an element of chirality, then it would be capable of discriminating between enantiotopic reactants. An especially successful example of stereoselective homogeneous catalysis is the use of a rhodium catalyst in the asymmetric hydrogenation of *N*-acetyldehydroamino acids. This rhodium catalyst is currently used by Monsanto in the synthesis of L-DOPA, **7.23**, a drug used in the treatment of Parkinson's disease.

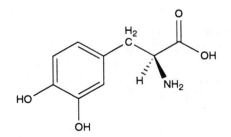

7.23

The Monsanto synthesis of L-DOPA starts with vanillin, **7.24**, which is elaborated to an enamine intermediate, **7.25**, see Eq. (7.12) (Ac-acetyl).

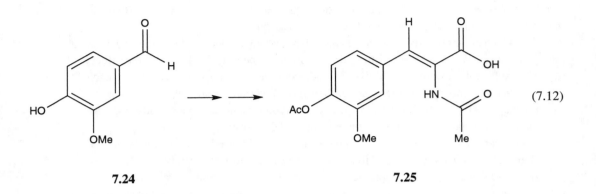

(7.12)

7.24   **7.25**

The enamine is then stereoselectively hydrogenated in the presence of a chiral rhodium(I) catalyst to the reduced, protected amino acid, **7.26**, see Eq. (7.13).

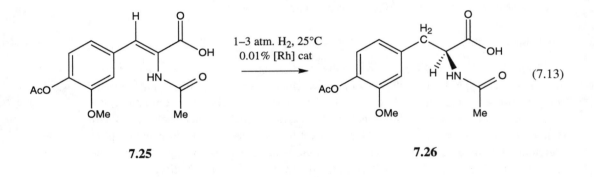

(7.13)

7.25   **7.26**

The protecting groups are then hydrolyzed off yielding L-Dopa, **7.23**.

The mechanism of the rhodium catalyzed hydrogenation reaction, Eq. (7.13), has been thoroughly studied (Landis and Halpern, 1987). The reaction sequence begins by an initial coordination of both the acetamidate and alkene functional groups of the chelating alkene to a Rh(I) center, Eq. (7.14).

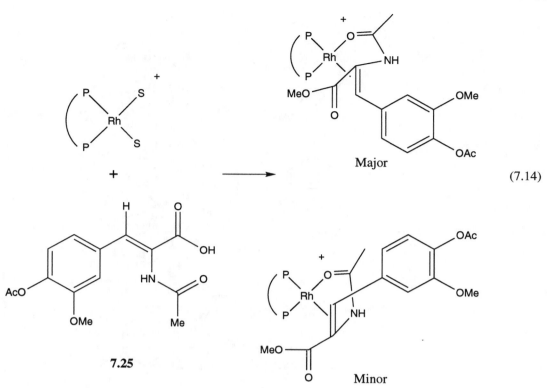

(7.14)

7.25

Major

Minor

In Eq. (7.14) S indicates a solvent molecule and $\left(\begin{smallmatrix} P \\ \\ P \end{smallmatrix}\right.$ indicates a chelating phosphine. Three examples of chelating phosphines are given in Figure 7.19. The complexation step of Eq. (7.14) is followed by the turnover-limiting reaction with dihydrogen, Eq. (7.15) (L′ indicates the acetamidate functional group).

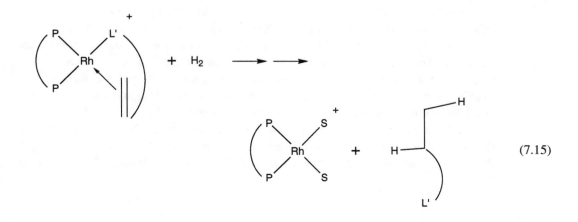

(7.15)

352

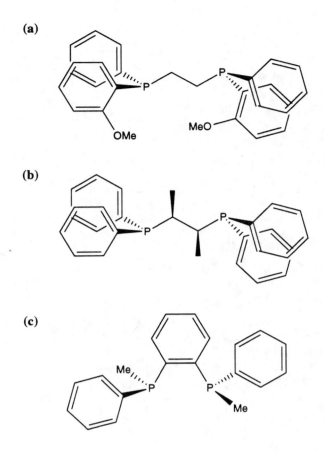

FIGURE 7.19.
Three representative chelating, chiral phosphine ligands, (a) (R, R)-1,2-bis[(o-
methoxyphenyl)phenylphosphino]ethane (DIPAMP), (b) $(2S,3S)$-bis(diphenylphosphino)butane
(CHIRAPHOS), and (c) (R^*,R^*)-($\pm$)-1,2-phenylenebis(methylphenylphosphine) (DIPH).

Since reaction with dihydrogen, Eq. (7.15), is turnover limiting there is no mechanistic
evidence for the subsequent steps leading to the reduced organic product.

Alkenes, because they are flat, are generally not chiral: The two alkene π clouds are
equivalent. However, when unsymmetrically substituted the π faces are enantiotopic.
When an alkene such as **7.25** is reacted with a chiral metal complex the diastereomers
resulting from complexation by each of the two alkene π clouds are not energetically the
same. Both diastereomeric alkene–Rh(I) complexes indicated in Eq. (7.14) have been
observed and the major diastereomer has been characterized by X-ray diffraction. The
alkene facial preference has been attributed, based on X-ray diffraction data, to a rigid,
chiral "edge–face–edge–face" array of the phosphorus aryl substituents, **7.27** (Koenig et
al., 1980).

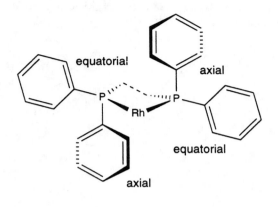

7.27

In this model, the two pseudoaxial phenyl rings are oriented with C−H bond "edges" pointed toward the Rh center. The two equatorial phenyl rings are oriented with π cloud "faces" pointed toward the Rh center. This arrangement of phenyl rings creates a stereochemically distinct "C_2" active site analogous to the C_2 symmetric active sites created by the *ansa*-metallocene catalysts discussed in Section 7.5.

The intriguing and mechanistically challenging feature of this catalytic process is that the major diastereomeric complex does not lead to the observed product! The minor isomer must proceed to product through a lower barrier pathway.

7.6.2 Steric Differentiation During the Complexation Event

Giovannetti et al. (1993) of the University of Wisconsin used molecular mechanics and NOE NMR spectroscopy to understand how the chirality of the bisphosphine ligand gives rise to enantiodifferentiating interactions and whether or not the solution structures of the catalysts are similar to the crystallographically determined solid state structures. The four-coordinate Rh(I) center found in the present catalyst poses a simulation problem because it is square planar. Without metal-centered angle terms the four-coordinate square planar structure would revert to a tetrahedral structure upon minimization, as tetrahedral coordination places the ligands as far apart as possible. The SHAPES force field does use a metal-centered angle term that permits square planar coordination. This angle term is reviewed in Box 7.3.

METHODS

Molecular mechanics calculations were carried out with the SHAPES force field (Allured et al., 1991) for the metal center and the CHARMM force field (Brooks et al., 1983) was used for the ligands, with phosphine CHARMM parameters as previously developed (Allured et al., 1991).

For both the molecular mechanics and NMR studies, the simpler methyl or isopropyl(Z)-α-acetamidocinnamate alkenes **7.28** and **7.29** (MAC and PRAC, respectively) were used to model the precursor to L-DOPA, **7.23**.

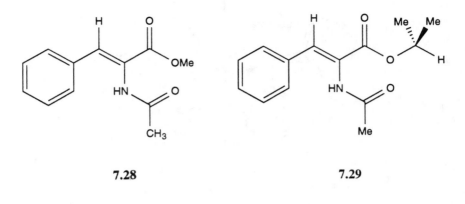

7.28   **7.29**

BOX 7.3   Shapes Force Field

The SHAPES force field includes angle bend terms (Fourier expansion) at Rh for a proper treatment of angular distortions at the metal center (Allured et al., 1991). This approach is discussed below.

For organic and biological molecules, angle bend interactions have typically been described as an expansion in θ, as either harmonic, Eq. (7.16), as proposed by Hill (1946)

$$V_\theta = \frac{1}{2}K_{IJK}(\theta_0 - \theta_{IJK})^2 \tag{7.16}$$

or as a higher order series in θ, Eqs. (7.17) and (7.18), used in MM2 and MM3 (Allinger, 1977; Allinger et al., 1989).

$$V_\theta = \frac{1}{2}K_{IJK}(\theta_{IJK} - \theta_0)^2\{1 + C(\theta_{IJK} - \theta_0)^4\} \tag{7.17}$$

$$V_\theta = \frac{1}{2}K_{IJK}(\theta_0 - \theta_{IJK})^2\{1 - B(\theta_{IJK} - \theta_0) + C(\theta_{IJK} - \theta_0)^2 +$$
$$D(\theta_{IJK} - \theta_0)^3 + E(\theta_{IJK} - \theta_0)^4\} \tag{7.18}$$

These expansions describe the distortions found in organic molecules but are not useful for the angular variations in metal complexes.

Allured et al. (1991) of the University of Wisconsin presented the SHAPES force field. The angular overlap model (AOM) of chemical bonding (reviewed in Larsen and La Mar, 1974) was used to motivate a Fourier expansion representation of the angle bend. The general form used in the SHAPES potential is given by Eqs. (7.19–7.21).

$$V_\theta = K_{IJK}[1 + \cos(p\theta + \psi)] \tag{7.19}$$

where

$$p = \frac{\pi}{\pi - \theta_0} \tag{7.20}$$

The alkene functional groups were attached to the metal through a pseudoatom as discussed in Section 7.5.

The SHAPES/CHARMM force field was structurally validated by comparison of the molecular mechanics minimized structures with the X-ray diffraction structures for a set of Rh chelating phosphine dialkene complexes. The structures of the major diastereomeric complexes of **7.29**, with two chelating chiral phospines CHIRAPHOS and DIPAMP, were compared as well. The central structural features of [Rh(CHIRAPHOS)(MAC)]$^+$ and [Rh-(DIPAMP)(MAC)]$^+$ are collected in Table 7.11.

As with the other case studies described in this text that use NOE data, Allured et al. (1991) began by obtaining a complete assignment of the ^{1}H spectrum. Chemical shift, correlation spectroscopy (COSY), and total correlation spectroscopy (TOCSY) experiments were used in this assignment. In order to enhance the magnitude of the observed nuclear Overhaurer effects (NOEs), ethylene glycol was used as the solvent. This more

and

$$\psi = \pi - p\theta_0 \tag{7.21}$$

For square planar and octahedral coordination environments, p is 4 and the potential is fourfold periodic with minima for both 90° and 180°. There is no need for multiple angles, because the SHAPES form treats these structures without difficulty. In analogy to the use of an inversion term for flat sp^2 centers, such as the C atoms of ethylene in the force fields discussed in Sections 2.2 and 3.2, the SHAPES force field uses an out-of-plane bending function to maintain planarity, Eq. (7.22).

$$V_\phi = K_{IJK}[1 + \cos(2\phi)] \tag{7.22}$$

The two angles, θ and ϕ, are defined by a spherical polar coordinate system, see **7.30**.

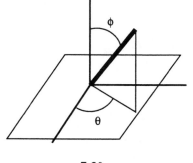

7.30

The angle ϕ is defined with respect to the polar axis, which is an axis perpendicular to the molecular square plane.

TABLE 7.11. Structural Parameters of Rh(MAC) Complexes

Distance/Angle	Calc		X-Ray
	Major	Minor	
[Rh(DIPAMP)(MAC)]⁺			
Rh−P1	2.25	2.26	2.27
Rh−P2	2.24	2.24	2.24
Rh−D1	2.12	2.12	1.92
Rh−O	2.11	2.11	2.11
P1−Rh−P2	83.0	83.0	83.0
P1−Rh−D1	176.0	175.0	150.0
P1−Rh−O	93.0	93.0	93.0
P2−Rh−D1	102.0	101.0	101.0
P2−Rh−O	173.0	174.0	173.0
$\phi1,\alpha$	69,115	1,128	48,115
$\phi1,\alpha$	22,128	107,114	52,133
$\phi2,\alpha$	−1,97	109,135	15,107
$\phi2,\alpha$	60,137	14,99	61,129
[Rh(CHIRAPHOS)(MAC)]⁺			
Rh−P2	2.24	2.24	2.29
Rh−P3	2.24	2.24	2.23
Rh−D1	2.11	2.12	2.07
Rh−O4	2.11	2.11	2.13
P2−Rh−P3	83.0	84.0	83.0
P2−Rh−D1	175.0	174.0	163.0
P2−Rh−O	92.0	92.0	90.0
P3−Rh−D1	101.0	102.0	101.0
P3−Rh−O	175.0	173.0	168.0
$\phi1,\alpha$	87,119	85,133	22,106
$\phi1,\alpha$	−12,125	9,144	84,133
$\phi2,\alpha$	14,99	−28,114	19,101
$\phi2,\alpha$	106,136	98,122	82,135
[Rh(DIPH)(MAC)]⁺			
Rh−P1	2.25	2.26	
Rh−P2	2.23	2.23	
Rh−D1	2.12	2.12	
Rh−O	2.11	2.11	
P1−Rh−P2	85.0	85.0	
P1−Rh−O	92.0	92.0	
P1−Rh−D1	175.0	176.0	
P2−Rh−D1	99.0	199.0	
P2−Rh−O	169.0	165.0	
$\phi1$	110	172	
$\phi2$	−19	−36	

viscous solvent lengthened the rotational correlation times of the relatively small molecules of interest in this study. The NOE intensities were collected at five mixing times.
Rather than use the NOE derived distances in a restrained MD simulation, a technique
more suited to molecular systems with multiple conformations was used. As you may
remember in Section 2.3.4, substantial errors were found for $^1H-^1H$ distances involving
phenyl rings with multiple conformations. Giovannetti et al. (1993) anticipated near free
rotation about the P-phenyl bonds, and hence that there would be more than one low-
energy conformation. The approach used, confomer population analysis (CPA), begins
with an ensemble of low-energy conformations generated through a combined systematic,
Monte Carlo conformational searching procedure (Landis and Allured, 1991). The weights
of these conformations are then fit to best reproduce observed NOE intensities.

RESULTS

In order to understand the differential interactions between the chiral bisphosphine
ligand and the two diastereomeric alkene complexes due to complexation using opposite
alkene faces, one must assess the conformational preferences of the phosphine ligand
bound to a metal center. This metal–ligand system will likely form an "active site,"
or pocket, for alkene complexation. For a chelating bisphosphine with aryl phosphine
substituents, there are two major elements of stereochemistry. As with the Co(III) ethylenediamine complexes discussed in Section 7.4, there are two conformational possibilities
for the five-membered chelate ring (δ and λ) (for the structurally constrained CHIRA-
PHOS and DIPH phosphines these possibilities have been removed). In addition, the arene
rings can be oriented either perpendicular (edge on) or parallel to (face on) the attached
chelating alkene. The orientational preferences of the phenyl substituents can be expressed
in terms of the $Rh-P-C_{ph}-C_{ph}$ torsional angles, ϕ and ψ. The puckered five membered
chelate ring can lead to either equivalent phosphines or to axial and equatorial phosphines.
The axial–equatorial nature of the phenyl substituents can be described by $P-Rh-P-C_{ar}$
torsional angles, α. These structural and conformational elements are summarized in Figure 7.20. For $[Rh(CHIRAPHOS)(MAC)]^+$ the arene ring orientational preferences were

(a)   **(b)**

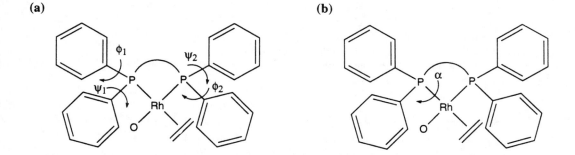

FIGURE 7.20.
Definition of conformational characteristics of metal bound bis(diarylphosphino)ethane ligands.

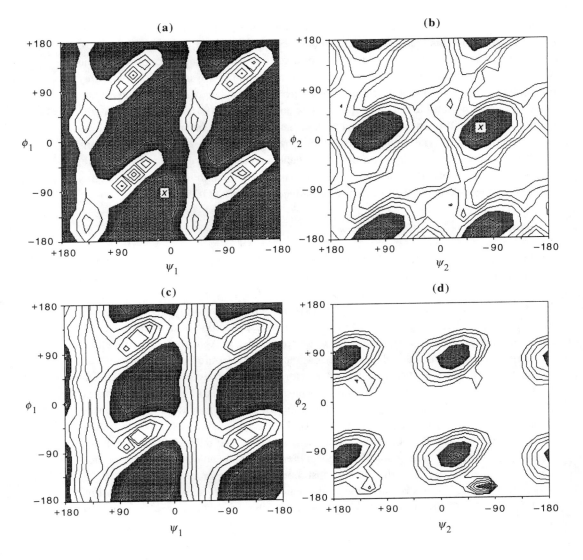

FIGURE 7.21.

The φ-ψ maps for the major (a and b) and minor (c and d) diastereomers for [Rh(CHIRAPHOS)(MAC)]$^+$. Contours are drawn and 2-kcal/mol intervals. The lowest contour is darkened as well as the next lowest contour. The x markings indicate crystallographic results for [Rh(CHIRAPHOS)(MAC)]$^+$. [Reproduced with permission from Giovannetti, J.; Kelly, C. M.; Landis, C. R. (1993), *J. Am. Chem. Soc.* **115**, 4040. Copyright © 1993 American Chemical Society.]

obtained by generating 2-D (φ–ψ) contour maps for both the major and minor diastereomers. The results are summarized in Figure 7.21. Considerable conformational flexibility was observed for the aryl groups attached to the phosphine trans to the alkene (labeled 1) for both the major and minor diastereomers. The aryls cis to the alkene (labeled 2) were substantially more conformationally constrained. The aryl ring orientations found in the crystal structure of the major diastereomer of the ethyl analog of [Rh(CHIRAPHOS) (MAC)]$^+$ denoted by x's in Figure 7.21 fell within the lowest contour of the molecular mechanics calculation. The conformational flexibility observed in the φ–ψ plots suggested

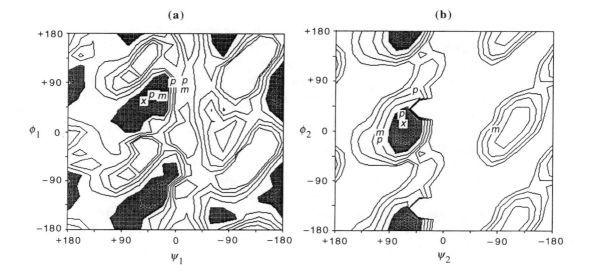

(a) **(b)**

FIGURE 7.22.
The φ-ψ maps for the major diastereomer for [Rh(DIPAMP)(MAC)]⁺. Contours are drawn at 2-kcal/mol intervals. The lowest contour is darkened as well as the next lowest contour. The x markings indicate crystallographic results for [Rh(DIPAMP)((Z)-β-propyl-α-acetamidoacrylate)]⁺ (the phenyl group of MAC is replaced with a propyl group). The p and m markings represent the positions for conformers giving the best fit to NOE data for the PRAC- and MAC- containing complexes, respectively. [Reproduced with permission from Giovannetti, J.; Kelly, C. M.; Landis, C. R. (1993), *J. Am. Chem. Soc.* **115**, 4040. Copyright © 1993 American Chemical Society.]

that a rigid "face–edge–face–edge" model as proposed by Koening was overly restrictive. The dominant steric interaction between the alkene and the chelating phosphine was found to be the interaction involving the complexed enamide ester group and the nearest arene ring. The enamide ester group and the nearby arene ring were always found to be coplanar. In order to minimize the ester–arene steric contacts, the chelate ring changes from a half-chair conformation for the major diastereomer to a higher energy envelope conformation for the minor diastereomer.

For [Rh(DIPAMP)(MAC)]⁺ the five-membered chelate ring can adopt two enantiomeric possibilities (δ and λ), so φ–ψ conformational searches were carried out for both alkene diastereomers along with both chelate possibilities. The lowest energy conformation found was the δ conformation of the major diastereomer. The 2-D (φ–ψ) contour map for the arene ring orientational preferences of the major diastereomer is given in Figure 7.22. As with the CHIRAPHOS system, considerable conformational flexibility was observed for the aryl groups attached to the phosphine trans to the alkene (labeled 1). The aryls cis to the alkene (labeled 2) were substantially more conformationally constrained. The aryl ring orientations found in the crystal structure of the major diastereomer of the (Z)-β-propyl-α-acetamidoacrylate)]⁺ analog of [Rh(DIPAMP)(MAC)]⁺ (the phenyl group of MAC is replaced with a propyl group) denoted by x's in Figure 7.22 fell within the lowest contour of the molecular mechanics calculation. As with the CHIRAPHOS system the conformational flexibility observed in the φ–ψ plots suggested that a rigid "face–edge–face–edge" model was overly restrictive. Here, the dominant steric interaction between the alkene and the chelating phosphine was also found to be the interaction involving the complexed enamide ester group and the nearest arene ring. In order to minimize the ester–arene steric contacts the chelate ring changes, adopting the λ chelate conformation for the minor diastereomer, rather than distorting to the higher energy envelope conformation. This δ to λ conformational change does cause the slightly larger anisyl ring to adopt the higher energy pseudoaxial configuration. Axial placement is disfavored due to 1,3-diaxial interactions.

The force field was found to reproduce the crystal structures for related CHIRAPHOS and DIPAMP complexes. Molecular mechanics derived energy differences between the major and minor diastereomers for the three systems CHIRAPHOS, DIPAMP, and DIPH of 2.3, 0.4, and 1.3 kcal/mol, respectively, were within 1 kcal/mol of the experimental free energy differences for these systems (> 2.4, 1.4, and ± 0.3 kcal/mol).

In order to determine whether or not the solid state structure was retained in solution Allured et al. (1991) carried out a detailed NOE study on the [Rh(DIPAMP)(MAC)]$^+$ and [Rh(DIPAMP)(PRAC)]$^+$ systems. The structures that the CPA analysis populated for the major diastereomers of both alkenes were consistent with the molecular mechanics analysis. The NOE-selected structures displayed the enamide ester–arene interaction dictated by the conformational preferences for both systems. This multiple conformation analysis implicated more than one conformational preference for the anisyl rings. For [Rh(DIPAMP)(MAC)]$^+$ the conformation where the methoxy group of the equatorial anisyl ring cis to the alkene was pointed in was populated 65%, and the conformation where it was pointed out was populated 35%. For [Rh(DIPAMP)(PRAC)]$^+$, only the conformation where the methoxy group was pointed in was populated. The essential features of the crystal structure appear to be retained in solution, though the anticipated conformational flexibility of the aryl rings was obtained in solution.

In summary, the SHAPES force field reproduced the solid state conformational preferences of a series of Rh(I) chiral phosphine chelating alkene complexes. That is, the experimentally observed conformations were found to be low energy in 2-D torsional space. The configurational energetic preferences for the major diastereomeric complexes were explained in terms of alkene–phosphine nonbonded interactions. Allured et al. (1991) were also able to confirm that the conformations in solution were those found in the solid state, with the addition of a set of anticipated aryl ring rotation conformers.

7.7. CONFORMATIONAL FLEXIBILITY

The coordination geometries of the metal complexes discussed in Sections 7.4–7.6 were easy to describe as being either octahedral, square planar, or tetrahedral. For a number of classes of complexes this assignment cannot be made unambiguously. A classic case is the coordination geometry of four-coordinate Cu(II) complexes, where structures ranging from near idealized square planar coordination to near tetrahedral coordination have been observed (reviewed in Hathaway, 1982).

Another class of complexes that are difficult to assign to idealized forms are the complexes of hexadentate amine cages. A systematic analysis of the X-ray structures for a series of such complexes found that the degree of distortion from an idealized octahedron was minimal for d^6 Co(III) complexes but rather substantial (rotation of the two triangular octahedrals faces by up to 36°) for complexes with little electronic preference for octahedral coordination (Comba et al., 1985).

The variety and underlying complexity of metal coordination environments can be partially understood by realizing that there are three major, and at times competing, contributions to the shapes of metal center coordination environments. These are (1) the directional bond hybridization component of organic chemistry, (2) a perturbation on this description due to the presence of either filled nonbonding d orbitals or low-lying empty d orbitals, and (3) the propensity of compounds to distort in order to relieve degeneracies. These degeneracy removing distortions are called first- and second-order Jahn–Teller effects and are described in more detail below (Jahn and Teller, 1937; Ammeter et al., 1979). In Section 7.2, we used orbital correlation diagrams to describe the coordination geometries mandated for six-coordinate d^6 complexes and four-coordinate d^8 complexes, respectively, by the presence of nonbonding d orbitals (contribution 2). For hexamine cage complexes, this analysis suggests some complexes will have a substantial preference for octahedral coordination but for other electronic configurations, such as high spin d^5, there is little effect on the coordination geometry from contribution 2. For d^9 Cu(II) complexes there is no clear predilection because, independent of geometry, antibonding orbitals are occupied.

7.7.1 Simulation of the Plasticity of Copper(II)

For a number of years beginning in 1982, Raos and co-workers of the University of Zagreb, Yugoslavia, attempted to develop force field models for the plasticity of Cu(II) (Raos et al., 1982; Raos and Simeon, 1983; Sabolovic and Raos, 1990a, 1990b, 1993). The primary focus of this research has been on the Cu(II) chelate complexes of N-alkylated amino acids, **7.31**.

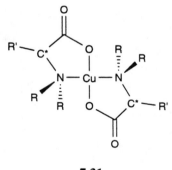

7.31

The coordination geometry of such complexes is of interest because the redox potentials and changes in redox potential of copper proteins (with amino acid ligands) are dependent on the coordination geometry at copper (Gould and Ehrenbeg, 1968). In addition, because there are two stereogenic centers in bis–chelate complexes (C* in **7.31**) such complexes are diastereomeric and possibly different in energy. Because of the diastereomeric nature of these complexes, Cu(II) has been used to separate racemic mixtures of amino acids (Davankov et al., 1974). Distortion of the bis amino acid complexes from square planarity is thought to play a role in the energetic differentiation between the diastereomers.

In addition to the difficulty of enforcing square planar coordination in a VSEPR-type model, wherein the coordination geometry is a consequence of ligand–ligand nonbonded interactions, Cu(II) complexes possess an additional electronic source for distortion, distortion to first-order Jahn–Teller effects. The first-order Jahn–Teller effect is discussed in Box 7.4.

Because a VSEPR-type model wherein the coordination geometry is a consequence of ligand–ligand nonbonded interactions cannot possibly work for four coordinate complexes, Raos and co-workers developed numerous distortional potentials for Cu(II) as well as a six center VSEPR-type model. The best models, reported by Sabolovic and Raos, are representative of the two extremes of inorganic modeling. The first model (Sabolovic and Raos, 1990a) included angle bend terms at copper as well as a distortional potential for out-of-plane bending (distortion from square planarity). We will call this a covalent model. The second model (Sabolovic and Raos, 1990b) abandoned copper centered bend terms and adopted a six-center VSEPR-type model. We will call this an ionic model. The results obtained with both models will be discussed below.

BOX 7.4   First-Order Jahn–Teller Effect

To see the role that the first-order Jahn–Teller effect plays in coordination geometries we will examine the square planar to tetrahedral distortional coordinate for Cu(II) complexes with four ligands. The frontier orbitals of square planar and tetrahedral complexes are collected in Figure 7.23. For d^9 Cu(II) complexes, nine electrons are placed in the frontier orbitals. For both square

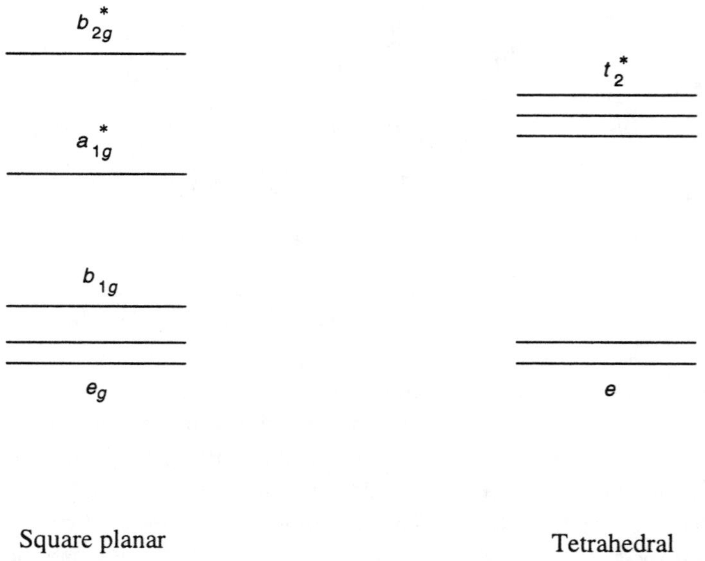

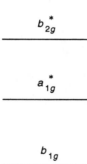

Square planar   Tetrahedral

FIGURE 7.23.
Frontier orbitals for square planar and tetrahedral metal complexes.

METHODS

For both the covalent and ionic models of Sabolovic and Raos, conventional harmonic stretch, harmonic bend, Fourier torsion, harmonic out-of-plane deformation (at sp^2 centers), and exponential-6 nonbond potential terms were used for the amino acid ligands. For the Cu$-$N and Cu$-$O bond stretches a harmonic potential was also used.

Since both tetrahedral and square planar minima are possible, a two-well angle bend potential was used for both the cis and trans ligand pairs in the covalent model (Sabolovic and Raos, 1990a). For the cis N$-$Cu$-$O angle, with minima at 90° and 109.471°, the product of a trigonometric function with a minimum at 90°, $\cos^2\theta$, and a function with a minimum at 109.471°, $(3\cos\theta + 1)^2$ was used, Eq. (7.23).

$$V_\theta = \frac{1}{2}k_\theta[\cos^2\theta \times (3\cos\theta + 1)^2] \tag{7.23}$$

planar coordination and tetrahedral coordination, all five frontier orbitals are occupied. At this qualitative level of analysis there should be, and is, little electronic preference for one geometry over the other and both coordination geometries should be observable. However, the tetrahedral structure possesses an unsymmetrically populated degenerate set of orbitals (the t_2 set with five electrons). According to the Jahn–Teller theorem, unsymmetrically populated degenerate levels are unstable (Jahn and Teller, 1937). That is, there is an energetic gain associated with breaking the degeneracy. This can be seen by considering the case with two degenerate orbitals, see Eq. (7.24)

$$\tag{7.24}$$

When a structural distortion occurs that will break the degeneracy, one of the levels will rise in energy and the other will become more stable. Since two electrons are in the stabilized energy level and only a single electron is in the destabilized level, this structural distortion will likely lead to a lowering in the energy of the system. The degree of distortion is dependent on the relative magnitudes of the stabilization and destabilization.

The more complex, threefold degeneracy case prevails for Cu(II), as seen in Figure 7.23. The triply degenerate t_2 set of tetrahedral coordination separates into three nondegenerate levels upon distortion towards square planar coordination. So, overlaid on the frontier orbital analysis, which suggested little energetic differentiation between square planar coordination and tetrahedral coordination, a first-order Jahn–Teller analysis suggests that the tetrahedral limit should not be seen. This provides a small, qualitative bias away from tetrahedral structures though complexes can occur along the entire distortion coordinate between the two limits.

For the trans $N-Cu-N$ and $O-Cu-O$ angles, with minima at $109.481°$ and $180°$, the product of a function with a minimum at $180°$, $\sin^2 \theta$, and a function with a minium at $109.471°$, $\left[\dfrac{3}{2\sqrt{2}} \sin \theta - 1 \right]^2$ was used, Eq. (7.25).

$$V_\theta = \frac{1}{2}k_\theta\left[\sin^2 \theta \times \left(\frac{3}{2\sqrt{2}} \sin \theta - 1 \right)^2 \right] \tag{7.25}$$

In addition, an out-of-plane deformation term was used to maintain planarity, Eq. (7.26)

$$V_\beta = \frac{1}{2}k_\beta(\beta - \beta_0)^2 \tag{7.26}$$

where β was defined as the angle between the bisectors of the two opposite $N-Cu-O$ and $N'-Cu-O'$ angles.

For the ionic model, all copper-centered angle terms were ignored (Sabolovic and Raos, 1990b). The copper coordination geometry was described in terms of the four atoms bonded to copper and two fictitious atoms, X and X', to make up a pseudooctahedral structure. For the two fictitious atoms, only bonding to the copper and electrostatic interactions with the four atoms bonded to copper were considered. The natural bond distance between X or X' and copper was set to zero and the harmonic force constant for the $Cu-X$ stretch term was determined so that the attractive interaction between X or X' and Cu was precisely balanced by the repulsive electrostatic interactions between X and N or O. The charge on X was -0.75 and on N and O it was -0.375. These charges were adjusted to achieve best reproduction of the coordination geometry for a set of Cu(II) amino acid and N-alkylated amino acid compounds.

RESULTS

For a series of N-alkylated amino acid Cu(II) complexes, the ionic and covalent models yielded structures of nearly the same accuracy. For example, for bis(L-*N,N*-dimethyli-soleucinato)copper(II), the experimental, covalent, and ionic structural parameters are collected in Table 7.12 (Kaitner et al., 1991). The molecular structure and atomic labeling scheme is presented in Figure 7.24. The rms errors in $Cu-L$ distances were 0.019 and 0.04 Å for the covalent and ionic models, respectively. The rms errors in the intraligand distances were 0.005 and 0.005 Å for the covalent and ionic models, respectively. The rms errors in $L-Cu-L$ angles were $1.56°$ and $1.54°$ for the covalent and ionic models, respectively.

The difficulty with this rms analysis of errors is that it does not take into account errors in shape. Since the copper complex, experimentally, can undergo a tetrahedral distortion from the idealized square planar structure, Eq. (7.27), or a pyramidal distortion, Eq. (7.28), the error analysis needs to differentiate between these different modes of distortion.

$$\tag{7.27}$$

$$(7.28)$$

Sabolovic and Raos (1993) reported a procedure for comparing computed structures with experimental structures in terms of a degree of distortion away from idealized limits. For a four-coordinate complex the idealized limits considered were the square planar limit, the tetrahedral limit, and a pyramidal limit (defined to be as far away from the square planar limit as the tetrahedral structure). The deviations were reported in terms of rms differences between idealized angles and experimental or calculated angles. The angles considered were the valence $N-Cu-N'$, $O-Cu-O'$, $N-Cu-O$, $N-Cu-O'$, $N'-Cu-O$, and $N'-Cu-O'$ angles, the $O-N-N'-O'$ dihedral angle, and the distortional coordinate β of Eq. (7.25). The values for the idealized structures are collected in Table 7.13. The rms distortions away from these idealized limits for a set of seven N-alkylated amino acid Cu(II) structures are collected in Table 7.14. For each system, except bis(L-N,N-dimethylisoleucinato)copper(II) (Entry 2 in Table 7.14), the covalent model underestimated the tetrahedral distortion. This result is likely due to the trans angles of the covalent model being too stiff. Each of the seven structures was found to be closest to the square planar limit, but each substantially distorted away from the square planar limit.

For each of the seven structures, except bis(L-N,N-dimethylisoleucinato)copper(II) (Entry 2 in Table 7.14), the ionic model mimicked the X-ray distortion better than the

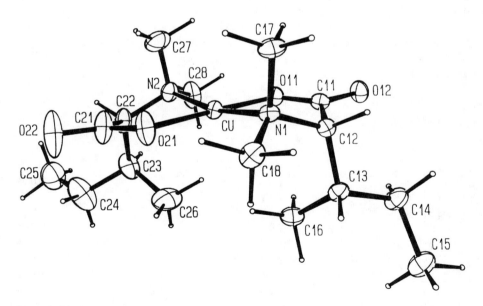

FIGURE 7.24.
Molecular structure of bis(L-N,N-dimethylisoleucinato)copper(II). [Reproduced with permission from Kaitner, B.; Paulic, N.; Raos, N. (1991), *J. Coord. Chem.* **22**, 269–279. Copyright © 1991 Gordon and Breach Science Publishers.]

TABLE 7.12. Selected Bond Distances and Valence Angles for Bis(L-*N*,*N*-dimethylisoleucinato)copper(II)

Internal Coordinate	Experiment	Covalent	Ionic
Cu−O11	1.911	1.944	2.005
Cu−O21	1.886	1.946	2.007
Cu−N1	2.009	1.988	2.004
Cu−N2	2.007	1.987	2.043
O11−C11	1.280	1.296	1.290
O12−C11	1.223	1.241	1.241
O21−C21	1.326	1.296	1.291
O22−C21	1.202	1.241	1.241
N1−C12	1.496	1.490	1.487
N1−C17	1.496	1.482	1.482
N1−C18	1.487	1.484	1.483
N2−C22	1.504	1.490	1.486
N2−C27	1.465	1.482	1.483
N2−C28	1.492	1.484	1.483
C11−C12	1.548	1.539	1.526
C12−C13	1.544	1.563	1.563
C13−C14	1.519	1.555	1.555
C13−C16	1.516	1.550	1.551
C14−C15	1.565	1.547	1.547
C21−C22	1.506	1.538	1.525
C22−C23	1.546	1.564	1.564
C23−C24	1.547	1.555	1.555
C23−C26	1.527	1.550	1.550
C24−C25	1.315	1.547	1.547
N1−Cu−N2	166.0	161.0	160.2
O21−Cu−N2	83.9	87.6	82.1
O21−Cu−N1	95.8	93.6	98.9
O11−Cu−N2	96.3	93.4	98.5
O11−Cu−N1	84.4	87.7	82.2
O11−Cu−O21	178.3	173.3	175.5
Cu−O11−C11	116.1	110.9	114.6
Cu−O21−C21	116.7	110.8	114.5
Cu−N1−C18	117.6	115.1	112.3
Cu−N1−C17	102.0	103.3	103.6
Cu−N1−C12	106.5	104.5	107.3
C17−N1−C18	107.7	109.1	109.0
C12−N1−C18	113.9	114.2	114.0
C12−N1−C17	108.2	110.0	110.1
Cu−N2−C28	116.4	115.6	112.9
Cu−N2−C27	102.7	103.1	103.5
Cu−N2−C22	105.7	104.3	107.1
C27−N2−C28	107.7	108.9	108.9
C22−N2−C28	114.0	114.2	113.9
C22−N2−C27	109.7	110.1	110.2
O11−C11−O12	124.7	120.5	121.1
O12−C11−C12	119.2	122.2	122.8
O11−C11−C12	116.1	117.2	116.1

TABLE 7.12. (*continued*)

Internal Coordinate	Experiment	Covalent	Ionic
N1−C12−C11	106.9	108.2	107.0
C11−C12−C13	110.5	109.1	109.4
N1−C12−C13	114.7	114.7	114.8
C12−C13−C16	113.6	111.9	112.0
C12−C13−C14	110.3	109.6	109.6
C14−C13−C16	112.0	110.3	110.3
C13−C14−C15	11.6	111.9	111.9
O21−C21−922	123.4	120.6	121.2
O22−C21−C22	122.5	122.3	122.9
O21−C21−C22	114.1	117.1	115.9
N2−C22−C21	107.8	108.1	107.0
C21−C22−C23	111.7	109.1	109.4
N2−C22−C23	113.6	114.5	114.6
C22−C23−C26	112.3	112.3	112.4
C22−C23−C24	113.5	112.0	111.9
C24−C23−C26	107.7	108.4	108.4
C23−C24−C25	117.6	112.3	112.3
O21−Cu−N2−C27	88.3	92.4	90.9
O11−Cu−N1−C17	88.4	93.2	91.8
N1−Cu−O11−C11	10.6	3.5	5.9
N2−Cu−O21−C21	11.9	4.0	6.5
Cu−O11−C11−C12	7.2	16.8	15.0
Cu−O21−C21−C22	7.2	16.7	14.8
Cu−N1−C12−C13	−90.2	−88.3	−84.8
Cu−N2−C22−C23	−88.7	−87.2	−83.6
O11−C11−C12−N1	−27.6	−35.7	−35.4
N1−C12−C13−C16	60.7	63.2	62.1
C12−C13−C14−C15	162.3	170.9	170.9
O21−C21−C22−N2	−29.4	−36.2	−35.9
N2−C22−C12−C26	61.7	61.4	60.2
C22−C23−C24−C25	70.3	60.9	60.9
β	171.8	171.5	169.8
O11−N1−N2−O21	163.5	154.3	155.5

TABLE 7.13. Valence Angles for Idealized Four-Coordinate Structures

Angle	Square Planar	Tetrahedral	Pydramidal
N−Cu−N′	180.0	109.471	102.8
O−Cu−O′	180.0	109.471	102.8
N−Cu−O	90.0	109.471	67.1
N−Cu−O′	90.0	109.471	67.1
N′−Cu−O	90.0	109.471	67.1
N′−Cu−O′	90.0	109.471	67.1
O−N−N′−O′	180.0	−70.529	180.0
β	180.0	180.0	102.8

TABLE 7.14. Distortions away from Idealized Structures

Model	Square Planar	Tetrahedral	Pyramidal
Bis(L-N,N-dimethylvalinato)copper(II)			
X-ray	34.1	126.8	138.8
Covalent	33.1	125.0	138.4
Ionic	37.0	127.3	138.4
Bis(L-N,N-dimethylisoleucinato)copper(II)			
X-ray	26.1	135.2	141.8
Covalent	34.3	123.6	138.2
Ionic	37.3	126.5	138.5
Bis(L-leucinato)copper(II)			
X-ray	14.4	147.3	148.4
Covalent	3.6	153.1	153.2
Ionic	15.4	151.8	153.0
Bis(L-alaninato)copper(II)			
X-ray	13.7	148.2	148.9
Covalent	6.3	152.8	154.0
Ionic	14.1	153.4	153.6
Bis(L-phenylalaninato)copper(II)			
X-ray	13.4	149.4	146.8
Covalent	3.8	152.4	153.1
Ionic	24.5	145.8	151.8
L-Serinato)clycinato)copper(II)			
X-ray	17.8	148.9	147.9
Covalent	3.6	152.9	152.7
Ionic	14.4	152.7	151.9
Bis(L-N-benzylprolinato)copper(II)			
X-ray	19.9	158.1	148.3
Covalent	11.3	144.4	152.1
Ionic	18.0	143.9	153.2

covalent model. However, due to the unphysical nature of the ionic model it is not surprising that Sabolovic and Raos (1990b) reported that it possesses unphysical minima, wherein the two fictitious atoms collapse on top of each other. Furthermore, there is nothing in the ionic model to prevent the trans fictitous atoms from becoming cis, yielding a family of unphysical distorted structures.

In summary, both the covalent and ionic models were able to reproduce the major structural features of the bis-N-alkyl amino acid Cu(II) complexes. However, each model had limitations. The covalent model underestimated the degree of distortion away from the square planar limit and the ionic model possessed unphysical minimia.

A possible way around the excessive stiffness of the covalent model can be found in an alternative, promising approach used by Vedani et al. (1989) of the University of Kan-

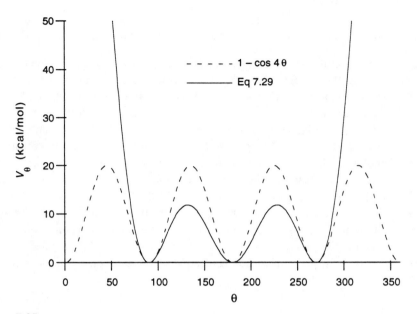

FIGURE 7.25.
Energy as a function of angle bend for a fourfold periodic potential $[1 + \cos(4\theta)]$ and a product potential [Eq. (7.30)]. A force constant of 80 kcal/mol-rad^2 and a natural angle of 90° is used for both potentials.

sas for their Zn(II) human carbonic anhydrase I (HCA I) molecular mechanics study. This approach is discussed in Chapter 3. In order to obtain the multiple minima associated with metal complexes (e.g., the present tetrahedral or square planar structures of four-coordinate metals) Vedani et al. (1986, 1988, 1990) suggested taking products of conventional terms, Eq. (7.29).

$$V_\theta = \frac{1}{2}K_{IJK}\{[\cos\theta_{tet} - \cos\theta_{IJK}]^2\} \times \{[\cos\theta_{sq-pl} - \cos\theta_{IJK}]^2\} \qquad (7.29)$$

Since this composite potential function goes to zero at both θ_{tet} and θ_{sq-pl}, this function will have two minima. The potential of Eq. (7.29) is the logical extension of the potentials described in Eqs. (7.24) and (7.25), and it provides a mechanism for selectively making the 180° minimum of the square planar structure softer. In the Sabolovic and Raos potential [Eqs. (7.24) and (7.25)], the 90° and 180° minima were described by $\cos^2\theta$. If the square planar potential were developed as a product of SHAPES type (Allured et al., 1991) cosine potentials, Eq. (7.30),

$$V_\theta = \frac{1}{2}k_\theta[1 + \cos\theta] \times [1 + \cos 2\theta] \qquad (7.30)$$

the minimum at 180° would be softer (have a smaller force constant) than the minimum at 90°, perhaps improving the Sabolovic and Raos covalent potential. A comparison of a $1 + \cos(4\theta)$ potential with the product potential of Eq. (7.30) is provided in Figure 7.25.

7.7.2 Simulation of the Coordination Geometries of Hexaamine Cage Complexes

In addition to the use of transition metal ions in catalysis, transition metal ion complexes have long served as the subjects of electronic spectroscopic investigations due to the presence of low-lying empty d orbitals. The excited states of transition metal complexes have potential to serve as the key elements in optoelectronic devices (Hopfield et al., 1989). Realization of this potential requires that the excited states persist under ambient conditions. The metal ion as well as the ligands play important roles in dictating excited state lifetime. Complexes where the ligand set plays a dramatic role in dictating excited-state lifetime are the cage or near cage hexamine complexes of the first transition series. For example, the lifetimes of the lowest energy excited state of trigonally strained octahedral (hexamine)chromium(III) complexes span a range of 10^8 s (Perkovic et al., 1991). The excited-state lifetime is thought to be dependent on a structural distortion in the ligand.

Cage hexamine complexes of the first transition series such as **7.32** are also of interest because the degree of trigonal distortion is dictated by the electronic configuration of the metal.

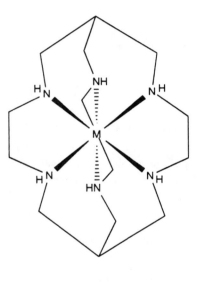

7.32

For d^6 Co(III) the distortion was found to be minimal, but the distortion was found to be rather substantial for complexes with little electronic preference for octahedral coordination (Comba et al., 1985). For example, the rotation of the two triangular faces away from an octahedral coordination geometry was 32.4° for a high spin d^5 Mn(II) complex. The degree of trigonal prismatic distortion has been shown to correlate with the d electronic configuration for a series of hexamine cage complexes (Comba et al., 1985). The electronic basis for geometric preferences in six-coordinate complexes is reviewed in Box 7.5.

BOX 7.5   Six-Coordination Revisited

To understand the geometric preferences for d^n ($n = 0–10$) six-coordinate complexes, let us consider a metal center with six equivalent ligands in its coordination environment. Each metal–ligand bond consists of two electrons. To begin with, let us consider the case without additional valence electrons. As suggested by VSEPR, the six bond pairs are mutually repelled and the ligands are as far apart as possible while retaining equivalent bond distances (Gillespie and Hargettai, 1991). For this six-coordinate case, there are two nearly equal energy configurations, the conventional octahedral arrangement **7.33a**, and a trigonal prismatic arrangement **7.33b.**

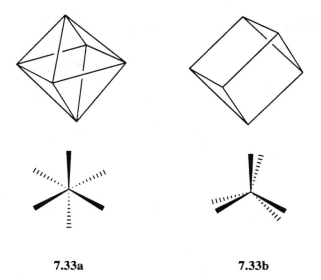

7.33a   **7.33b**

For this d^0 case, there is little intrinsic geometric preference for **7.33a** over **7.33b.**

As seen in Section 7.2, if nonbonding d electrons are added, a substantial preference for **7.33a** develops. The more detailed extended hückel orbital correlation diagram of Figure 7.26 shows that there is little net difference in the M−L bonding orbitals between an octahedral configuration and a trigonal prismatic configuration (Kang et al., 1993). The e_g level rises upon distortion from **7.33a** to **7.33b** but when the t_{1u} set splits, the e' component is stabilized. The major difference occurs in the t_{2g} nonbonding d set; it is no longer nonbonding, as there are now ligand orbitals of the same symmetry. The e' component of the t_{2g} rises because of mixing with the lower e' set. The a_1' orbital changes little because its cone shaped nodal surface provides substantial orthogonality to the ligand orbitals. As discussed in Section 7.3, for a d^6 configuration such as Co(III) the t_2 nonbonding set is full, distortion from an octahedral structure to a trigonal prismatic structure will be unfavorable. As we will see below, the degree of preference for octahedral coordination depends on the occupation of the t_{2g} and e_g frontier orbitals with each electronic configuration requiring a unique molecular mechanics description.

Beginning with the d^0 case we see a complication (Kang et al., 1993). The VSEPR picture, and our MO analysis both have ignored the presence of the low-lying empty (for a d^0 complex) t_{2g} orbital set. Components of this orbital set will mix (interact) with the filled t_{1u} M−L bonding set when a distortion from an octahedral structure to a trigonal prismatic structure occurs, stabilizing the molecule. This filled orbital-empty orbital interaction brought about by molecular distortion is

(continued)

BOX 7.5   Six-Coordination Revisited *(continued)*

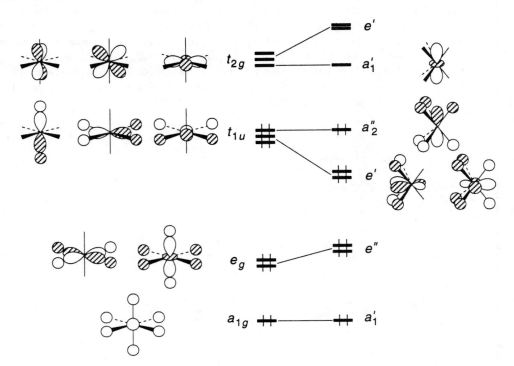

FIGURE 7.26.
Orbital energy correlation diagram for ML_6 for the distortion from octahedral to trigonal prismatic coordination. [Reproduced with permission from Kang, S. K.; Tang, H.; Albright, T. A. (1993), *J. Am. Chem. Soc.* **115**, 1971. Copyright © American Chemical Society.]

referred to as a second-order Jahn–Teller effect. The magnitude of the interaction between the e' components of the t_{2g} and t_{1u} sets depends on the energy difference between the t_{2g} and t_{1u} levels. Favorable mixing prevails for hexamethyl complexes such as WMe_6 where an electron diffraction structural study suggests a trigonal prismatic structure (Haaland et al., 1990), and $ZrMe_6^{2-}$, where the coordination environment in the X-ray crystal structure is trigonal prismatic (Morse and Giro-lami, 1989).

For the d^1–d^{10} configurations the frontier orbital contribution to the structural preferences of octahedral complexes can be developed by placing electrons in the t_{2g} and e_g levels. Only the lowest energy (most stable) configurations will be considered and the general rule that electrons in degenerate levels prefer to be spin aligned will be obeyed. The results are collected in Figure 7.27. For d^4–d^7 both high-spin and low-spin configurations are listed; the preferred configuration is dependent on the nature of the metal ion and the ligands involved.

As with the occupation diagrams of Figure 7.27 indicate, d^3 complexes and low-spin d^6 complexes prefer octahedral coordination. As discussed in Section 7.3, distortion away from octahedral coordination causes antibonding orbitals to be occupied. High-spin d^5 complexes and d^{10} complexes should have little geometric preference, as all five d levels are equally populated. Other occupations, with partially full degenerate levels, are subject to first-order Jahn–Teller distortional effects away from octahedral coordination.

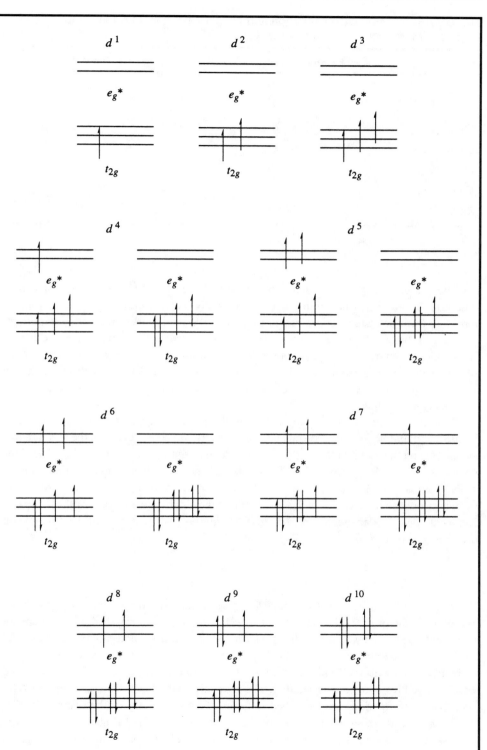

FIGURE 7.27.
Orbital occupanies of the octahedral metal frontier orbitals for d^1–d^{10} configurations.

TABLE 7.15. Electronic Configuration and Degree of Trigonal Prismatic Distortion for a Series of Hexamine Cage Complexes, **7.32**

Metal Ion	Likely Electronic Configuration	Degree of Trigonal Distortion (°)
Mg(II)	d^0	32
Cr(III)	d^3	11
Mn(II)	High-spin d^5	32
Co(III)	Low-spin d^6	3
Ni(II)	d^8	12
Cu(II)	d^9	30
Zn(II)	d^{10}	31

The degree of trigonal prismatic distortion has been related to the d electronic configuration for a series of hexamine cage complexes (Comba, 1985). The degree of distortion away from octahedral coordination as well as the likely electronic configuration are collected in Table 7.15 for a subset of the complexes studied. For reference, the metal-free ligand is distorted 36° away from an octahedron. The complexes with a preference for octahedral coordination show little distortion away from an octahedron, but those complexes with little electronic preference for coordination geometry (d^0, d^5, d^9, and d^{10}) adopt a geometry dictated by the ligand.

Two factors left out of this analysis were the preferred geometry of the ligand as a function of metal–ligand bond distance (as one progresses across the row, the metal radius decreases) and the fact that the cage structure is composed of five-membered chelate rings which, as discussed in Section 7.4, are not flat and have intrinsic conformational preferences. Comba (1989) of the Universität Basel reported a molecular mechanics study on the distortional characteristics of **7.32** to examine the significance of both of these factors. Comba studied the dependence of the strain energy on M-N distance for each of several low-energy cage conformations.

METHODS

The force field used by Comba (1989) was a modified (angle force constants decreased by a factor of 2) (Hambley et al., 1981) version of the classic Co(III) hexamine force field due to Snow (Snow, 1970). Harmonic stretch and bend, Fourier torsion, and exponential-6 nonbond terms were used. The L$-$M$-$L angle terms were replaced by 1,3 nonbonded interactions. Comba studied a set of six conformations of **7.32**, each having been observed for one metal or another. The six conformations are defined in Figure 7.28, where the conformations are described in terms of the symmetry of the complex, either D_3, C_3, or C_2 and the stereochemistry of the three five-membered rings, either lel or ob. For a D_3lel$_3$ arrangement two conformations were considered, one with a small trigonal distortion and one with a large trigonal distortion. The conformation with a large trigonal distortion was denoted D_3lel$'_3$. For each conformation for a range of M$-$N distances the

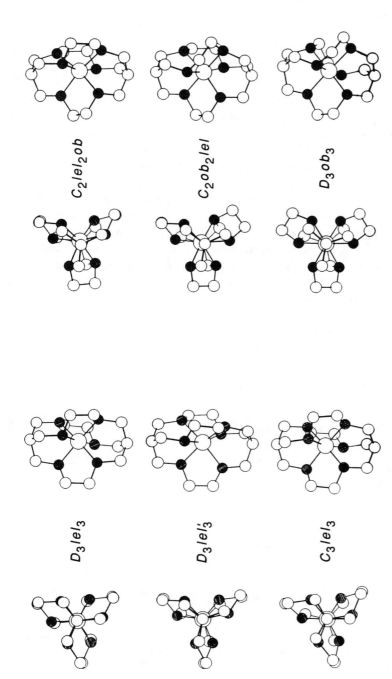

C_2lel_2ob

C_2ob_2lel

D_3ob_3

D_3lel_3

$D_3lel'_3$

C_3lel_3

FIGURE 7.28.
Low-lying conformations of **7.32**. The conformations are described in terms of the symmetry of the complex, either D_3, C_3, or C_2 and the stereochemistry of the three five-membered rings, either lel or ob. [Reproduced with permission from Comba, P. (1989), *Inorg. Chem.* **28**, 426. Copyright © 1989 American Chemical Society.]

M—N distance was fixed and the remaining geometric variables were optimized using a modified Newton–Raphson procedure. The M—N distances studied ranged from 1.90 to 2.36 Å.

RESULTS

In order to understand the distortional characteristics of hexamine cages, Comba minimized the energy of each of a set of six conformations of **7.32** with fixed M—N distance. Plots of energy verus M—N distance are presented in Figure 7.29 for the six conformations shown in Figure 7.28. For short M—N distance (< 2.1 Å) all of the conformations were nearly degenerate, but the C_2lel$_2$ob conformation was lowest in energy. For M—N distances greater than 2.1 Å the D_3lel$'_3$ conformation was lowest. Experimentally, the Co(III) complex, with a M—N distance of 1.974 Å adopts a D_3lel$_3$ conformation. The Cr(III) complex, with a M—N distance of 2.073 Å adopts a C_3lel$_3$ conformation. All of the observed structures with M—N distances above 2.17 Å adopt D_3lel$'_3$ conformations.

An interesting conformational lability of the three lel$_3$ conformations was found in the region of 2.07 Å. For M—N distances above 2.07 Å the D_3lel$_3$ conformation distorted into the C_3lel$_3$ conformation. For M—N distances above 2.14 Å the C_3lel$_3$ conformation distorted into the D_3lel$'_3$ conformation.

The distance preferences of the various conformations is of importance because the different conformations support different trigonal distortions. The D_3lel$_3$ conformation,

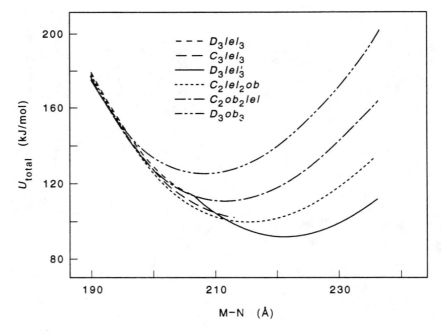

FIGURE 7.29.
Plots of total energy as a function of M—N distance. Minimized strain energies of M (sar)$^{m+}$ as a function of the M—N bond length: $(--)$ D_3lel$_3$; $(-\cdot-)$ C_3lel$_3$; $(-)$ D_3lel$_3'$; $(--)$ C_2lel$_2$ob; $(-\cdot-\cdot-)$ C_2ob$_2$lel; $(-\_\_\_\_-)$ D_3ob$_3$. [Reproduced with permission from Comba, P. (1989), *Inorg. Chem.* **28**, 426. Copyright © 1989 American Chemical Society.]

observed for small M−N distances, is consistent with a nearly octahedral structure, adopting a twist angle of less than 10°. The D_3lel'$_3$ conformation, observed for large M−N distances, adopts a structure with a twist angle of greater than 30°.

Since N−M−N angle terms were ignored in this molecular mechanics study, the computed lack of a trigonal distortion for Co(III) was not due an electronic preference of low-spin d^6 Co(III) complexes for octahedral coordination. The electronic preference of low-spin d^6 Co(III) complexes for octahedral coordination would have had to have been mimicked by an angle bend potential. Because both the molecular mechanics study (without angle bend terms) and the MO analysis did explain the trends in trigonal distortion as a function of metal ion, Comba concluded that the electronic ground states of transition metal cage complexes are the result of the structural environment and not the reverse (the structural environment being determined by the electronic ground state of the complex).

An alternative explanation is that complexes that would differentiate between a model that ignores angle bends and a model based on a significant electronic preference for octahedral coordination simply have not yet been prepared. For example, a Re(I) complex would have an electronic preference for octahedral coordination but the Re−N bond distances (~2.2 Å) would place the complex into the group that would prefer a distorted structure according to Comba's analysis.

Homework

7.1. Angle bend terms at square planar and octahedral centers require double-welled potentials because there are minima for both 90° and 180° bond angles. To see possible solutions to this problem generate V versus θ curves for the functions given in Eqs. (7.31)–(7.33). Plot the data from $\theta = 0°$ to $360°$ in $10°$ increments.

$$V_\theta = 1 + \cos 4\theta \tag{7.31}$$

$$V_\theta = [1 + \cos \theta] \times [1 + \cos 2\theta] \tag{7.32}$$

$$V_\theta = [(\cos 180° - \cos \theta)^2] \times [(\cos 90° - \cos \theta)^2] \tag{7.33}$$

7.2. Use your molecular modeling software to sketch in and minimize the Λ and Δ enantiomers of the lel$_3$ configuration of tris(ethylenediamine)cobalt(III), **7.8**. Use the docking facility of your molecular modeling software to convince yourself that the molecules are not superimposable.

7.3. Use your molecular modeling software to sketch in and minimize the four Λ diastereomers (lel$_3$, lel$_2$ob, lelob$_2$, and ob$_3$) of tris(ethylenediamine)cobalt(III), **7.8**. How large are your calculated energy differences?

7.4. As discussed in Section 7.5, modeling π ligands is not particularly straightforward in molecular mechanics. How does the software you are using handle π ligands? Sketch in and minimize the structure for Cp$_2$TiCl$_2$. Replace the two Cl ligands by an ethylene to make a metallacyclopropane. Replace one of the Cl atoms of Cp$_2$TiCl$_2$ with a Me, remove the other Cl to make a cation species, and coordinate an ethylene. For this complex the ethylene cannot be a metallacyclopropane because no d electrons are left.

7.5. Construct **7.14a** and **7.14b** and convince yourself that they have the same energy.

7.6. In asymmetric catalysis a stereoselectivity, described in terms of enantiomeric excess (%ee) of 95% is synthetically useful. Assuming the reaction is run at $-90°C$, how large of a transition state energy difference does a 95%ee correspond to? Use Eq. (7.34) for %ee, and Eq. (7.11) ($R = 1.98$ cal/mol-K), to obtain your answer.

$$\%ee = |\%R - \%S| \tag{7.34}$$

7.7. Use your molecular modeling software to sketch in and minimize the fragment **7.27** to see the edge-face arrangement of arene rings.

7.8. Dock **7.29** to **7.27** using each of the two alkene faces, attach the carboxylate and alkene functional groups, and minimize each structure. How different are the energies?

7.9. Sketch in and minimize the structure for *mer*-tribromotris(dimethylphenylphosphine) rhodium(III) (the three bromines are coplanar in this *mer* octahedral structure). Experimentally, the P$-$Rh distances are 2.392, 2.392, and 2.296 Å and the three Br$-$Rh distances are 2.498, 2.498, and 2.568 Å. Why are all three P$-$Rh distances not the same? How do your molecular mechanics determined distances compare?

7.10. Use your molecular modeling software to sketch in and minimize **7.31** with R = R' = Me (L-*N*-methyl alanine). Is the Cu ring flat? Is the Cu tetrahedral? What angle term is being used in your molecular modeling software package? Construct the two possible diastereomers and determine the energy difference.

7.11. Use your molecular modeling software to sketch in and minimize the lel$_3$ configuration of **7.32** with and without a Co(III) ion in the middle. Does the metal ion make a difference in the geometry of the ligand?

7.12. For use in electronic devices, a molecule should undergo a large geometric change upon electronic excitation. Can you use your molecular modeling software to evaluate the geometry of the electronic excited state resulting from a $t_{2g}{}^5 e_g{}^1 \leftarrow t_{2g}{}^6$ excitation?

7.13. Sketch in and minimize VCl_5 and WCl_4O. Do they both have trigonal bipyramidal geometries after minimization? (Experimentally WCl_4O is square pyramidal.) What angle term is being used in your molecular modeling software package?

7.14. Sketch in and minimize $W(Me)_6$. Is it octahedral after minimization? (Experimentally it is trigonal prismatic.) What angle term is being used in your molecular modeling software package?

7.15. Read in and navigate through the crystal structure of faujasite. The unit cell and unique atom coordinates are given below (Czjzek et al., 1992). Find the four unique oxygen sites in the material. Replace a single Si in the unit cell with an Al, add the corresponding proton to each of the four oxygen sites and calculate the relative energies for each.

Space group: $Fd3m$, $a = b = c = 24.7390$ ($\alpha = \beta = \gamma = 90°$)

Atom	X	Y	Z
Si(T)	-0.0524	0.0361	0.1248
O1	0.0	-0.10525	0.10525
O2	-0.0035	-0.0035	0.1428
O3	0.1789	0.1789	-0.0353
O4	0.1755	0.1755	0.3199

References

Albright, T. A.; Burdett, J. K; Whangbo, M.-H. (1985), *Orbital Interactions in Chemistry*, Wiley, New York.

Allen, F. H.; Davies, J. E.; Galloy, J. J.; Johnson, D.; Kennard, O.; Macrae, C. F.; Mitchell, E. M.; Smith, J. M.; Watson, D. G. (1991), The development of versions 3 and 4 of the Cambridge structural database system, *J. Chem. Inf, Comp. Sci.* **31**, 187.

Allinger, N. L. (1977), Conformational Analysis. 130. MM2. A hydrocarbon force field utilizing V_1 and V_2 torsional terms, *J. Am. Chem. Soc.* **99**, 8127.

Allinger, N. L.; Yuh, Y. H.; Lii, J.-H. (1989), Molecular mechanics. The MM3 force field for hdyrocarbons. 1, *J. Am. Chem. Soc.* **111**, 8551.

Allured, V. S.; Kelly, C. M.; Landis, C. R. (1991), SHAPES empirical force field: New treatment of angular potentials and its application to square-planar transition-metal complexes, *J. Am. Chem. Soc.* **113**, 1.

Ammeter, J. H.; Burgi, H. B.; Gamp, E.; Meyer-Sandrin, V.; Jensen, W. P. (1979), Static and dynamic Jahn–Teller distortions in CuN_6 complexes. Crystal structures and EPR spectra of complexes between copper(II) and rigid, tridentate *cis,cis*-1,3,5-triaminocyclohexane (tach): $Cu(tach)_2(ClO_4)_2$, $Cu(tach)_2(NO_3)_2$. Crystal structure of $Ni(tach)_2(NO_3)_2$, *Inorg. Chem.* **18**, 733.

Arlman, E. J.; Cossee, P. J. (1964), Ziegler–Natta catalysis III. Stereospecific polymerization of propene with the catalyst system $TiCl_3 - AlEt_3$, *J. Catal.*, **3**, 99.

Battaglia, L. P.; Bonamartini Corradi, A. B.; Marcotrigiano, G.; Menabue, L.; Pellacani, G. C. (1982), *N*-(2-ammonioethyl)morpholinium tetrachlorocuprates(II). The first instance of two forms, one green and one yellow, both stable at room temperature, *Inorg. Chem.* **21**, 3919.

Bender, B. R.; Norton, J. R; Miller, M. M.; Anderson, O. P.; Rappé, A. K. (1992), Structure of $Os(CO)_4(C_2H_4)$, an osmacyclopropane, *Organometallics* **11**, 3427.

Brooks, B. R.; Bruccoleri, R. E.; Olafson, B. D.; States, D. J.; Swaminathan, S.; Karplus, M. (1983), CHARMM: A program for macromolecular energy, minimization, and dynamics calculations, *J. Comp. Chem.* **4**, 187.

Castonguay, L.A.; Rappé, A. K. (1992), Ziegler–Natta catalysis. A theoretical study of the isotactic polymerization of propylene, *J. Am. Chem. Soc.*, **114**, 5832.

Chatt, J.; Duncanson, L. A. (1953), Olefin co-ordination compounds. Part III. Infra-red spectra and structure: attempted preparation of acetylene complexes, *J. Chem. Soc.* 2939.

Comba, P. (1989), Coordination geometries of hexaamine cage complexes, *Inorg. Chem.* **28**, 426.

Comba, P.; Sargeson, A. M.; Englehardt, L. M.; Harrowfield, J. M.; White, A. H.; Horn, E.; Snow, M. R. (1985), Analysis of trigonal-prismatic and octahedral preferences in hexaamine cage complexes. *Inorg. Chem.* **24**, 2325.

Corey, E. J.; Bailar, J. C. (1959), The stereochemistry of complex inorganic compounds. XXII. Stereospecific effects in complex ions, *J. Am. Chem. Soc.* **81**, 2620.

Cossee, P. J. (1964), Ziegler–Natta Catalysis I. Mechanism of polymerization of α-olefins with Ziegler–Natta catalysts. *J. Catal.*, **3**, 80.

Cotton, F. A.; Wilkinson, G (1988), *Advanced Inorganic Chemistry*, Wiley, New York, p. 283.

Czjzek, M.; Jobic, H.; Fitch, A. N.; Vogt, T. (1992), Direct determination of proton positions in D-Y and H-Y zeolite samples by neutron powder diffraction, *J. Phys. Chem.* **96**, 1535.

Davankov, V. A.; Rogozhin, S. V.; Semechkin, A. V. (1974), Ligand-exchange chromatography of racemates resolution of α-amino acids, *J. Chromatogr.* **91**, 493.

Dewar, M. J. S. (1951), A review of the π-complex theory, *Bull. Soc. Chim. Fr,* **18**, C71.

Doman, T. N.; Landis, C. R.; Bosnich, B. (1992), Molecular mechanics force fields for linear metallocenes, *J. Am. Chem. Soc.* **114**, 7264.

Dyer, A. (1988), *An Introduction to Zeolite Molecular Sieves*, Wiley, New York.

Erker, G.; Nolte, R.; Aul, R.; Wilker, S.; Krüger, C.; Noe, R. (1991), Cp-substituent additivity effects controlling the stereochemistry of the propene polymerization reaction at conformationally unrestricted $(Cp-CHR^1R^2)_2ZrCl_2$/methylalumoxane catalysts, *J. Am. Chem Soc.* **113**, 7594.

Ewen, J. A.; Elder, M. J.; Jones, R. L.; Haspeslagh, L.; Atwood, J. L.; Bott, S. G.; Robinson, K. (1991), Metallocene/polypropylene structural relationships: implications on polymerization and stereochemical control mechanics, *Makromol. Chem., Macromol. Symp.,* **48/49**, 253.

Gates, B.C.; Katzer, J. R.; Schuit, G. C. A. (1979), *Chemistry of Catalytic Processes,* McGraw-Hill, New York.

Gillespie, R. J.; Hargittai, I (1991), *The VSEPR Model of Molecular Geometry,* Allyn and Bacon, Boston, MA.

Giovannetti, J.; Kelly, C. M.; Landis, C. R. (1993), Molecular mechanics and NOE investigations of the solution structures of intermediates in the [Rh(chiral bisphosphine)]$^+$-catalyzed hydrogenation of prochiral enamides, *J. Am. Chem. Soc.* **115**, 4040.

Gould, D. G.; Ehrenberg, A. (1968), Cu^{2+} in non axial-field: A model for Cu^{2+} in copper enzymes, *Eur. J. Biochem.* **5**, 451.

Haaland, A.; Hammel, A.; Rypdal, K. Volden. H. V. (1990) Coordination geometry of gaseous hexamethyltungsten: not octahedral, *J. Am. Chem. Soc.* **112**, 4547.

Hald, N. C. P.; Rasmussen, K. (1978), Conformational analysis of coordination compounds. VI. Force field calculation of thermodynamic properties of Tris(diamine)cobalt(III) coordination complexes, *Acta Chem. Scand. A* **32**, 879.

Hambley, T. W.; Hawkins, C. J.; Palmer, J. A.; Snow, M. R. (1981) Conformational analysis of coordination compounds. XI Molecular structure of tetraammine-{(±)-pentane-2,4-diamine }cobalt(III) dithionate, *Aust. J. Chem.* **34**, 45.

Harnung, S. E.; Kallesoe, S.; Sargeson, A. M.; Schaffer, C. E. (1974), The tris[(±)-1,2-propanediamine]cobalt(III) system, *Acta Chem. Scand., A* **28**, 385.

Hart, J. R.; Rappé, A. K. (1992), van der Waals functional forms for molecular simulations, *J. Chem. Phys.* **97**, 1109.

Hart, J. R.; Rappé, A. K. (1993), Predicted structure selectivity trends: propylene polymerization with substituted *rac*-(1,2-Ethylenebis(η^5-indenyl))zirconium(IV) catalyts, *J. Am. Chem. Soc.* **115**, 6159.

Hathaway, B. J. (1982) Copper, *Coord. Chem. Rev.* **41**, 423.

Hill, T. L (1946), On Steric Effects, *J. Chem. Phys.* **14**, 465.

Hopfield, J. J.; Onuchic, J. N.; Beratan, D. N. (1989), Electronic shift register memory based on molecular electron-transfer reactions, *J. Phys. Chem.* **93**, 6350.

Hückel, W. (1925), Zur stereochemie bicyclischer ringsysteme. I. Die stereoisomerie des dekahydronaphthalins und seiner derivate, *Justus Liebigs Ann. Chem.* **441**, 1.

Jahn, H. A.; Teller, E. (1937), Stability of polyatomic molecules in degenerate electronic states. I- Orbital degeneracy, *Proc. R. Soc., London* **161**, 220.

Kaitner, B.; Paulic, N.; Raos, N. (1991), Stereochemistry of complexes with N-alkylated amino acids. III. Crystal structure and conformational analysis of bis(L-*N,N*-dimethylisoleucinato)-copper(II), *J. Coord. Chem.,* **22**, 269.

Kaminsky, W.; Külper, K.; Brintzinger, H. H.; Wild, F. R. W. P. (1985), Polymerization of propene and butene with a chiral zirconocene and methylalumoxane as cocatalyst, *Angew. Chem., Int. Ed. Engl.* **24**, 507.

Kang, S. K.; Tang, H.; Albright, T. A. (1993) Structures for d^0 ML$_6$ and ML$_5$ complexes, *J. Am. Chem. Soc.* **115**, 1971.

Kobayashi, M (1943), Optical rotatory power and dichroism III. Optical rotatory dispersion of aqeous d[Coen$_3$]Br$_3$, *J. Chem. Soc. Jpn.* **64**, 648.

Koenig, K. E.; Sabacky, M. J.; Bachman, G. L.; Christopfel, W. C.; Barnstorff, H. D.; Friedman, R. B.; Knowles, W. S.; Stultz, B. R.; Vineyard, B. D.; Weinkauff, D. J. (1980), Asymmetric hydrogenation with rhodium chiral phosphine catalysts, *Ann. N. Y. Acad. Sci.* **333**, 16.

Kramer, G. J.; van Santen, R. A. (1993), Theoretical determination of proton affinity differences in zeolites. *J. Am. Chem. Soc.* **115**, 2887.

Landis, C. R.; Allured, V. S. (1991) Elucidation of solution structures by conformer population analysis of NOE data, *J. Am. Chem. Soc.* **113**, 9493.

Landis, C. R.; Halpern, J. (1987), Asymmetric hydrogenation of methyl-(Z)-α-acetamidocinnamate catalyzed by {1,2-bis((phenyl-*o*-anisoyl)phosphino)ethane}rhodium(I): kinetics, mechanism, and origin of enantioselection, *J. Am. Chem. Soc.* **109**, 1746.

Larsen, E.; La Mar, G. N. (1974), The angular overlap model. How to use it and why, *J. Chem. Educ.* **51**, 633.

Lee, I.-M.; Gauthier, W. J.; Ball, J. M.; Iyengar, B.; Collins, S. (1992), Electronic effects in Ziegler–Natta polymerization of propylene and ethylene using soluble metallocene catalysts. *Organometallics* **11**, 2115.

Lepard, D. W.; Shaw, D. E.; Welsh, H. L. (1966), *Can J. Phys.* **44**, 2353.

Mayo, S. L.; Olafson, B. D.; Goddard III, W. A. (1990), DREIDING: A generic force field for molecular simulations, *J. Phys. Chem.*, **94**, 8897.

Morrison, J. D., Ed. (1985), *Asymmetric Synthesis*, Academic, New York.

Morse, P. M.; Girolami, G. S. (1989), Are d^0 ML$_6$ complexes always octahedral? The X-ray structure of trigonal-prismatic [Li(med)]$_2$[ZrMe$_6$], *J. Am. Chem. Soc.* **111**, 4114.

Niketic, S. R.; Rasmussen, K. (1978), Conformational analysis of coordination compounds. IV. tris(1,2-ethanediamine)- and tris(2,3-butanediamine)cobalt(III) complexes, *Acta Chem. Scand. A.* **32**, 391.

Parshall, G. W.; Ittel, S. D. (1992), *Homogeneous Catalysis: The applications and Chemistry of Catalysis by Soluble Transition Metal Complexes*, Wiley-Interscience, New York.

Pauling, L. (1960), *The Nature of the Chemical Bond*, Cornell University Press, Ithaca, NY.

Perkovic, M. W.; Heeg, M. J.; Endicott, J. F. (1991), Stereochemical perturbations of the relaxation behavior of (^{2}ECr(III). Ground-state X-ray crystal structure, phophysics, and molecular mechanics simulations of the quasi-cage complex [4,4',4'',-ethylidynetris(3-azabutan-1-amine)]chromium tribromide, *Inorg. Chem.* **30**, 3140.

Pino, P.; Cioni, P.; Wei, J. (1987), Asymmetric hydrooligomerization of propylene, *J. Am. Chem. Soc.* **109**, 6189.

Pino, P.; Cioni, P.; Wei, J.; Rotzinger, B.; Arizzi, S. (1988), Recent develpments in basic research on the stereospecific polimerization of α-olefins, in *Transition Metal Catalyzed Polymerizations; Ziegler–Natta and Metathesis Polymerizations*, Quirk, R. P., Ed., Cambridge University Press, Cambridge, p. 1.

Raos, N.; Niketic, S. R.; Simeon, V. (1982), Conformational analysis of copper(II) chelates with epimeric amino acids, *J. Inorg. Biochem.*, **16**, 1.

Raos, N.; Simeon, V. (1983), Conformational analysis and hydration models of copper threoninates and isoleucinates, *J. Inorg. Biochem.*, **18**, 133.

Rappé, A. K.; Goddard, W. A. (1991), Charge equilibration for molecular dynamics simulations, *J. Phys. Chem.*, **95**, 3358.

Rosenblatt, F.; Schleede, A. (1933), Der räumliche bau der platin-tetrammin-salze, *Justus Liebigs Ann. Chem.* **505**, 51.

Sabolovic, J.; Raos, N. (1990a), Critical evaluation of empirical force-field models for simulation of plasticity of copper(II) coordination, *Polyhedron* **9**, 1277.

Sabolovic, J.; Raos, N. (1990b), Simulation of plasticity of a copper(II) coordination polyhedron with a force field based on Coulombic interactions: conformational analysis of copper(II) chelates with .alpha.-amino acids, *Polyhedron*, **9**, 2419.

Sabolovic, J.; Raos, N. (1993), The shapes of the copper(II) coordination polyhedra: an attempt at a comparison, *THEOCHEM*, **100**, 101–5.

Saito, Y. (1985), Conformations of the transition metal complexes, *J. Mol. Struct.* **126**, 461.

Schröder, K. P.; Sauer, J.; Leslie, M.; Catlow, C. R. A. (1992a), Siting of Al and bridging hydroxyl groups in ZSM-5: A computer simulation study, *Zeolites* **12**, 20.

Schröder, K. P.; Sauer, J.; Leslie, M.; Catlow, C. R. A.; Thomas, J. M. (1992b), Bridging hydroxyl groups in zeolite catalysts: a computer simulation of their structure, vibrational properties and acidity in protonated faujasites (H-Y zeolites), *Chem. Phys. Lett.* **188**, 320.

Shriver, D. F.; Atkins, P. W.; Langford, C. H. (1990), *Inorganic Chemistry*, Freeman, New York.

Snow, M. R. (1970), Structure and conformational analysis of coordination complexes. The αα isomer of chlorotetraethylenepentaminecobalt(III), *J. Am. Chem. Soc.* **92**, 3610.

Sudmeier, J. L.; Blackmer, G. L.; Bradley, C. H.; Anet, F. A. L. (1972), Conformational analysis of trisethylenediamine and tris-*R*-propylenediamine complexes of cobalt(III) by 251-MHz proton magnetic resonance with Cobalt-59 decoupling, *J. Am. Chem. Soc.* **94**, 757.

Theilacker, W. (1937), **ILL**, *Z. Anorg. Allgem. Chem.* **234**, 161.

Vedani, A. (1988), YETI: an interactive molecular mechanics program for small-molecule protein complexes, *J. Comput. Chem.* **9**, 269.

Vedani, A.; Dobler, M.; Dunitz, J. D. (1986), An empirical potential function for metal centers: application to molecular mechanics calculations on metalloproteins, *J. Comput. Chem.* **7**, 701.

Vedani, A.; Huhta, D. W. (1990), A new force field for modeling metalloproteins, *J. Am. Chem. Soc.* **112**, 4759.

Vedani, A.; Huhta, D. W.; and Jacober, S. P. (1989), Metal coordination, H-bond network formation, and protein–solvent interactions in native and complexed human carbonic anhydrase I: a molecular mechanics study, *J. Am. Chem. Soc.* **111**, 4075.

Wielers, A. F. H.; Vaarkamp, M.; Post, M. F. M. (1991), Relation between properties and performance of zeolites in paraffin cracking, *J. Catal.* **127**, 51.

Further Reading

Inorganic Chemistry

Cotton, F. A.; Wilkinson, G. (1988), *Advanced Inorganic Chemistry*, Wiley, New York, p. 283.

Shriver, D. F.; Atkins, P. W.; Langford, C. H. (1990), *Inorganic Chemistry*, Freeman.

Organometallic Chemistry

Collman, J. P.; Hegedus, L. S.; Norton, J. R.; Finke, R. G. (1987), *Principles and Applications of Organotransition Metal Chemistry*, University Science, Mill Valley, CA.

Parshall, G. W.; Ittel, S. D. (1992), *Homogeneous Catalysis*, Wiley, New York.

Asymmetric Catalysis

Morrison, J. D., Ed. (1985), *Asymmetric Synthesis*, Academic, New York.

Zeolites

Catlow, C. R. A., Ed. (1992), *Modeling of Structure and Reactivity in Zeolites*, Academic, London.

Dyer, A. (1988), *An Introduction to Zeolite Molecular Sieves*, Wiley, New York.

Molecular Orbital Theory

Albright, T. A.; Burdett, J. K; Whangbo, M.-H. (1985), *Orbital Interactions in Chemistry*, Wiley, New York.

Molecular Mechanics of Inorganic Compounds

Boeyens, J. C. A. (1985), Molecular mechanics and the structure hypothesis, *Structure Bonding* **63**, 67.

Brubaker, G. R.; Johnson, D. W. (1984), Molecular mechanics calculations in coordination chemistry, *Coord. Chem. Rev.* **53**, 1.

Comba, P. (1993), The relationship between ligand structures, coordination stereochemistry, and electronic and thermodynamic properties, *Coord. Chem. Rev.* **123**, 1.

Hancock. R. D. (1990), Molecular mechanics calculations and metal ion recognition, *Acc. Chem. Res.* **23**, 253.

Hancock, R. D.; Martell, A. E. (1989), Ligand design for selective complexation of metal ions in aqueous solution, *Chem. Rev.* **89**, 1875.

Hay, B. P. (1993), Methods for molecular mechanics modeling of coordination compounds, *Coord. Chem. Rev.* **126**, 177.

Landis, C. R.; Root, D. M.; Cleveland, T (1995), Molecular mechanics force fields for modeling inorganic and organometallic compounds. In *Reviews in Computational Chemistry*, Vol. 6, Lipkowitz, K. B.; Boyd, D. B., Eds., VCH, New York. p. xx.

Supplemental Case Studies

Bogdan, P. L.; Irwin, J. J.; Bosnich, B. (1989), Asymmetric synthesis. Molecular graphics and enantioselection in asymmetric catalytic hydrogenation, *Organometallics* **8**, 1450.

Bond, A. M.; Hambley, T. W.; Mann, D. R.; Snow, M. R. (1987), Theoretical analysis of the cobalt(III)-cobalt(II) tris[($\pm$)-1,2-propanediamine] electron-transfer reaction using molecular mechanics modeling of the configurational isomer distribution in both oxidation states, *Inorg. Chem.* **26**, 2257.

Charles, R.; Ganly-Cunningham, M.; Warren, R.; Zimmer, M. (1992), A modified MM2 force field for bleomycin analysis, *J. Mol. Structure* **265**, 385.

DiMeglio, C. M.; Ahmed, K. J.; Luck, L. A.; Weltin, E. E.; Rheingold, A. L.; Bushweller, C. H. (1992), Stereodynamics of sterically crowded metal-phosphine complexes: *trans*-[(t-Bu)$_2$P(i-Pr)]$_2$MCl$_2$ [M = Pt(II) and Pd(II)]. One-dimensional dynamic and two-dimensional chemical exchange NMR studies, X-ray crystallographic studies, molecular conformation trapping, and molecular mechanics calculations, *J. Phys. Chem.* **96**, 8765.

du Plooy, K. E.; Marais, C. F.; Carlton, L.; Hunter, R.; Boeyens, J. C. A.; Coville, N. J. (1989), Steric effects associated with monosubstituted cyclopentadienyl transition-metal complexes. Synthesis and NMR spectroscopic and molecular mechanics study of [(η^5-C$_5$H$_4$Bu^t)Fe(CO)(L)I] complexes and crystal structure determination of [(η^5-C$_5$H$_4$Bu^t)Fe(CO)(PPh$_3$)I], *Inorg. Chem.* **28**, 3855.

Fossheim, R.; Dahl, S. G. (1990), Molecular structure and dynamics of aminopolycarboxylates and their lanthanide ion complexes, *Acta Chem. Scand.* **44**, 698.

Hambley, T. W. (1991), Molecular mechanics analysis of the stereochemical factors influencing monofunctional and bifunctional binding of *cis*-diamminedichloroplatinum(II) to adenine and guanine nucleobases in the sequences d(GpApGpG)·d(CpCpTpC) and d(GpGpApG)·d(CpTp-CpC) of A- and B-DNA, *Inorg. Chem.* **30**, 937.

Hambley, T. W.; Snow, M. R. (1986), Crystal structure of (azacapten)cobalt(II) diperchlorate and strain energy minimization analyses of the (azacapten)cobalt(III) and (azacapten)cobalt(II) cations, *Inorg. Chem.* **25**, 1378.

Mustafi, D.; Telser, J.; Makinen, M. W. (1992), Molecular geometry of vanadyl–adenine nucleotide complexes determined by EPR, ENDOR, and molecular modeling, *J. Am. Chem. Soc.* **114**, 6219.

CHAPTER 8

Force Fields

8.1. INTRODUCTION

Underpinning all of the applications discussed in the previous chapters is the energy expression. The energy expression is the sum of potential energy terms, and is often referred to as a force field because it provides a mathematical representation for how each atom would move under the influence of the motions of all the other atoms in the system. That is, it is a parametric or functional representation for the forces that each atom experiences. In general, a force field is defined by an energy expression (set of functions), the choice of functional forms used, and the parameters that are used for specific combinations of atoms.

Molecular mechanics, in contrast to quantum mechanics, is a strictly empirical, or deductive, technique for describing the potential energy of a molecule. The potential energy is expressed in terms of discrete localized interactions such as bond stretches, bond angle bends, and torsional motions and interactions that extend over the entire system, such as van der Waals forces and electrostatics. That the terms in molecular mechanics are deduced can be seen by examining origins of the bond stretch term. Bjerrum (1914) proposed a linear restoring force for the interactions responsible for infrared (IR) spectroscopy. This led to a harmonic interaction potential; a bond stretch harmonic potential is given in Eq. (8.1).

$$V_r = \frac{1}{2}k_{IJ}(r - r_{IJ})^2 \tag{8.1}$$

In Eq. (8.1) k_{IJ} is the spring or force constant for the stretch between I and J, and r_{IJ} is the unstrained or natural bond distance. The observation of a compression in the IR overtone spectrum of diatomic molecules, which could not be explained by the harmonic potential of Bjerrum, led Morse (1929) to propose a new function to describe how the energy of a diatomic molecule would change upon increasing the internuclear separation, Eq. (8.2).

$$V_r = D_{IJ}(e^{-\alpha(r - r_{IJ})} - 1)^2 \tag{8.2}$$

385

The parameter D_{IJ} is the bond energy for the bond between centers I and J, and α is defined in terms of k_{IJ} and D_{IJ}, Eq. (8.3).

$$\alpha = \left[\frac{k_{IJ}}{2D_{IJ}} \right]^{1/2} \tag{8.3}$$

Support for this improved shape for the potential energy as a function of bond distance was provided by quantum mechanical plots of the total energy as a function of bond distance. Whether based on quantum mechanics or interpretation of experimental observation, the function was chosen based on an observation; it was not derived in advance, and hence it was a deduced quantity rather than an induced, predicted, or derived quantity.

First, Teller and Topley (1935), and then Kemp and Pitzer (1936) suggested a threefold torsional potential for ethane, Eq. (8.4).

$$V_\phi = \frac{1}{2} V_{IJKL} (1 - \cos 3\phi) \tag{8.4}$$

As with the bond stretch, the torsional potential was based on observation; in order to explain the entropy of ethane hindered rotation about the C−C bond had to be invoked.

Hill (1946) suggested the use of harmonic bond stretch, Eq. (8.1), and harmonic bond angle bend potentials, Eq. (8.5), for molecular deformations due to steric interactions

$$V_\theta = \frac{1}{2} k_{IJK} (\theta - \theta_{IJK})^2 \tag{8.5}$$

and the use of a Lennard-Jones van der Waals potential (Lennard-Jones, 1924), Eq. (8.6),

$$V = \frac{\lambda_m}{\rho^m} - \frac{\lambda_n}{\rho^n} \tag{8.6}$$

with an attractive exponent, n, of 6 and a repulsive exponent, m, of 12 for nonbond or steric interactions.

The currently used functions and functional forms of molecular mechanics were established during the 1940s and were also based on the available theoretical and experimental data. What remained to be established by Kemp and Pitzer and Hill was the particular choice of parameters, for example, the numerical values of k_{IJ} and r_{IJ} in Eqs. (8.1) and (8.2), the value of V_{IJKL} in Eq. (8.4), the values of k_{IJK} and θ_{IJK} in Eq. (8.5), and the values of λ_n and λ_m in Eq. (8.6). The approach used to obtaining parameters was to choose parameters to fit a particular set of experimental observations. For example, Kemp and Pitzer (1936) found that a torsional barrier [V_{IJKL} in Eq. (8.4)] of 3.15 kcal/mol gave the best reproduction of the entropy for ethane. The establishment of force field parameters continues as an issue today.

Implicit in the development of force field parameters is the concept of transferability. If all of the natural distances, bond angles, torsions, and inversions in a particular molecule are chosen to reproduce a particular molecular structure, then this structure will be reproduced exactly (assuming the force constants used are sufficiently large to pin the atoms in place), but these parameters are not likely to reproduce the structures of related molecules.

As evidenced by the applications discussed in the earlier chapters, one usually wants to model molecular systems where the 3-D structures are not known in advance. For molecular mechanics to be useful, more than a limited degree of transferability is mandated. One must choose functional forms and force constants that permit physically realistic distortions to occur so that parameters can be transferred from one structure to another. Fortunately, this is possible and several successful, transferable, force fields have been developed.

In this chapter, we discuss sources of data for parameterization and present a description of force field augmentation and design. Most of the common force fields in use today use the same basic set of functions in the energy expression. They are stretch, bend, torsion, inversion, van der Waals, and electrostatics. A number of the common force fields also use precisely the same equations and differ only in choice of parameters. In Section 3.2 a set of force fields that are used primarily for biological applications and have similar functional forms were summarized and compared. This set of force fields can be generically called harmonic force fields because harmonic stretch and harmonic bend functions are used. In Section 2.2 the premier force fields for organic applications, Allinger's MM2 and MM3 force fields, were summarized. These MM2 and MM3 force fields use the same set of functions as the harmonic force fields but also include anharmonic effects.

Often a molecular mechanics force field is used to study a molecule containing a particular combination of atoms for which the parameters do not exist. It is then up to the researcher to provide them. By understanding the underlying philosophy of that particular force field, new parameters for the specific combination of atoms of interest can be estimated. Techniques for parameter estimation will be provided in this chapter. Since force field parameters are not derived, data must be used in a fitting process. In Section 8.2 experimental and computational sources of data are discussed, and an assessment of the intrinsic error in both the theoretical and experimental data is given. Procedures for utilizing data in parameter estimation are overviewed in Section 8.3.

8.2. SOURCES OF PARAMETERIZATION DATA

Molecular structures to be fit or reproduced can be obtained from either experiment, or the more expensive semiempirical or ab initio electronic structure techniques. Experimental sources include X-ray (or neutron) diffraction studies on crystals, electron diffraction studies of gas-phase samples, microwave spectroscopy of gas-phase samples, or NMR spectroscopy on solution samples. Vibrational frequencies (to aid in the development of force constants) can be obtained experimentally or from electronic structure theoretical techniques. Molecular conformational energy differences, internal rotation barriers, and inversion barriers can be obtained experimentally from NMR coalescence data, vibrational spectroscopy, and the temperature dependence of thermodynamic data. These approaches can provide energy differences and barrier heights but cannot provide information about the detailed shapes of the potential energy curves. Energy differences, barrier heights, and the detailed shapes of the potential energy curves can be obtained from electronic structure theoretical techniques.

All types of data are not equally available. Structural data is most plentiful, with tabulations of X-ray structural data and gas-phase structural data being readily available. Gas-phase vibrational data is less common, though somewhat available. Solid state energetic data is relatively rare. Heats of sublimation are generally not available. Elastic moduli and phonon modes have also not been measured or tabulated for most materials. Thus, it is most straightforward to fit parameters to or compare the results of a force field calculation with gas-phase or solid state molecular structural data. Vibrational frequencies can be compared, though often one is comparing a condensed state experimental spectrum with a set of isolated molecule molecular mechanics vibrational frequencies. There is, however, precious little data for comparison of a van der Waals parameterization. Furthermore, the van der Waals parameters selected are not independent of the electrostatic parameters (partial charges). Electrostatic parameterization is itself difficult, being dependent on the same limited set of solid state data and the dubious extraction from molecular electric moments. It is unfortunate that the two major determinants of intermolecular interactions, van der Waals interactions, and electrostatics, are the least well defined from experimental data.

The question of accuracy of the data used to develop molecular mechanics force field parameters must be addressed. Whether the data comes from experiment or electronic structure calculation there are intrinsic limitations to the accuracy of the data, and hence limits to how well one can reproduce the data with a force field. A summary of the intrinsic accuracy of both experimental and theoretical sources of data is presented below.

8.2.1 Experimental Sources

There are three generally available experimental techniques for determining accurate molecular structure: X-ray (or neutron) diffraction, electron diffraction, and microwave spectroscopy. X-ray diffraction is carried out on crystalline samples of a material and has a well-defined estimation of error. The presence of systematic deviations and random error can be underscored by an analysis of the sample standard deviations in analogous bond distances collected in the Cambridge Structural Database (CSD). Allen et al. (1987) and Orpen et al. (1989) used the CSD to generate mean bond distances, Eq. (8.7), and sample standard deviations, Eq. (8.8), for a wide variety of combinations of atoms.

$$d = \sum_{i=1}^{n} \frac{d_i}{n} \tag{8.7}$$

$$\sigma = \sum_{i=1}^{n} \left[\frac{(d_i - d)^2}{(n - 1)} \right]^{1/2} \tag{8.8}$$

A sampling of this data for bonds involving C, N, and O is collected in Table 8.1. The distribution of the experimental data, $\sigma \geq 0.01$ Å, suggests that 0.01 Å represents a floor in the error in computed bond distances when they are compared to a set of X-ray data. Errors in a force field of less than 0.01 Å, relative to X-ray data, make little sense as one is inside the experimental sample standard deviation for virtually all bond distances. Furthermore, since X-rays are scattered by electron densities, X-ray diffraction determines the position of electron densities rather than nuclei. For elements other than hydrogen or helium this is not an issue, but X$-$H bond distances obtained by X-ray diffraction are underestimated (Ebsworth et al., 1987).

In addition to random experimental error, structures obtained from gas-phase spectroscopy are subject to systematic uncertainties and are not all equivalently useful for

389

TABLE 8.1. Mean Bond Distances d (Å), Standard Deviations σ (Å), and Sample Sizes n

Bond	Substructure	d	σ	n
Csp^3–Csp^3	C–CH$_2$Me	1.513	0.014	192
	C$_2$–CH–Me	1.524	0.015	226
	C$_3$–C–Me	1.534	0.011	825
	C–CH$_2$–CH$_2$–C	1.524	0.014	2459
	C–C overall	1.530	0.015	5777
Csp^3–Csp^2	Me–C=C	1.503	0.011	215
	C–CH$_2$–C=C	1.502	0.016	193
	C–C=C overall	1.507	0.015	1456
Csp^3–Car	Me–Car	1.506	0.011	454
	C–Car overall	1.513	0.014	1813
Csp^2–Csp^2	C=C–C=C overall	1.460	0.015	38
Car–Car	C–C overall	1.384	0.013	3264
Csp^3–Nsp^3	C–NH$_2$	1.469	0.010	19
	C$_2$NH	1.469	0.012	152
	C$_3$N	1.469	0.014	1042
	C–N overall	1.469	0.014	1201
Csp^2–Nsp^2	NH$_2$–C=O	1.325	0.009	32
	C–NH–C=O	1.334	0.011	78
	C^2–N–C=O	1.346	0.011	5
Csp^3–Osp^3	Me–OH	1.413	0.018	17
	C–CH$_2$–OH	1.426	0.011	75
	C$_2$–CH–OH	1.432	0.011	266
	C$_3$–C–OH	1.440	0.012	106
	C–OH overall	1.432	0.013	464
	C–O–C overall	1.426	0.019	236

comparison with the results of isolated molecule theoretical studies. For example, electron diffraction measures the average distances between nuclear positions in the gas phase; as such, electron diffraction is quite subject to vibrational effects such as shrinkage from bond angle bending and bond elongation due to anharmonicity in the bond stretch (Harmony et al., 1979). Bond distances obtained directly from the electron diffraction experiment or vibrationally corrected are different by a few thousandths of an angstrom. Microwave spectroscopy determines molecular moments of inertia from which (for simple molecules) distances can be derived (Harmony et al., 1979), again vibrational corrections need to be applied.

The electric moments (dipole, quadrapole, octapole, etc.) of a molecule can be used to fit atomic partial charges [the q_I values of Eq. (1.10)]. For example, for a gas-phase diatomic molecule such as NaCl, the dipole moment μ, in debye, can be expressed in terms of the bond distance r_{IJ}, in angstroms, and a partial charge q, in electron units (the same magnitude for each center), see Eq. (8.9).

$$\mu = 4.80324\, r_{IJ}\, q \tag{8.9}$$

For NaCl, a partial charge of $+0.7912$ on Na is obtained using an experimental dipole moment of 8.97141 D and an experimental bond distance of 2.360795 Å (Huber and Herzberg, 1979). The process becomes problematic for larger molecules, molecules with limited symmetry, and molecules with lone pairs of electrons. Because the dipole moment is a Cartesian vector it only contains three degrees of information. This means that dipole moments cannot be used to assign partial charges for molecules with even three unique atoms. For example, the dipole moment of HCN, 2.98 D, can be fit with charges of $+0.28$, 0, and -0.28 on H, C, and N, respectively, or with charges of 0, $+0.54$, and -0.54 on H, C, and N, respectively. Higher order moments can be included in the analysis for larger molecules but the effect of lone pairs is still a problem. For example, the dipole moment of carbon monoxide [r_e = 1.128323 Å, μ = 0.1222 D (Huber and Herzberg, 1979)] would suggest a transfer of 0.0225 electrons from oxygen to carbon, even though oxygen is substantially more electronegative than carbon. The culprit here is the lone pair on carbon, which extends out the back of the molecule, and this asymmetric electron distribution contributes significantly to the dipole moment (Billingley and Krauss, 1974).

8.2.2 Theoretical Sources

For small organic molecules (H_mABH_n) Hehre et al. (1986), reported average absolute errors for an ab initio Hartree–Fock wave funtion with the 6-31G* basis of 0.030 Å for AB single bonds, 0.018 Å for AB multiple bonds, 0.014 Å for AH bonds, and 1.5° for bond angles. Smaller basis sets gave correspondingly larger errors and inclusion of electron correlation decreased the error. Average absolute errors in distance of 0.054, 0.050, and 0.036 Å for modified intermediate neglect of diffractional overlap (MNDO), Austin model 1 (AM1), and Parameter model 3 (PM3), respectively, have been reported for the atomic associations that have been parameterized (Stewart, 1989). Bond angle errors of 4.3°, 3.3°, and 3.9° have also been reported for MNDO, AM1, and PM3, respectively. Reported dihedral angle absolute average errors are 21.6°, 12.5°, and 14.9° for MNDO, AM1, and PM3, respectively.

A comparison between experimental and theoretical vibrational frequencies has been reported (Hehre et al., 1986). Mean absolute percentage deviations of theoretical vibrational frequencies from experimental frequencies (uncorrected for anharmonicity effects) of 12.8% for an ab initio Hartree–Fock wave funtion with the 4-31G basis, 13.0% for an ab initio Hartree–Fock wave funtion with the 6-31*G basis, and 7.5% for an ab initio MP2 wave funtion with the 6-31*G basis were found. The theoretical frequencies obtained with a Hartree–Fock wave function are systematically high so a viable empirical correction can be applied to the theoretical frequencies. The "corrected" vibrational frequencies or underlying force constants can be used in force field parameterization.

Errors in internal rotation barriers (for an ab initio MP2 wave funtion with the 6-31*G basis) of 1.0 kcal/mol for BH_3-NH_3, 0.2 kcal/mol for ethane, 0.6 kcal/mol for methyl amine, and 2.4 kcal/mol for the cis barrier in hydrogen peroxide have been reported (Hehre et al., 1986). Errors in inversion barriers (for an ab initio MP2 wave funtion with the 6-31*G basis) of 0.8 kcal/mol for ammonia and 1.7 kcal/mol for methyl amine have also been recorded (Hehre et al., 1986). The shapes of the potential curves are likely reliable, but the magnitudes of the barrier heights are not of sufficient accuracy to permit direct force field parameterization.

Electronic structure techniques can also be used to obtain the partial charge parameters of a point charge model of electrostatics, the q values of Eq. (1.10). The simplest approach is to use the Mulliken partitioning (Mulliken, 1962) of the electron density, see Eq. (8.10),

$$n_a = \sum_{i=1}^{m} f_i \sum_{k=1}^{p_a} c^i_k \sum_{j=1}^{p_a} S_{jk} c^i_j \qquad (8.10)$$

where n_a is the number of electrons on center a, f_i is the occupation of orbital i (either 0, 1, or 2). The i summation runs over the m occupied orbitals in the wave function, and the j and k summations run over the p_a basis functions on center a in the wave function c^i_j is the coefficient for the jth basis function in the ith orbital, and S_{jk} is the overlap between basis functions j and k. The charges obtained are dependent on the basis set used, and the equal partitioning of electron density implicit in Eq. (8.10) has its limitations.

The wave function can also be used to evaluate the electrostatic potential, Eq. (8.11),

$$V(r_i) = \langle \Psi | \frac{1}{r_i} | \Psi \rangle \qquad (8.11)$$

over a grid of points r_i (Cox and Williams, 1981; Singh and Kollman, 1984; Chirlian and Frand, 1987; Besler et al., 1990; Breneman and Wiberg, 1990). This grid representation of the molecular electrostatic potential can be fit to a set of partial charges. Since the partial charge approximation is only valid outside the range of the electron density (inside the electron cloud penetration effects decrease the magnitude of the electron density seen by a point charge probe), the usual approach is to only use grid points outside the van der Waals radius of each center. For most molecules, using only those points that are outside the van der Waals radii of all the atoms causes some centers to be buried. For example, in carbon tetrachloride (CCl_4) virtually all of the points in a grid that is set at 1.5 times the van der Waals radii arise from the Cl atoms. Because there are very few grid points associated with the carbon, the carbon is not adequately represented in the data set of grid points (Stouch and Williams, 1993). This leads to the the carbon charge being ill-determined or arbitrary. Thus, the electrostatic potential approach has its own limitations.

8.3. FORCE FIELD AUGMENTATION AND DESIGN

It is rather rare that a person designs a force field from scratch. It is more common to add parameters for a particular functional group or atom type to an existing force field. In order to blend in with the existing parameterization of a given force field one must understand the underlying philosophy of that parameterization. A force field is, in a sense, a living interdependent structure, the precise form of which is determined by all of the types of functions, all of the functional forms, and all of the parameters. For example, since 1,4 nonbonded interactions are included for most force fields, the magnitudes of torsional barriers for each force field are not only dependent on the torsional parameters but also on the van der Waals parameters and the electrostatic terms. Changing a van der Waals radius or well depth will influence the torsional parameters. In this section we begin with descriptions of four different parameterization philosophies. This is followed by a discussion of augmentation procedures. This section concludes with a discussion of force field design strategy.

8.3.1 Parameterization Approaches

Most force fields use the same basic set of functions (stretch, bend, torsion, inversion, van der Waals, and electrostatics) (see Section 3.2). Furthermore, most force fields use the same basic functional forms (harmonic stretch and bend, cosine torsion, 6–12 or exp-6 van der Waals). The distinction between force fields then comes at the level of parameterization procedure. Here we discuss four different approaches to parameterization.

FITTING EXPERIMENTAL DATA

The first parameterization approach, and by far the most common, is to adjust force field parameters to reproduce experimental observables such as molecular structure, vibrational frequencies, internal rotation and inversion barriers, conformational energy differences, and occasionally heats of formation. The force fields MM2, MM3, AMBER, CHARMM, GROMOS, and Tripos 5.2 all use this approach. The primary difficulty with this method is the accuracy and availability of experimental data.

FITTING ELECTRONIC STRUCTURE DATA

To circumvent the problem of availability of experimental data, a second parameterization approach has been developed (Maple et al., 1988; Dasgupta and Goddard, 1989; Dinur and Hagler, 1990). With the ready availability of ab initio electronic structure codes, which have facilities for generating first and second derivatives of the energy with respect to molecular structure, it is possible to quickly generate a tremendous amount of theoretically derived structural, vibrational, and energetic data to be used in the fitting process. This method suffers from one of the same flaws that fitting experimental data does— accuracy of the data base. The electronic structure method that is generally used to obtain theoretically derived data (Hartree–Fock with a modest basis set) does not yield molecular structure, vibrational frequencies, barriers, or conformational energy differences of sufficient accuracy to be directly used in parameterization. The Discover force field (Maple et al., 1988; Dinur and Hagler, 1990) and the MSXX force field used by Karasawa et al. (1991) and Karasawa and Goddard (1992) for studies on polymers in Chapter 6 are examples of this parameterization approach. In both approaches the theoretically derived data was empirically corrected.

There is hope that gradient corrected local density functional (LDF) methods will help in the economic generation of force field parameters from electronic structure technologies (Sim et al., 1991; Ziegler, 1991; Fan and Ziegler, 1992; Johnson et al., 1993; Andzelm et al., 1993). Work validating LDF methods is progressing rapidly.

RULE-BASED PARAMETERIZATION

The third approach to parameterization uses chemical rules to develop parameters. The method is far older than either the purely empirical or purely theoretical methods. The rule-based approach forms one of the cornerstones of chemical "theory"—though it only recently has gained popularity in force field development. Radii have been used as a tool for understanding what "normal" bond distances were beginning with the formulation of a set of atomic radii by Bragg, the development of ionic radii by Landé, Wasastjerna, Goldschmidt, and Pauling, and the development of covalent radii (reviewed in Pauling, 1960b). These radii were developed as tools for understanding what "normal" bond distances were. These theorists obtained bond distances by the simple addition of the radii (ionic or covalent). Deviations from "normal" bond distances then revealed important concepts about bonding. For example, electronegativity corrections to "normal" bond distances were suggested by Schomaker and Stevenson (1941) to account for partial ionicity in covalent bonds, Eq. (8.12),

$$r_{IJ} = r_I + r_J - \beta \, | \chi_I - \chi_J | \tag{8.12}$$

where β is an adjustable parameter (set to 0.09 in the original paper) and χ_I is the electronegativity for center I. An early discussion of "carbonyl back-bonding" was based on a comparison of ideal covalent $M-C$ single-bond distances (based on covalent radii) with the substantially shorter experimental bond distances in metal carbonyls (Cotton and Wing, 1965).

In the 1929 paper, wherein the Morse bond stretch function was introduced, Morse (1929) also presented a relationship between bond distance and stretching force constant ($r \propto k^{1/3}$). This relationship was extended in 1934 by Badger (1934, 1935).

Beginning with the crystal energy method of Skorczyk (1976) and the DREIDING force field (Mayo et al., 1990) for general organic and biological systems, efforts have been directed at using chemical rules to devise force field parameters. In the universal force field (UFF), developed for the entire periodic table, bond distances were obtained from a combination of covalent radii, r_I, an electronegativity correction, r_{EN}, and a bond-order correction, r_{BO}, as given in Eq. (8.13) (Rappé et al., 1992; Casewit et al., 1992).

$$r_{IJ} = r_I + r_J + r_{BO} + r_{EN} \tag{8.13}$$

The electronegativity correction of O'Keeffe and Brese (1991), Eq. (8.14), was used

$$r_{EN} = \frac{r_I r_J (\sqrt{\chi_I} - \sqrt{\chi_J})^2}{(\chi_I r_I + \chi_J r_J)} \tag{8.14}$$

where χ_I is the electronegativity of center I. The bond-order correction, Eq. (8.15) was a Pauling-type term (Pauling, 1960a)

$$r_{BO} = -\lambda (r_I + r_J) \ln (n) \tag{8.15}$$

where n was the bond order between centers I and J, and λ an adjustable parameter.

In the VALBOND force field developed by Landis and co-workers, valence bond ideas were used to construct a set of rules to reduce the number of unique angle bend parameters for the entire p block of the periodic table from 427,500 force constants and natural angles (assuming three hybridizations per element, sp, sp^2, and sp^3) to 650 weighting factors and constants (a 660-fold reduction in the number of possible parameters) (Root et al., 1993).

The advantage of rule-based parameterizations rests in the reduction of the number of parameters to be determined. The strictly empirical and theoretical parameterization approaches hinge on the availability of data and the quality of that data. If the number of parameters to be determined is decreased, the dependence on data is decreased. Furthermore, for a rule-based force field, new atom types can be added and integrated in with minimal effort.

SIMPLE ASSIGNMENT

The fourth approach, typically used for force constants, is to assume that force constants are indeed constant and independent of atom type. Andrews (1930) suggested that stretching force constants were all approximately 600 kcal/mol-Å^2 and that variation in Raman frequencies was due to variations in mass. Gollogly and Hawkins (1968) suggested a bend force constant of 100 kcal/mol-rad^2. The DREIDING force field assigned all stretch terms a force constant of 700 kcal/mol-Å^2 and all bend terms a force constants 100 kcal/mol-rad^2 (Mayo et al., 1990). The presence of single digit parameters in the parameter listings for each of the harmonic force fields of Section 3.2 suggests a degree of "assignment" in all of these force fields.

Simple assignment is not wrong; it merely is an admission there is a limited amount of data, there is a limited amount of time, and some parameters are thought to be more important than others in determining the properties of interest. For example, the DREIDING force field used assigned force constants for stretch and bend, but developed an elaborate set of rules for torsional barriers, and used van der Waals parameters that were fit to experimental data. The underlying assumption was that the low-energy interactions such as torsional motions and nonbonded terms dominated intermolecular steric interactions, and hence were more important. Thus these interactions needed to be determined more precisely than stretch or bend force constants.

8.3.2 Augmentation Strategies

At this point, we can address the scenario posed in Section 8.1. You, the researcher, are using an established force field to study an important problem that contains a particular combination of atoms for which parameters do not exist. How are these parameters to be selected? The proper approach would be to blend into the force field. By knowing the procedures used to generate the existing parameters one will know how to generate the needed parameters. For example, if the parameterization of the force field being augmented was based on fitting experimental data, the proper approach would be to collect the appropriate experimental data and adjust the augmented parameters to reproduce the collected data. If the parameterization was based on fitting theoretical data, then the proper approach would be to fire up an electronic structure code to generate the appropriate data. In practice, augmentations are rarely this elaborate. When one begins a study on a new system one is usually interested in a "quick and dirty" solution to see if molecular mechanics can address the question of interest. If one always spent the time necessary to carry out a correct parameterization each time force field parameters were missing, one would soon be looking for alternate sources of employment.

The pragmatic approach to force field augmentation is to first scan the set of published parameters for the given force field looking for similar combinations of atoms. If a similar combination of atoms (same hybridization, analogous functional groups, etc.) is found, the parameters could be used as is. If a similar combination of atoms is not found, then simple assignment of force constants is often used in conjunction with either setting or fitting the geometric parameter (distance or angle) to an experimentally observed value. Fitting the geometric parameter would take more time than simply setting the geometric parameter to the experimental value, but fitting would reflect the nonbonded interactions present in the experimentally observed molecule.

If the preliminary molecular mechanics study, carried out with first guess parameters, did lead to an interesting result, then one could test the sensitivity of the result to the choice of parameters by arbitrarily modifying the parameters by some amount, say 10%, and redoing the calculation. If the result is sensitive to the choice of parameters, then it might be necessary to carry out the "correct" augmentation.

8.3.3 Design Strategies

A force field is defined by a general energy expression (set of functions), functional forms, and parameters. In developing a force field, the first and most global decision is which functions should be included. There is almost universal agreement that bond stretch terms are needed and that van der Waals interactions must be included, but what about the remaining terms? For example, should angle bend terms be used at metal centers, is an explicit hydrogen-bonding term needed, does one need to have an explicit inversion term, and should electrostatic interactions be included?

Once the types of functions to be included are chosen, the next most general decision is the precise choice of functional forms for each of the terms in the force field. Given the underlying concern over computational efficiency (expense) this decision boils down to a desire for either a precise description of the potential surface near the minimum (near-exact reproduction of molecular structure), or a need for a more global description of the potential surface suitable for MD or molecular structures far removed from equilibrium. Factored into this decision for each functional form is the number of terms to be calculated for this function and the percentage of the total effort involved for this term. For example, the number of stretch terms in a molecule grows as roughly $2n$, where n is the number of atoms. This can be contrasted to the $\dfrac{n(n-1)}{2}$ dependence of the nonbond terms. Practically anything short of a large basis ab initio electronic structure calculation could be carried out for the stretch terms and still take less time than the nonbond terms.

The remaining decision is how to obtain the numerical values of the parameters dictated by the choice of functional forms. Underlying this decision is the question of degree of transferability and the acceptable level of error in the computed structures. Any of the parameterization approaches discussed above could be used, though it worth remembering the difficulties associated with determining van der Waals and electrostatic parameters.

Homework

8.1. Calculate the dipole moment based partial charges for NO. Use a bond distance of 1.15077 Å and a dipole moment of 0.1587 D (N^+O^-) and Eq. (8.9). How do these charges compare with those in Section 8.2 for carbon monoxide and do they agree with your electronegativity-based intuition?

8.2. Technetium is a radioactive element for which realtively little structural information is available. Nonetheless it is important in the radiopharmaceutical industry. Which of the parameter augmentation methods would allow you to directly develop parameters for technetium? Which ones allow development through interpolation using parameters from other metals?

8.3. Use the following published MM2 parameters to interpolate C–Ge distance and force constant parameters.

Atom Pair	r_{ij} (Å)	k_{ij} (kcal/mol-A²)
C–Si	1.880	427.3
C–Sn	2.147	305.6
C–Pb	2.2417	273.4

8.4. For diatomic molecules a harmonic estimate of vibrational frequencies can be obtained from Eq. (8.16), where k is the force constant in kilocalories per mole per square angstrom, μ is the reduced mass [$\mu = \dfrac{m_1 m_2}{m_1 + m_2}$], and 108.593 converts to frequencies in reciprocal centimiters.

$$\omega = 108.593 \sqrt{\frac{k}{\mu}} \tag{8.16}$$

Estimate force constants for the following diatomics:

Molecule	ω
H_2	4401.213
F_2	916.64
Cl_2	559.72
Br_2	325.321
I_2	214.502
HF	4138.32
HCl	2990.9463
HBr	2648.975
HI	2309.014

Atomic weights: $^1H = 1.0078$, $^{19}F = 18.9984$, $^{35}Cl = 34.9688$, $^{79}Br = 78.9183$, $I = 126.9044$.

8.5. Use the data you generated in Problem 8.4 and the following distance data:

Molecule	R (Å)
H_2	0.74144
F_2	1.41193
Cl_2	1.9879
Br_2	2.28105
I_2	2.6663

to obtain the homonuclear proportionality constants of Badgers rules ($r \propto k^{1/3}$). Use these constants to estimate the heteronuclear frequencies of Problem 8.4 using the following distance data:

Molecule	R (Å)
HF	0.916808
HCl	1.274552
HBr	1.414435
HI	1.60916

References

Allen, F. H.; Kennard, O.; Watson, D. G.; Brammer, L.; Orpen, A. G.; Taylor, R. (1987), Tables of bond lengths determined by X-ray and neutron diffraction. Part 1. Bond lengths in organic compounds, *J. Chem. Soc. Perkin Trans. 2*. S1.

Andrews, D. H. (1930), The relation between the Raman spectra and the structure of organic molecules, *Phys. Rev.* **30**, 544.

Andzelm, J.; Sosa, C.; Eades, R. A. (1993), Theoretical study of chemical reactions using density functional methods with nonlocal corrections, *J. Phys. Chem.* **97**, 4664.

Badger, R. M. (1934), A relationship between internuclear distances and bond force constants, *J. Chem. Phys.* **2**, 2128.

Badger, R. M. (1935), The relation between the internuclear distances and force constants of molecules and its applications to polyatomic molecules, *J. Chem. Phys.* **3**, 710.

Besler, B. H.; Merz, K. M.; Kollman, P. A. (1990), Atomic charges derived from semiempirical methods, *J. Comp. Chem.* **11**, 431.

Billingsley II, F. P.; Krauss, M. (1974), Mulitconfiguration self-consistent-field calculation of the dipole moment function of CO ($X\ ^1\Sigma^+$), *J. Chem. Phys.* **60**, 4130.

Bjerrum, N. (1914), Uber die ultraroten spektren der gase. III. Die konfiguration des kohlendioxyd-molekuls und de gestze der intramolekularen krafte. *Verhandl. Deut. Physik. Ges.* **16**, 737.

Breneman, C. M.; Wiberg, K. B. (1990), Determining atom-centered monopoles from molecular electrostatic potentials. The need for high sampling density in formamide conformational analysis, *J. Comp. Chem.* **11**, 361.

Casewit, C. J.; Colwell, K. S.; Rappé, A. K. (1992), Application of a universal force field to organic molecules, *J. Am. Chem. Soc.* **114**, 10035.

Chirlian, L. E.; Francl, M. M. (1987), Atomic charges derived from electostatic potentials: a detailed study, *J. Comp. Chem.* **8**, 894.

Cotton, F. A.; Wing, R. M. (1965), The crystal and molecular structure of *cis*-(diethylenetriamine) molybdenum tricarbonyl; the dependence of Mo−C bond length on bond order, *Inorg. Chem.* **4**, 314.

Cox, S. R.; Williams, D. E. (1981), Representation of the molecular electrostatic potential by a net atomic charge model, *J. Comp. Chem.* **2**, 304.

Dasgupta, S.; Goddard III, W. A. (1989), Hessian-biased force fields from combining theory and experiment, *J. Chem. Phys.* **90**, 7207.

Dinur, U.; Hagler, A. T. (1990), A novel decomposition of torsional potentials into pairwise interactions. A study of energy second derivatives, *J. Comp. Chem.* **11**, 1234.

Ebsworth, E. A. V.; Rankin, D. W. H.; Cradock S. (1987), *Structural Methods in Inorganic Chemistry*, Blackwell Scientific, Boston, p. 331.

Fan, L.; Ziegler, T. (1992), Application of density functional theory to infrared absorption intensity calculations on main group molecules, *J. Chem. Phys.*, **96**, 9005.

Gollogly, J. R.; Hawkins, C. J. (1968), Conformational analysis of coordination compounds. II. Substituted five-membered diamine chelate rings, *Inorg. Chem.* **8**, 1168.

Harmony, M. D.; Laurie, V. W.; Kuczkowski, R. L.; Schwendeman, R. H.; Ramsay, D. A.; Lovas, F. J.; Lafferty, W. J.; Maki, A. G. (1979), Molecular structures of gas-phase polyatomic molecules determined by spectroscopic methods, *J. Phys. Chem. Ref. Data* **8**, 619.

Hehre, W. J.; Radom, L.; Schleyer, P. v. R.; Pople, J. A. (1986), *Ab Initio Molecular Orbital Theory*, Wiley, New York.

Hill, T. L (1946), On steric effects, *J. Chem. Phys.* **14**, 465.

Huber, K.; Herzberg, G. K. (1979) *Constants of Diatomic Molecules*, Van Nostrand-Reinhold, New York.

Johnson, B. G.; Gill P. M. W.; Pople, J. A. (1993), The performance of a family of density functional methods, *J. Chem. Phys.* **98**, 5612.

Karasawa, N.; Dasgupta, S.; Goddard III, W. A. (1991), Mechanical properties and force field parameters for polyethylene crystal, *J. Phys. Chem.* **95**, 2260.

Karasawa, N.; Goddard III, W. A. (1992), Force fields, structures, and properties of poly(vinylidene fluoride) crystals, *Macromolecules* **25**, 7268.

Kemp, J. D.; Pitzer, K. S. (1936), Hindered rotation of the methyl groups in ethane, *J. Chem. Phys.* **4**, 749.

Lennard-Jones, J. E. (1924), On the determination of molecular fields.-II. From the equation of state of a gas *Proc. R. Soc. London, Ser. A* **106**, 463.

Maple, J. R.; Dinur, U.; Hagler, A. T. (1988), Derivation of force fields from molecular mechanics and dynamics from *ab initio* energy surfaces, *Proc. Natl. Acad. Sci., U.S.A.* **85**, 5350.

Mayo, S. L.; Olafson, B. D.; Goddard III, W. A. (1990), DREIDING: A generic force field for molecular simulations, *J. Phys. Chem.* **94**, 8897.

Morse, P. M. (1929), Diatomic molecules according to the wave mechanics. II. Vibrational levels, *Phys. Rev.* **34**, 57.

Mulliken, R. S. (1962), Critera for the construction of good self-consistent-field molecular orbital wave functions, and the significance of LCAO–MO population analysis, *J. Chem. Phys.* **36**, 3428.

O'Keefe, M.; Brese, N. E. (1991), Atom sizes and bond lengths in molecules and crystals, *J. Am. Chem. Soc.* **113**, 3226.

Orpen, A. G.; Brammer, L.; Allen, F. H.; Kennard, O.; Watson, D. G.; Taylor, R. (1989), Tables of bond lengths determined by X-ray and neutron diffraction. Part 2. Organometallic compounds and co-ordination complexes of the *d*- and *f*-block metals, *J. Chem. Soc. Dalton Trans.* S1.

Pauling, L (1960a), *The Nature of the Chemical Bond*, Cornell University Press, Ithaca, NY, p. 239.

Pauling, L (1960b), The sizes of ions and the structure of ionic crystals. In *The Nature of the Chemical Bond*, Cornell University Press, Ithaca, NY, pp. 511–532.

Rappé, A. K.; Casewit, C. J.; Colwell, K. S.; Goddard III, W. A.; Skiff, W.M. (1992), UFF, a full periodic table force field for molecular mechanics and molecular dynamics simulations, *J. Am. Chem. Soc.* **114**, 10024.

Root, D. M.; Landis, C. R.; Cleveland, T. (1993), Valence bond concepts applied to the molecular mechanics description of molecular shapes. 1. Application to nonhypervalent molecules of the *p*-block, *J. Am. Chem. Soc.* **115**, 4201.

Sim, F.; Salahub, D. R.; Chin, S.; Dupuis, M. (1991), Gaussian density functional calculations on the allyl and polyene radicals: C_3H_5 to $C_{11}H_{13}$, *J. Chem. Phys.* **95**, 4317.

Singh, U. C.; Kollman, P. (1984), An approach to computing electrostatic charges for molecules, *J. Comp.Chem.* **5**, 129.

Schomaker, V.; Stevenson, D. P. (1941), Some revisions of the covalent radii and the additivity rule for the lengths of partially ionic single covalent bonds, *J. Am. Chem. Soc.* **63**, 37.

Skorczyk, R. (1976), The calculation of crystal energies as an aid in structural chemistry. I. A semiempirical potential-field model with atomic constants as parameters, *Acta Crystallogr. Sect. A* **32**, 447.

Stewart, J. J. P. (1989), Optimization of parameters for semiempirical methods II. applications, *J. Comput. Chem.* **10**, 221.

Stouch, T. R.; Williams, D. E. (1993), Conformational dependence of electrostatic potential-derived charges: Studies of the fitting procedure, *J. Comp. Comput. Chem.* **14**, 858.

Teller, E.; Topley, B. (1935), On the equilibrium and the heat of reaction $C_2H_4 + H_2 \rightleftharpoons C_2H_6$, *J. Chem. Soc.* 876.

Ziegler, T. (1991), Approximate density functional theory as a practical tool in molecular energetics and dynamics, *Chem. Rev.* **91**, 651.

Further Reading

Bowen, J. P.; Allinger, N. L. (1991), Molecular mechanics: The art and science of parameterization. In *Reviews in Computational Chemistry*, Vol. 2, Lipkowitz, K. B.; Boyd, D. B., Eds., VCH, New York, p. 81.

Dinur, U.; Hagler, A. T. (1991), New approaches to empirical force fields. In *Reviews in Computational Chemistry*, Vol. 2, Lipkowitz, K. B.; Boyd, D. B., Eds., VCH, New York, p. 99.

Lipkowitz, K. B.; Osawa, E. (1995), Published force field parameters. In *Reviews in Computational Chemistry*, Vol. 6, Lipkowitz, K. B.; Boyd, D. B., Eds., VCH., New York, p. 1.

Murrell, J. N.; Carter, S.; Farantos, S. C.; Huxley, P.; Varandas, A. J. C. (1984), *Molecular Potential Energy Functions*, Wiley, New York.

Rigby, M.; Smith, E. B.; Wakeman, W. A.; Maitland G. C. (1986), *The Forces Between Molecules*, Oxford University Press, Oxford, UK.

Stereochemical Terms

Molecular structure can be described by the four "C's": composition, connectivity, configuration, and conformation. This Appendix focuses on the language of the third "C," configuration. Because sp^3 carbon makes four bonds and has a tetrahedral geometry, the interchange of substituents in 3-D space can lead to molecules that share connectivities but are not spatially the same or superimposable. These molecules are configurational isomers or stereoisomers. The atoms from which stereoisomers are produced by the interchange of two of their substituents are called stereogenic atoms (Mislow and Siegel, 1984).

CHIRAL MOLECULES. A molecule is *chiral* if it cannot be superimposed on its mirror image and *achiral* if it can be. A molecule can be established as being chiral or achiral from the symmetry elements present in the molecule. A molecule is achiral if an S_n symmetry operation transforms the molecule into itself and chiral otherwise. The most common S_n symmetry operations are S_1, reflection through a mirror plane and S_2, inversion through a point. Even though the terms chiral and *asymmetric* are used somewhat interchangably they are not the same. Asymmetric molecules lack any symmetry. Chiral molecules are *dissymmetric* because they lack an S_n symmetry operation but they may still possess symmetry, for example, they may have C_2 symmetry. For example, the puckered amine **A.1a** is C_2 symmetric but cannot be superimposed on its mirror-image **A.1b**.

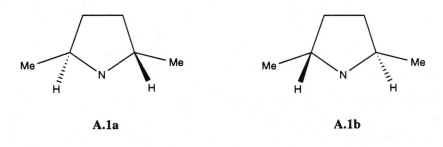

A.1a	**A.1b**

ENANTIOMERS. When substituents connected to a stereogenic center are interchanged, the new stereoisomer is an *enantiomer* of the original molecule if the molecule contains only one stereogenic center.

401

DIASTEREOMERS. When substituents connected to a stereogenic center are interchanged, the new stereoisomer is a *diasteoromer* of the unperturbed molecule if the structure contains two or more stereogenic centers.

ACHIRAL CONFIGURATIONAL ISOMERS. Configurational isomers are interesting even when chirality is not involved. Examples include the isotactic or syndiotactic polymers discussed in Sections 6.3.4 and 7.5. Isotactic or syndiotactic polymers are not chiral, but the pairwise relative stereochemistries of the side chains are distinguishable by NMR and the long-range repetition of stereochemistry,or stereoregularity, determines the physical properties of the material.

PROSTEREOGENIC CENTERS. Substituents attached to centers that become stereogenic upon single substitution are called *prostereogenic* or "prochiral." If one of the two methylene hydrogen atoms in proprionic acid is substituted by a halogen, the methylene center becomes stereogenic and the molecule chiral, see Eq. (A.1). The stereogenic center is indicated by a star. The methylene center in the reactant is prostereogenic.

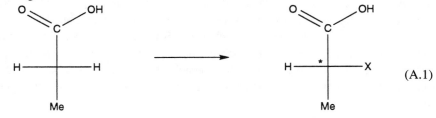

(A.1)

The methylene protons are called *enantiotopic*. If one of the two methylene protons in 3-chlorobutanoic acid is substituted by a halogen the product has two stereogenic centers and is a member of a diastereomer pair , see Eq. (A.2). The methylene protons in the reactant are called *diastereotopic*.

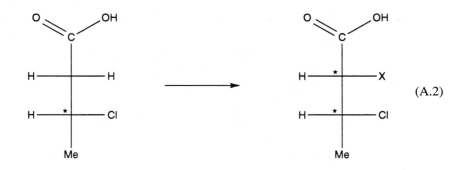

(A.2)

PROSTEREOGENIC FACES. When reagents are added across C=C and C=O π bond stereogenic centers can be created. The C=C or C=O π faces are called prostereogenic faces or "prochiral" faces. When the addition reaction gives a molecule with a single stereogenic center, the π face is called *enantiotopic*, see Eqs. (A.3) and (A.4).

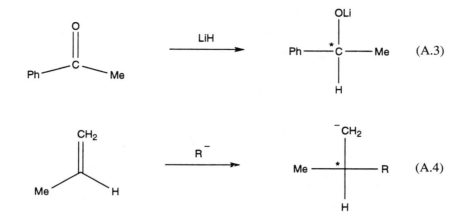

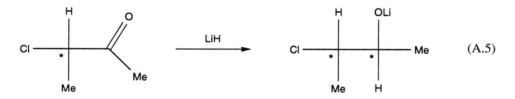

When the addition reaction yields a molecule with two or more stereogenic centers, the π face is called *diastereotopic*, see Eq. (A.5).

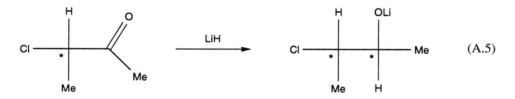

Enantiotopic substituents or π faces react differently and provide chiral products only in chiral environments. The chiral environment can be provided by chiral reagents, catalysts, or solvents. The transition state modeling case studes in Sections 2.4 and 7.5 discussed additions across enantiotopic carbonyl and alkene π faces, respectively. Diastereotopic substituents or π faces show differing reactivity patterns under almost all conditions. They do not require a chiral environment.

References

Mislow, K.; Siegel, J. (1984), Stereoisomerism and local chirality, *J. Am. Chem. Soc.* **106**, 3319.

General References

Juaristi, E. (1991), *Introduction to Stereochemistry and Conformational Analysis*, Wiley, New York.
Eliel, E. L.; Wilen, S. H. (1994), *Stereochemistry of Organic Compounds*, Wiley, New York.

Thermodynamic Corrections

The energetic results of a molecular mechanics study often require manipulation to bring the theoretical results to the same set of thermodynamic conditions, primarily temperature, under which the experiments were carried out. The two major corrections are a correction for zero-point motion and the temperature-dependent population of rotational, vibrational, and electronic states.

ZERO-POINT CORRECTION. The most basic correction is for zero-point vibrational motion. As shown in Figure B.1, the lowest energy vibrational level, the zeroth vibrational level, is well above the bottom of the potential well. Due to quan-

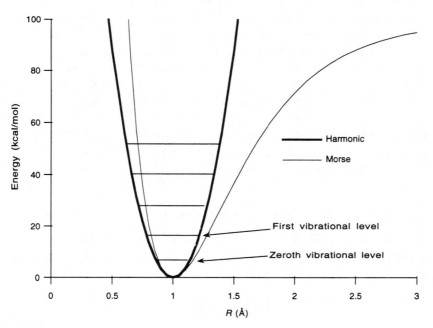

FIGURE B.1.
Harmonic potential curve and harmonic vibrational progression superimposed on a Morse potential curve.

TABLE B.1. Harmonic Zero-Point Vibrational Energies[a]

Molecule	Vibational Frequency (cm^{-1})	Zero-Point Energy (kcal/mol)
H_2	4401.21	6.3
D_2	3115.50	4.5
CO	2169.81	3.1
N_2	2358.57	3.4

[a] Zero-point energy (kcal/mol) $= \dfrac{1}{699.5} \times \nu$ (cm^{-1}).

tum mechanics this zeroth vibrational level is as low in energy as the system can get experimentally. The minima of molecular mechanics correspond to the bottom of the potential well. To compare theoretical energy differences to experimental energy differences, the zero-point energy must either be added to the theoretical results or subtracted from the experimental results. As the data in Table B.1 demonstrate, zero-point effects can be quite large. For a polyatomic molecule the harmonic approximation to the zero-point energy can be obtained from Eq. (B.1) where the zero-point energy is expressed in kilocalories per mole, the frequencies in reciprocal centimeters, and 699.5 is a conversion factor.

$$\text{Zero point energy} = \frac{1}{699.5} \sum_i \nu_i \tag{B.1}$$

IDEAL GAS STATISTICAL THERMODYNAMICS. Energetic differences between molecular conformations are of interest in molecular mechanics. Experimentally, conformational energy differences are obtained by measuring the difference in a molecular property that can be equated to equilibrium population differences among the set of conformations at one or more temperatures. For example, for butane the *trans* and the two *gauche* conformations (g^+ and g^-) have different vibrational frequencies. The ratios of the intensities of comparable bands for each conformation can be used to find the ratios of the populations of the two conformations. Murphy et al. (1991) obtained a mole fraction of 0.62 $\pm$ 0.04 for the *trans* conformation of butane at room temperature by fitting the relative vibrational intensities. Assuming that only the *trans* and the two *gauche* conformations are populated at room temperature, each of the *gauche* conformations would have a mole fraction of 0.19. From this ratio $\left(\dfrac{0.19}{0.62}\right)$, the difference in free energies of the two conformations can be obtained using Eq. (B.2), where RT is 0.592 kcal/mol at 298 K [$R = Nk = 1.9872$ cal K^{-1}, where N is Avogadro's number (6.02252×10^{23}) and k is the Boltzmann constant (3.29957×10^{-24} cal K^{-1})].

$$\Delta G = -RT \ln \frac{p_1}{p_2} \tag{B.2}$$

The free energy difference of 0.70 kcal/mol obtained from Eq. (B.2) can be decomposed into enthalpic and entropic contributions by Eq. (B.3) and the enthalpy difference can be related to the internal energy ΔU by Eq. (B.4), where $\Delta(PV)$ is associated with pressure–volume work.

$$\Delta G = \Delta H - T \, \Delta S \tag{B.3}$$

$$\Delta H = \Delta U + \Delta(PV) \tag{B.4}$$

For isolated molecules or ideal gas systems, the PV term can be ignored and changes in the enthalpy of the system directly associated with changes in internal energy, ΔU. For our purposes ΔU is the same as differences in molecular mechanics energies, ΔE.

The temperature dependence of the internal energy of a system can be found by taking an average of the energies of all possible states or configurations of the system, see Eq. (B.5), where the energy E_i of each configuration is weighted by its probability of occurring, P_i, at the particular temperature of interest, T.

$$\langle E \rangle + PV = \Delta H = \sum_i P_i E_i + PV \tag{B.5}$$

The parameter P_i is given by the Boltzmann equation, Eq. (B.6), again we will usually ignore the PV term.

$$P_i = \frac{f_i e^{\frac{-E_i}{kT}}}{\sum_j f_j e^{\frac{-E_j}{kT}}} \tag{B.6}$$

In Eq. (B.6), P_i is the probability or population of conformation i, f_i is the number of states or conformations of energy E_i (or the degeneracy of conformation i), and the j summation is over all the conformations. Note the summations over states should include rotational states, vibrational states, electronic states, and molecular conformations.

At 0 K only the lowest energy state is populated, but as the temperature rises higher energy states will be populated and the internal energy, $\langle E \rangle$, will rise. Furthermore, as the number of populated states increases the number of distinguishable combinations of populated states increases. This leads to an entropic contribution to the free energy of the system. In Section 6.5 the number of distinguishable, equivalent arrangements of a system, W, was related to the entropy of a system by Eq. (B.7).

$$S = k \ln W \tag{B.7}$$

In general, the free energy can be expressed as in Eq. (B.8), where Q is the partition function of the system, given by Eq. (B.9), the energies E_i have an arbitrary zero, and the i summation is over all the energy states of the sytem.

$$G = -kT \ln Q + PV \tag{B.8}$$

$$Q = \sum_i f_i e^{-E_i/kT} \tag{B.9}$$

If the energies for all of the states, i, are the same and the PV term is ignored, then Eq. (B.9) will reduce to Eq. (B.7) because the exponentials would all be 1.

If Eq. (B.8) is differentiated with respect to T and the E_i values are assumed to be temperature independent $\left[\text{remember } d \ln (x) = \frac{1}{x} dx \right]$, then comparison of this derivative with Eq. (B.5) leads to an expression for $\langle E \rangle$ or ΔH in terms of Q, given by Eq. (B.10).

$$\Delta H = kT^2 \left(\frac{\partial \ln Q}{\partial T} \right)_{E_i} \tag{B.10}$$

The entropy of the system, S, can also be related to the partition function, Q [by Eq. (B.11)].

$$S = kT \left(\frac{\partial \ln Q}{\partial T} \right)_{E_i} + k \ln Q \tag{B.11}$$

Since the free energy, ΔG, the enthalpy, ΔH, and the entropy of the system, S, can all be obtained from the partition function, Q, a discussion of the expressions for the partition function in terms of quantities obtained from molecular mechanics is in order.

It is not a bad approximation to assume that the translational, rotational, and vibrational motions of a molecule are independent and the energies associated with each can be considered separately. This assumption, along with the multiplicative nature of probabilities, leads to Eq. (B.12).

$$Q = Q_{\text{trans}} Q_{\text{rot}} Q_{\text{vib}} \tag{B.12}$$

Combining Eq. (B.12) with Eq. (B.8) yields Eq. (B.13), suggesting that the translational, rotational, and vibrational contributions to the free energy can be separately determined.

$$G = -kT (\ln Q_{\text{trans}} + \ln Q_{\text{rot}} + \ln Q_{\text{vib}}) \tag{B.13}$$

The translational partition function is given by Eq. (B.14).

$$Q_{\text{trans}} = \frac{V^N}{N!} \left(\frac{2\pi MkT}{h^2} \right)^{3N/2} \tag{B.14}$$

The rotational partition function for nonlinear polyatomic molecules is given by Eq. (B.15).

$$Q_{\text{rot}} = \frac{\sqrt{\pi}}{\sigma_\epsilon} \left(\frac{8\pi^2 I_m^3 kT}{h^2} \right)^{3N/2} \tag{B.15}$$

In Eq. (B.15), I_m^3 is the product of the three principal moments of inertia and σ_ϵ is the external symmetry number of the molecule. The vibrational partition function for nonlinear polyatomic molecules is given by Eq. (B.16).

$$Q_{\text{vib}} = \prod_i \frac{1}{(1 - e^{-h\nu_i/kT})} \tag{B.16}$$

For systems with low-energy excited states or molecular conformations, the partition function is not separable into the product form given in Eq. (B.12) because the rotational and vibrational frequencies (or energy levels) are not independent of the electronic excited state or molecular conformation. However, the free energy can be taken as a Boltzmann weighted combination of individual free energies. This can be seen from a bit of mathematical manipulation. For a system with more than one low-energy state the more general partition function expression given in Eq. (B.17) must be used where the summation is over electronic states or molecular conformations.

$$Q = Q_{trans} \sum_i f_i e^{-E_i/kT} Q_{i \, rot} Q_{i \, vib} \tag{B.17}$$

The enthalpic contribution to free energy is straightfoward. Insertion of Eq. (B.17) into Eq. (B.10) leads to the recovery of Eq. (B.5), that is, the enthalpy of a molecule with multiple conformations is simply the Boltzmann weighted average of enthalpies of the individual conformations. To obtain the entropy for this type of system, let us place the Q_{trans} term of Eq. (B.17) inside the summation, yielding Eq. (B.18).

$$Q = \sum_i Q_i \tag{B.18}$$

Inserting Eq. (B.18) into Eq. (B.11), differentiation of $\ln Q$, and insertion of the identity for Q arising from Eq. (B.9) yields Eq. (B.19) (ignoring the f_i term).

$$S = k \left\{ \frac{1}{kT} \frac{\sum_i E_i e^{-E_i/kT}}{Q} + \ln Q \frac{\sum_i e^{-E_i/kT}}{Q} \right\} \tag{B.19}$$

By using Eq. (B.6), the two summations of Eq. (B.19) can be combined, yielding Eq. (B.20).

$$S = k \sum_i P_i \left(\frac{E_i}{kT} + \ln Q \right) \tag{B.20}$$

Taking the log of Eq. (B.6) ($\ln P_i = -\frac{E_i}{kT} - \ln Q$) and insertion into Eq. (B.20) yields Eq. (B.21).

$$S = -k \sum_i P_i \ln P_i \tag{B.21}$$

Equation (B.21) indicates that the entropy of a set of conformations can be obtained from the probabilities of the individual conformations.

References

Benson, S. W (1976), *Thermochemical Kinetics*, Wiley-Interscience, New York.

Chandler, D. (1987), *Introduction to Modern Statistical Mechanics*, Oxford University Press, New York.

Hill, T. L. (1960), *An introduction to Statistical Thermodynamics*, Addison-Wesley, Reading MA.

Knox, J. H. (1978), *Molecular Thermodynamics*, Wiley, New York.

McQuarrie, D. A. (1976), *Statistical Mechanics*, Harper Collins, New York.

Murphy, W. T.; Fernandez-Sanchez, J. M.; Raghavachari, K. (1991), Harmonic force field and raman scattering intensity parameters of *n*-butane. *J. Phys. Chem.* **95**, 1124.

Whalen, J. W. (1991), *Molecular Thermodynamics: A Statistical Approach*, Wiley, New York.

Molecular Dynamics

Molecular motion is most easily discussed in terms of the positions of the particles in a system, denoted $\mathbf{r}$, and the time dependence of the positions of the particles, the velocities, denoted $\dot{\mathbf{r}}$, where the dot denotes differentiation with respect to time and the boldface lettering denotes a multidimensionaal variable. (McCammon and Harvey 1987, van Gursteven and Berendson 1990) The time dependence of the velocities, the accelerations, denoted $\ddot{\mathbf{r}}$, are obtained from the forces exerted upon a given atom by all the other atoms in the system (using Newtonian mechanics, force equals mass times acceleration). The forces are given by the negative of the gradient of the n particle potential energy function with respect to the positions of the particles, shown for particle i in Eq. (C.1).

$$\mathbf{F}_i = -\frac{\partial V}{\partial \mathbf{r}_i} \tag{C.1}$$

The Newtonian equation of motion for particle i is given by Eq. (C.2), where m_i is the mass of particle i.

$$m_i \ddot{\mathbf{r}}_i = \mathbf{F}_i \tag{C.2}$$

The trick is turning the gradient information (also used in an energy minimization) into velocities.

Consider the 1-D case. Here the acceleration is given by Eq. (C.3):

$$\ddot{x} = \frac{F}{m} \tag{C.3}$$

where m is atomic mass and F is the force acting on the particle. For motion along x, if we know the value of x at time t, $x(t)$, then the position after a short-time interval can be estimated by a standard Taylor series, Eq. (C.4) and derivatives of position with respect to time can be estimated by finite differences.

$$x(t + \Delta t) = x(t) + \dot{x}(t)\Delta t + \ddot{x}(t)\Delta t^2/2 + \cdots \tag{C.4}$$

411

The derivative of position with respect to time can be obtained from the present position and the position at Δt, see Eq. (C.5).

$$\dot{x}(t)_+ \approx \frac{x(t + \Delta t) - x(t)}{\Delta t} \qquad (C.5)$$

Alternatively, the present position and the one at $-\Delta t$ could be used, see Eq. (C.6).

$$\dot{x}(t)_- \approx \frac{x(t) - x(t - \Delta t)}{\Delta t} \qquad (C.6)$$

These two approximate estimates can be combined to give an improved, second-order center difference approximation, see Eq. (C.7).

$$\dot{x}(t) \approx \frac{1}{2}\left[\dot{x}(t)_+ + \dot{x}(t)_- \right] = \frac{x(t + \Delta t) - x(t - \Delta t)}{2\Delta t} \qquad (C.7)$$

The analogous center difference formula for second derivatives is given in Eq. (C.8).

$$\ddot{x}(t) \approx \frac{x(t + \Delta t) + x(t - \Delta t) - 2x(t)}{\Delta t^2} \qquad (C.8)$$

Equations (C.3) and (C.8) can be combined and rearranged to yield an estimate for x at $t + \Delta t$ in terms of the positions at t and $t - \Delta t$ and the force at time t, Eq. (C.9).

$$x(t + \Delta t) = 2x(t) - x(t - \Delta t) + \Delta t^2 F(t) \qquad (C.9)$$

The multidimensional analog of Eq. (C.9) is the Verlet equation (Verlet, 1967). Once the ball is rolling, so to speak, Eq. (C.9) can be used to obtain new positions. A remaining problem is how to start things off. Assigning $t = 0$ in Eq. (C.9) yields Eq. (C.10), which demonstrates that we need to know the position prior to the start of time $x(-\Delta t)$.

$$x(\Delta t) = 2x(0) - x(-\Delta t) + \Delta t^2 F(0) \qquad (C.10)$$

Plugging $t = 0$ into Eq. (C.7) yields Eq. (C.11) and a solution to our problem.

$$\dot{x}(0) \approx \frac{x(\Delta t) - x(-\Delta t)}{2\Delta t} \qquad (C.11)$$

Substituting Eq. (C.11) into Eq. (C.10) gives the position at Δt in terms of the initial position, an initial velocity, and an initial force, see Eq. (C.12).

$$x(\Delta t) = x(0) + \Delta t\, \dot{x}(0) + \frac{1}{2}\Delta t^2 F(0) \qquad (C.12)$$

Generalizing to $3n$ coordinates for n atoms yields Eqs. (C.13) and (C.14).

$$\mathbf{r}(t + \Delta t) = 2\mathbf{r}(t) - \mathbf{r}(t - \Delta t) + \Delta t^2 \mathbf{F}(t) \tag{C.13}$$

$$\mathbf{r}(\Delta t) = r(0) + \Delta t\, \dot{r}(0) + \frac{1}{2}\Delta t^2 \mathbf{F}(0) \tag{C.14}$$

Given an initial set of velocities, which are usually taken as a random distribution, we can stride forward in time. The remaining issues are time step and temperature. The time step Δt must be small enough that the finite difference approximation is valid. This limits time steps in MD to the 1–2-fs range if hydrogen atoms are present in the molecule.

In MD, temperature is calculated from the atomic velocities using Eq. (C.15), where T is temperature, n is the number of atoms, k_B is Boltzmann's constant, m_i is the mass of atom i, and $\dot{\mathbf{r}}_i$ is the velocity of atom i.

$$3k_B T = \sum_{i=1}^{n} m_i \dot{\mathbf{r}}_i \cdot \dot{\mathbf{r}}_i / n \tag{C.15}$$

If one is interested in molecular motion at a particular temperature, T_{set}, the velocities in Eq. (C.15) can be scaled to achieve this temperature, see Eq. (C.16).

$$\dot{\mathbf{r}} T_{set} = \tau\, \dot{r} \tag{C.16}$$

The scale factor τ in Eq. (C.16) can be obtained from a rearranged Eq. (C.15), see Eq. (C.17).

$$\tau = \frac{T_{set}}{3k_B} \sum_{i=1}^{n} m_i\, \dot{\mathbf{r}}_i \cdot \dot{\mathbf{r}}_i / n \tag{C.17}$$

References

McCammon, J. A.; Harvey, S. C. (1987), *Dynamics of Proteins and Nucleic Acids*, Cambridge University Press, Cambridge UK.

van Gunsteren, W. F.; Berendsen, H. J. C. (1990), Computer simulation of molecular dynamics: methodology, applications, and perspectives in chemistry, *Angew. Chem. Int. Ed. Engl.* **29**, 992.

Verlet, L (1967), Computer "Experiments" on Classical Fluids. I. Thermodynamical Properties of Lennard-Jones Molecules, *Phys. Rev.* **159**, 98.

Monte Carlo Sampling

In addition to making changes in molecular coordinates by heading downhill (minimization), by heading forward in time (molecular dynamics), and by making a systematic change (grid search), a change in molecular coordinates can be generated randomly. Use of a random change is referred to as a Monte Carlo process (in reference to gambling) or Monte Carlo sampling (Wood and Parker, 1957). As discussed in numerial mathematics texts, Monte Carlo techniques are often used to numerically evaluate integrals (Fröberg, 1985).

A simple example is the evaluation of the area inside an irregularly shaped object, as shown in Figure D.1. If a square is drawn to encompass the object and a large number of

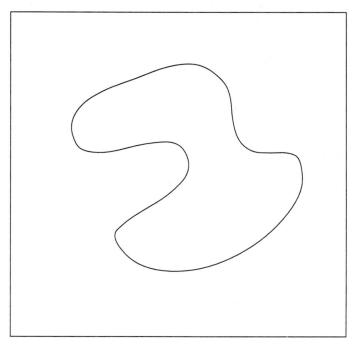

FIGURE D.1.
Irregularly shaped object and enclosing square.

random x and y pairs are generated inside the box, the area of the object can be obtained by comparing the number of x,y pairs inside the object to the total number of pairs and scaling by the area of the square, see Eq. (D.1).

$$\text{Area}_{\text{object}} = \text{Area}_{\text{square}} \times \frac{\text{number inside}}{\text{number total}} \tag{D.1}$$

The principal difficulty with the approach is its relative inefficiency. For a uniform spacing or distribution of random numbers the error in function evaluation decreases as $\sqrt{n}$. If a doubling of the accuracy is desired, then four times more throws of the dice are needed. This inefficiency can be improved upon within the realm of molecular simulation by using a nonuniform distribution of random numbers—this technique is referred to as importance sampling.

Importance sampling techniques choose random numbers from a distribution that causes the function evalualtions to be concentrated in the "important" regions of space. In one example of importance sampling, the Metropolis method, configuration space is biased towards low-energy conformations. As such, a new molecular configuration or conformation is generated by randomly selecting or changing one or more of the conformational degrees of freedom of interest, usually dihedral angles. The energy of this configuration is evaluated. If the energy of a new configuration is less than the previous energy, it is accepted and used as the new starting configuration. If the energy is greater than the previous configuration, then a random number (between 0 and 1) is generated. If the random number is less than or equal to $\exp(-\Delta E/RT)$ (ΔE being the energy difference between the old and new configurations), then the new configuration is included in the average. This procedure skews the population of configurations generated towards low-energy structures while still providing a statistically significant distribution of configurations.

References

Fröberg, C.-E. (1985), *Numerical Mathematics, Theory and Computer Applications*, Benjamin/Cummings, Menlo Park, CA.

Wood, W. W.; Parker, F. R. (1957), Monte Carlo equation of state of molecules interacting with the Lennard-Jones potential. I. A supercritical isotherm at about twice the critical temperature, *J. Chem. Phys.* **27**, 720.

Conformational Searching

Molecular mechanics studies on small molecules typically start with a reasonable guess of the molecular structure. The molecular mechanics energy of a molecule is then minimized by changing the coordinates of the atoms. This process is repeated for each of the likely conformations. The molecular mechanics project is completed by carrying out an analysis of why one particular conformation is favored over the others. Armed with a reasonable intuition as to what the most favorable conformations are likely to be for a given molecule, this approach was successfully used for most of the case studies described in this text.

However, the number of probable conformations grows rapidly as the complexity of the molecule increases. Furthermore, as molecules become more complex, the number of functional groups increases and the subtle interplay between electrostatics (allignment of dipole moments), van der Waals interactions (favorable *gauche* conformations), and torsional preferences clouds reasonable intuition. After all, molecular mechanics is used when the best conformation for a molecule cannot be reasonably guessed.

In addition to assessing individual molecular structures, molecular mechanics can be used to assess the number of minima and the energetic differences between these minima or molecular conformations. Determination of molecular shape or structure is thus a problem of finding the global or lowest energy minimum and the relatively small set of low-energy conformations present in a large family of viable structures. Insuring that one has obtained the best solution is, in fact, an insoluble problem. Though the global minimum problem has yet to be solved in mathematics, many conformational searching techniques have been developed and used successfully.

The most straightfoward and most rigorous approach (the only approach guaranteed to find the correct answer) is to simply find all of the minima on the potential surface for the molecule of interest, order them by energy, and select the lowest energy set of these conformations. To survey the shapes of potential energy surfaces, each valence or bonded term will be considered in sequence. To a first approximation, the bond stretch terms in a force field can be ignored because there are, or should be, single minima associated with each stretch, which can be determined through simple energy minimization. Angle bend terms can also be ignored because again there usually are single minima associated with each bend term (see Chapter 7 for exceptions). The same is true for out-of-plane bends.

Inversion terms such as in ammonia (NH_3), although important when present, are relatively rare and will be ignored in this discussion. The remaining valence terms are the torsions, each torsion supplying several minima to the potential energy surface. The global sampling of a molecular potential energy surface can be accomplished by generating a multidimensional grid consisting of combinations of the minima for each of the torsions while simultaneously minimizing the remaining molecular degrees of freedom. Each combination of torsions is called a configuration. An idealized representation of this is given in Figure E.1(a) where the dots indicate configurations included in the search. Unfortunately, low-energy conformations are known that are substantially removed from the bottoms of the torsional wells (Saunders, 1987) (~90° instead of 60° for sp^3–sp^3 torsions). In the idealized representation of Figure E.1(a) both the primary and secondary minima of the idealized system are indicated as being significantly removed from the bottoms of the individual torsional wells. The displacement of minima away from the bottoms of the torsional wells suggests that the resolution of the search should be increased; instead of using a spacing of 120° for sp^3–sp^3 torsions, a spacing of 30° might be more appropriate (so the 90° minima would be found). This would yield 12^t conformations for a set of t torsions (rather than 3^t for a 120° spacing). For molecules with 2, 3, 4, 5, or 6 torsions there would be 144, 1728, 20,736, 248,832, or 2,985,984 structures to be minimized with each structure requiring at least 100 steps of conjugate gradient minimization. Unfortunately, there are a lot of interesting molecules with more than six bonds (torsions).

An alternative approach would be to skip the minimization step and sample more points for each torsional coordinate. If a spacing of 10° were used, instead of the 30° spacing above, and the three maxima of the potential were not sampled, then 33 configurations would be included for each torsion. Again, conformational anlaysis of molecules with six torsions becomes unthinkable.

Conformational searching techniques for medium-to-large molecules must abandon the safety of searching the full torsional potential energy surface. Currently used techniques can be classified as being either ''smart'' grid searches, random searches, distance driven searches, or energy driven searches. Examples from each class of search technique are discussed below.

An additional complication arises when the molecule of interest contains one or more rings: the vast majority of the possible combinations of torsional minima do not correspond to low-energy structures, that is, they simply do not close the rings. For example, for a five-membered ring of sp^3 centers all of the low-energy conformations correspond to torsions near 0°. For this set of 5 torsions there are 243 possible configurations corresponding to torsional minima, none of which are associated with the low-energy conformations of the ring! For six-membered rings of sp^3 centers the low-energy conformations correspond to alternating $\pm60°$, $\mp60°$ combinations for the chair conformation and the sequence: $-60°$, 30°, 30° for the twist–boat conformation. This set of 5 low-energy conformations (2 chairs and 3 twist–boats) is out of a set of 729 torsional minima (when a spacing of 120° is used). A systematic torsional search of a six-membered ring would consist of 0.6% productive work! An analytic representation for the set of torsions that will close a geometric gap have been presented (Gō and Scheraga, 1970; Bruccoleri and Karplus, 1985). For up to six torsions the ring closure condition is completely defined (Gō and Scheraga, 1970) unless angle bend distortions are permitted (Bruccoleri and Karplus, 1985). Other approaches for surmounting this problem will be discussed below.

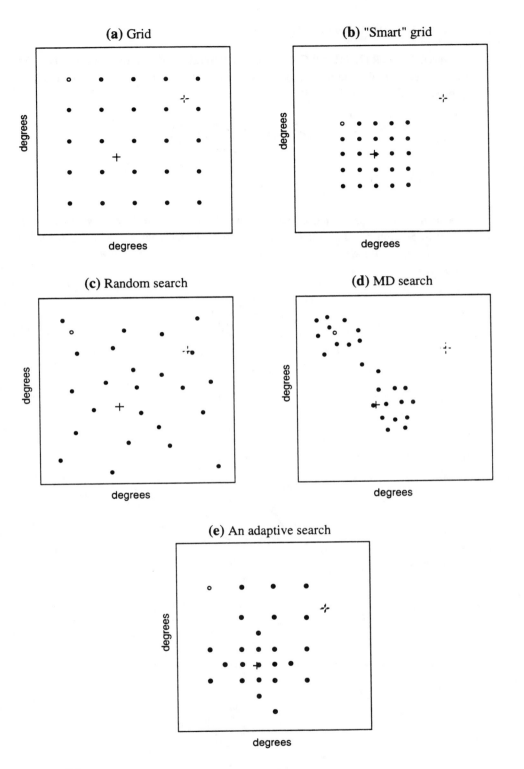

FIGURE E.1.
Two-dimensional, idealized representations of several conformational searching strategies: (a) a systematic grid search, (b) a "smart" systematic grid search, (c) a random search, (d) a molecular dynamics search, and (e) an adaptive search. The solid and dashed x's indicate a global and a secondary minimum, respectively. The open circle indicates the starting configuration for each search.

"SMART" GRID SEARCHING. Of the 729 conformations associated with the minima in a systematic search of six torsions, **E.1**, a number of torsions can be ignored due to intrinsic difficulties.

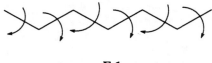

E.1

One such set of torsions that could be ignored is due to the "pentane effect" discussed in Section 6.4. If we consider rotation about the central two C–C bonds of pentane, we end up with four unique conformations, the *trans–trans* conformation (*tt*), **E.2**,

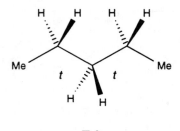

E.2

the *trans–gauche*$^+$ conformation (*tg*$^+$ or *tg*$^-$), **E.3**,

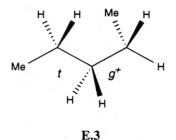

E.3

the *gauche*$^+$–*gauche*$^+$ conformation (*g*$^+$*g*$^+$ or *g*$^-$*g*$^-$), **E.4**,

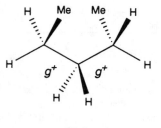

E.4

and the *gauche⁺–gauche⁻* conformation (g^+g^- or g^-g^+), **E.5**.

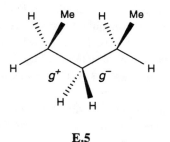

E.5

As should be evident from **E.5**, there will be substantial interaction between the terminal methyl groups in the g^+g^- conformation. This conformation (and the equivalent g^-g^+) will be substantially higher in energy than the tt, $tg^{\pm}$, $g^{\pm}t$, g^+g^+, or equivalent g^-g^- conformations. So out of the nine conformations for each pair of sp^3–sp^3 torsions two can be ignored. This is one example of a "smart" reduction in the torsional space to be sampled. The overall effect is as shown schematically in Figure E.1(b). By eliminating unfavorable configurations up front, the mesh in the grid can be made finer without increasing the work. As shown in Figure E.1(a) the original systematic search sampled points fairly near, but not at, the global minimum. The "smart" search shown in Figure E.1(b) used a finer mesh and found the global minimum, but missed the secondary minimum. This illustrates a limitation of any systematic search. By excluding conformations through intuition, unanticipated or novel structures can be missed. For example, if torsional distortion is allowed, low-energy conformations associated with the "forbidden" g^+g^- conformation **E.4** are observed to be low-energy structures within 1–2 kcal/mol of the global minimum (Saunders, 1987). One of the two *gauche* angles merely has to open up from 60° to about 90° for the structure to be low energy. The key is to use a fine enough grid to find these distorted structures but to use enough knowledge in inexpensive screening, prior to the energy evaluation stage, to eliminate whole families of conformations. Several techniques have been developed to recognize and eliminate, prior to creation, structures containing high-energy conformational features (Bruccoleri and Karplus, 1987; Lipton and Still, 1988; Dammkoehler et al., 1989). These techniques commonly use nonbonded distance as well as ring-closure criteria to screen conformations. For example, if a configuration is being generated for a ring that places a pair of atoms far enough apart that the sum of bond distances for the remaining bonds in the ring cannot possibly close the ring, then this conformation can safely be abandoned. The vast majority of the time associated with a conformational search (>96%) (Saunders et al., 1990) is used minimizing the energy of the molecular structures. Prescreening by calculating nonbonded distances and summing bond distances is certainly cost-effective relative to energy minimization.

Rather than defining the screening criteria in advance, knowledge can be built up as the search proceeds either through a heuristic approach (Leach and Prout, 1990) or by the use of genetic algorithms (a popular account is provided in Holland, 1992) as will be discussed below.

RANDOM SEARCHING. Rather than systematically generating all possible combinations of torsions, see Figure E.1(a), or generating a subset of configurations based on intuition or cutoff criteria, see Figure E.1(b), one could randomly choose the combinations of torsions to be sampled. As indicated schematically in Figure E.1.(c), if one is lucky enough, it is quite possible to find both the global and secondary minimum. In practice, if one were that lucky there would be many more profitable ways to use this luck than on conformational searching. This fully random approach simply will not work for ring systems; for a six-membered ring rolling dice with a success ratio of 5:729 is not efficient.

Rather than being confined to the torsional minima used in systematic searches or the large number of high-energy combinations of torsions found through a completely random sampling, the torsions can be selected as a set of perturbations from a starting, low-energy conformation (Chang et al., 1989). In this approach, an initial trial structure is built and minimized. This structure is perturbed by applying random increments to the torsional degrees of freedom, and the new structure is minimized. If the new structure is different from previously found minima and the new structure is lower in energy than previously found structures it is kept, and the perturbation, minimization sequence is repeated. If the new structure is not new or lower in energy than previous structures, then it is discarded and the random perturbation proceeds from the previous structure. Since, for ring systems the starting structure did satisify the ring closure condition, this approach does work for rings.

Given the coupling between torsional degrees of freedom and angular degrees of freedom previously discussed (Bruccoleri and Karplus, 1985), Ferguson and Raber (1989, 1990) and Saunders (1987, 1989) proposed using the full set of molecular coordinates rather than merely the torsional degrees of freedom in a random perturbation scheme. The Cartesian (x, y, z) coordinate space is used for both the Ferguson–Raber and Saunders approaches. As before, starting from an initial, minimized structure coordinate, perturbations are selected by applying random increments, here to the Cartesian coordinates. For the Ferguson–Raber procedure, the increments selected are relatively small, less than 1.7 Å per atom, but for the Saunders procedure, bigger kicks are given (~3 Å). The resulting structures are minimized, a uniqueness check is applied, and a lower energy criteria used. Use of the large step size in the Saunders approach places a substantial load on the minimization procedure. To circumvent this problem Saunders and Jarret (1986) proposed a simplified, inexpensive force field that can be used to "roughly" minimize the perturbed structure prior to subjecting it to a conventional force field minimization. As with the random perturbation of torsions approach, this method works for rings because the starting structure satisfied the ring closure condition.

Each of these random approaches, as described here, suffer from a removable flaw. A new structure might well be higher in energy than previous structures but it could be on a path towards a lower energy structure. These "passes" to lower energy structures can be explored by adding a Monte Carlo attribute to the procedure. Rather than only selecting lower energy structures, new structures can be selected by a Metropolis Monte Carlo sampling algorithm (Wood and Parker, 1957) (see Appendix D). Use of a high-temperature acceptance criterion in the Metropolis sampling will permit conformational barriers to be surmounted. Slowly lowering the temperature used in the acceptance scheme will cause the system to sample lower energy conformations more frequently. This procedure is called simulated annealing (Wilson et al., 1991).

DISTANCE DRIVEN SEARCHING. Rather than representing a molecular structure consisting of n atoms in terms of the usual $3n$ Cartesian coordinates the molecular structure can be described in terms of n^2 intramolecular distances, see Eq. (E.1).

$$d_{ij} = \sqrt{(x_i - x_j)^2 + (y_i - y_j)^2 + (z_i - z_j)^2} \tag{E.1}$$

Since we are accustomed to thinking about molecular structures in terms of easily guessed bond distances, bond angles (which can be expressed in terms of 1,3 non-bonded distances), and torsions (which can be expressed in terms of 1,4 nonbonded distances) distance geometry (Menger, 1931) has become a useful technique for searching conformational space (Crippen, 1977; Crippen and Havel, 1978), for building trial structures, and for incorporating nonbonded distance information from nuclear Overhauser effect (NOE) data into the large molecule structure determination process (Havel and Wüthrich, 1985).

While it is true that bond distances and bond angles are a well-defined part of a molecular structure the remaining distances of a distance geometry analysis are more uncertain. Torsions can take on several values and, as a consequence, 1,4 nonbonded distances can be ill-defined. For example, for a H−C−C−H unit the H−H distance will range from 2.3 Å for an eclipsed conformation to 3.1 Å for a trans arrangement of the hydrogen atoms. Nonbonded distances have a lower bound of roughly 2.8 Å, due to van der Waals repulsions but the upper bound can achieve rather large values limited by the all-*trans* or extended arrangement of the intervening bonds. Auxiliary data such as NOE distance or three bond scalar coupling data can be used to define the 1,4 distances. These uncertainties are a reflection of the fact that molecules do adopt numerous conformations. By assigning these ill-determined distances randomly (within bounds), a family of molecular conformations can be generated and subjected to minimization. In a sense, this is another variant of the random searching procedures into which external, experimental data can be easily incorporated.

Briefly, the process begins by developing lower and upper bound distance matrices from all available information (bond distances, bond angles, torsional ranges, van der Waals minima, NOE data, $^3J_{HH}$ data, etc.) Given these data, which is usually refined due to geometric constraints, a distance is assigned randomly (within the bounds) for each pair of atoms. This set of distances can be used, through a series of steps, to generate the least-squares best set of Cartesian coordinates (best matched to this set of distances). A real, symmetric matrix, **G**, is constructed whose elements are the scalar products of the respective distances from the origin, Eq. (E.2).

$$g_{ij} = x_i x_j + y_i y_j + z_i z_j \tag{E.2}$$

Since the coordinates of the atoms are not known (they are what we are trying to find) the elements of **G** can be constructed from the distance matrix, Eq. (E.3),

$$g_{ij} = \frac{1}{2}(d_{i0}^2 + d_{j0}^2 - d_{ij}^2) \tag{E.3}$$

where the terms d_{i0} and d_{j0} can be obtained from Eq. (E.4).

$$d_{i0}{}^2 = \frac{1}{n} \sum_{j=1}^{n} d_{ij}{}^2 - \frac{1}{n^2} \sum_{j=2}^{n} \sum_{k=1}^{j-1} d_{jk}{}^2 \qquad (E.4)$$

The largest three eigenvalues, λ_j, of $\mathbf{G}$ and the corresponding eigenvectors, w_{ij}, are used to construct the Cartesian coordinates, v_{ij}, see Eq. (E.5), where j runs from 1 to 3 and i runs from 1 to n.

$$v_{ij} = \sqrt{\lambda_j}\, w_{ij} \qquad (E.5)$$

Because the ring-closing distance is explicitly included in the set of distances used in distance geometry, distance geometry is particulary well suited to molecules containing rings.

ENERGY DRIVEN SEARCHING. In each of the previous conformational searching procedures the criteria of low energy was only applied near the end of the process. Without a "smart" prescreening of conformations each of the conformational searching techniques spend most of the time generating high-energy structures, which may or may not minimize to low-energy conformations. One conformational searching technique that always samples low-energy conformations is generically referred to as molecular dynamics (MD). An overview of MD is provided in Appendix C. In MD, heat is used as a means of surmounting potential energy barriers, and hence changing from one conformation or local minimum to another. This result is schematically represented in Figure E.1(d) where there are numerous dots near the starting configuration, a few dots are shown to represent a conformational barrier being surmounted and then numerous dots around the final configuration. The advantage of MD, that low-energy structures are always sampled, is also a disadvantage of the technique. The internal energy supplied, 0.894 kcal/mol per atom at 300 K, is usually not concentrated in modes associated with traversing potential energy barriers, so it takes quite a while for potential barriers to be surmounted in MD.

As discussed in Section 1.4, there are three common types of MD calculations used for conformational searching: (1) conventional microcanonical MD (2) quenched MD, and (3) annealed dynamics. In conventional microcanonical MD, the dynamical behavior of a molecular system is monitored as a function of time at a "constant" temperature. This MD procedure samples relatively high potential energy structures due to the presence of significant kinetic energy in the system. For a molecule consisting of 50 atoms, roughly 45 kcal/mol of "excess" energy is contained in the molecule in order to maintain a temperature of 300 K. In the second procedure, quenched MD, structures are periodically extracted from the microcanonical MD time progression and minimized. The configurations found in this set of minimized structures are analyzed for uniqueness and the low-energy subset of unique structures said to represent the "structure" of the molecular system. In annealed dynamics [as with Monte Carlo simulated annealing (Wilson et al., 1991)], the temperature of the system is incrementally increased and then decreased between low and high temperatures for a number of designated cycles with the lowest energy structures being saved for further minimization and analysis. The high-temperature excursions are used to facilitate crossing potential barriers and the low-temperature phases are used to "minimize" the structures. Although annealed dynamics does not actually minimize the energy, if the low temperature is small (i.e., 0 K) and the temperature step size is small, then the system will cool slowly enough to find a low-energy minimum on the potential energy surface without getting trapped in a high-energy local minimum. This process is the computational analog of annealing that is used in metallurgy for the formation of stable alloys.

Because vibrational energy is used as the driving force in MD conformational searching, MD does not have any difficulties with ring systems.

GENETIC ALGORITHMS. Each of the other searching techniques discussed in this Appendix proceeds to completion using information or a direction that was determined prior to starting the process. In "smart" systematic searching, the initial grid size and nonbond distance criteria were established, and then the search proceeded. In random searching, the dice were rolled for a prescribed number of times. In MD, a trial structure was created and the molecule set off to wiggle for a defined period of time. In this final search technique, genetic algorithms (Blommers et al., 1992; Judson, 1992; Judson et al., 1993; McGarrah and Judson, 1993), knowledge about the search is built up as the search proceeds. This is illustrated in Figure E.1(e), where the initial structures are far apart but (optimistically) rapidly converge to the global minimum.

This procedure is based on an evolution metaphor. Conformational information is encoded into chromosomes [just like protein amino acid sequence information is encoded in codons (sets of three bases in DNA)]. The energy of each chromosome is determined and the "better" chromosomes are chosen to generate new chromosomes. The process is repeated with random mutations until the conformational pool is exhausted (all members of the pool the same) and the best conformation is found.

Operationally, one begins with a population of conformations where the variable conformational information (typically torsional angles) is encoded into a bit string. For example, for an alkene, a single bit (0 or 1) can be used to distinguish between the cis and trans orientations. For sp^3–sp^3 bonds at least three minima need to be included. As with the base pairs of DNA, more than one bit can be used to encode each angle. If two bits are used to encode an angle, then four possible choices can be stored. In general, the number of possibilities for an angle goes as 2^c, where c is the number of bits used to encode the angle. The initial population is generated using a coin flip for each bit in each chromosome in the population. The survivability of each chromosome is determined by an energy evaluation of the corresponding molecular conformation (the lower the energy the better the chromosome). Once this is done the next generation of chromosomes is created through reproduction, crossover breeding, mutation, and so on, with the size of the pool remaining constant. This process is repeated until the population has converged to a single chromosome.

Given that finite choices of torsional angles are included in each chromosome, genetic algorithms serve best as a technique for generating a family of low-energy structures to be subjected to energy minimization.

Since the initial combinations of torsions in the population is generated randomly and the percentage of combinations of torsions that will satisify a ring closure condition is quite small, it is likely that genetic algorithms will have difficulty with ring systems.

References

Blommers, M. J. J.; Lucasius, C. B.; Kateman, G.; Kaptein, R. (1992), Conformational analysis of a dinucleotide photodimer with the aid of the genetic algorithm, *Biopolymers* **32**, 45.

Bruccoleri, R. E.; Karplus, M. (1985), Chain closure with bond angle variations, *Macromolecules* **18**, 2767.

Bruccoleri, R. E.; Karplus, M. (1987), Prediction of the folding of short polypeptide segments by uniform conformational sampling, *Biopolymers* **26**, 137.

Chang, G.; Guida, W. C.; Still, W. C. (1989), An internal coordinate monte carlo method for searching conformational space, *J. Am. Chem. Soc.* **111**, 4379.

Crippen, G. M. (1977), A novel approach to the calculation of conformation: distance geometry, *J. Comp. Phys.* **24**, 96.

Crippen, G. M. Havel, T. F. (1978), Stable calculation of coordinates from distance information, *Acta Cryst. Allogr., Sect. A* **34**, 282.

Dammkoehler, R. A.; Karasek, S. F.; Shands, E. F. B.; Marshall, G. R. (1989), Constrained search of conformational hyperspace, *J. Comput. Aided Mol. Design* **3**, 3.

Ferguson, D. M.; Raber, D. J. (1989), A new approach to probing conformational space with molecular mechanics: random incremental pluse search, *J. Am. Chem. Soc.* **109**, 4371.

Ferguson, D. M.; Raber, D. J. (1990), Molecular mechanics calculations of several Lanthanide complexes: An application of the random incremental pulse search, *J. Comp. Chem.* **11**, 1061.

Gō, N.; Scheraga, H. A. (1970), Ring closure and local conformational deformations of chain molecules, *Macromolecules* **3**, 178.

Havel, T. F.; Wüthrich, K. (1985), An evaluation of the combined use of nuclear magnetic resonance and distance geometry for the determination of protein conformations in solution, *J. Mol. Biol.* **182**, 281.

Holland, J. H. (1992), Genetic algorithms, *Sci. Am.* July, 66.

Judson, R. (1992), Teaching polymers to fold, *J. Phys. Chem.* **96**, 10102.

Judson, R. S.; Jaeger, E. P.; Treasurywala, A. M.; Peterson, M. L. (1993), Conformational searching methods for small molecules. II. Genetic algorithm approach, *J. Comp. Chem.* **14**, 1407.

Leach, A. R.; Prout, K. (1990), Automated conformational analysis: directed conformational search using the A* algorithm, *J. Comp. Chem.* **11**, 1193.

Lipton, M.; Still, W. C. (1988), The multiple minimum problem in molecular modeling. Tree searching internal coordinate conformational space, *J. Comp. Chem.* **9**, 343.

McGarrah, D. B.; Judson, R. S. (1993), Analysis of the genetic algorithm method of molecular conformation determination, *J. Comp. Chem.* **14**, 1385.

Menger, K. (1931), New foundation of euclidean geometry, *Am. J. Math.* **53**, 721.

Saunders, M.; Jarret, R. M. (1986), A new method for molecular mechanics, *J. Comp. Chem.* **7**, 578.

Saunders, M. (1987), Stochastic exploration of molecular mechanics energy surfaces. Hunting for the global minimum, *J. Am. Chem. Soc.* **109**, 3150.

Saunders, M. (1989), Stochastic search for the conformations of bicyclic hydrocarbons, *J. Comp. Chem.* **10**, 203.

Saunders, M.; Houk, K. N.; Wu, Y.-D.; Still, W. C.; Lipton, M.; Chang, G.; Guida, W. C. (1990), Conformations of cycloheptadecane. A comparison of methods for conformational searching, *J. Am. Chem. Soc.* **112**, 1419.

Wilson, S. R.; Cui, W.; Moskowitz, J. W.; Schmidt, K. E. (1991), Applications of simulated annealing to the conformational analysis of flexible molecules, *J. Comp. Chem.* **12**, 342.

Wood, W. W.; Parker, F. R. (1957), Monte Carlo Equation of State of Molecules Interacting with the Lennard-Jones Potential. I. A Supercritical Isotherm at about Twice the Critical Temperature, *J. Chem. Phys.* **27**, 720.

General References

Leach, A. R. (1991), A survey of methods for searching the conformational space of small and medium-sized molecules. In *Reviews in Computational Chemistry*, Vol. 2, Lipkowitz, K. B.; Boyd, D. B., Eds., VCH, New York, p. 1.

Answers to Homework

CHAPTER 1

1.1. (a) The same molecule, (b) stereoisomers, (c) the same molecule, (d) positional isomers, (e) conformers, (f) configurational isomers.

1.2. The energy associated with molecular distortions away from "natural" positions.

1.3. No, the natural bond distance is 1.509 Å, given in Table 1.3.

1.4. Yes, natural bond distances are adjustable parameters of the force field.

1.5. $V_r = \frac{1}{2}k_{IJ}(r - r_{IJ})^2 \Rightarrow V = \frac{1}{2}(700)\,(1.2 - 1.1)^2 \Rightarrow V = 350\,(0.1)^2 \Rightarrow$

$$V = 3.50 \text{ kcal/mol}$$

$$\Rightarrow V = \frac{1}{2}(700)\,(1.0 - 1.1)^2 \Rightarrow V = 350\,(0.1)^2 \Rightarrow$$

$$V = 3.50 \text{ kcal/mol}$$

since the potential is harmonic extension or compression leads to the same increase in energy

$$\Rightarrow V = \frac{1}{2}(1000)\,(1.2 - 1.1)^2 \Rightarrow V = 1000\,(0.1)^2 \Rightarrow$$

$$V = 5.00 \text{ kcal/mol}$$

increasing the force constant causes a proportional increase in the strain energy.

1.6. $V_r = D_{IJ}[e^{-\alpha(r - r_{IJ})} - 1]^2 \qquad \alpha = \left[\dfrac{k_{IJ}}{2D_{IJ}}\right]^{1/2} \Rightarrow \alpha = \left[\dfrac{700}{2 \times 100}\right]^{1/2} \Rightarrow \alpha = 1.87$

for $r = 1.2 \quad e^{-\alpha(r - r_{IJ})} = e^{-1.87(1.2 - 1.1)} = 0.829$
$$\Rightarrow V = 100\,(1.0 - 0.829)^2 = 2.92 \text{ kcal/mol}$$
for $r = 1.0 \quad e^{-\alpha(r - r_{IJ})} = e^{-1.87(1.0 - 1.1)} = 1.206$
$$\Rightarrow V = 100\,(1.0 - 1.206)^2 = 4.23 \text{ kcal/mol}$$

since the potential is anharmonic, compression or extension does not lead to the same increase in energy, for the Morse potential, compression causes a larger increase than extension. Furthermore, since the same force constant and distortion distance was used as in Problem 1.5, we can see that the Morse potential gives a greater energy rise than the harmonic potential for compression, and a smaller energy rise for extension.

427

1.7. In order to use Eq. (1.24):

$$x_0 - x_1 = \frac{1}{d^2V/dx^2} \frac{dV}{dx}$$

we need the first and second derivatives of V with respect to x.

$$\alpha = \left[\frac{k_{IJ}}{2D_{IJ}}\right]^{1/2} => \alpha = \left[\frac{720}{2 \times 110}\right]^{1/2} => \alpha = 1.809$$

$$V_r = D_{IJ}\{\exp[-\alpha(r - r_{IJ})] - 1]\}^2$$

$$= 110\{\exp[-1.809(r - 1.1)] - 1\}^2$$

$$\frac{dV}{dx} = 2D_{IJ}\alpha\{\exp[-\alpha(r - r_{IJ})] - \exp[-2\alpha(r - r_{IJ})]\}$$

$$= 2 \times 110 \times 1.809\{\exp[-1.809(r - 1.1)] - \exp[-2 \times 1.809(r - 1.1)]\}$$

$$= 398.0\{\exp[-1.809(r - 1.1)] - \exp[-2 \times 1.809(r - 1.1)]\}$$

$$\frac{d^2V}{dx^2} = 2D_{IJ}\alpha^2\{2\exp[-2\alpha(r - r_{IJ})] - \exp[-\alpha(r - r_{IJ})]\}$$

$$= 2 \times 110 \times 1.809^2\{2\exp[-2 \times 1.809(r - 1.1)] - \exp[-1.809(r - 1.1)]\}$$

$$= 719.9\,[2\exp(-2 \times 1.809(r - 1.1)] - \exp[-1.809(r - 1.1)]\}$$

First iteration

For $r = 1.3$ $\exp[-1.809(r - 1.1)] = \exp[-1.809(1.3 - 1.1)] = 0.6964;$
$\exp[-2 \times 1.809(1.3 - 1.1)] = 0.4850$

$V = 10.14$ kcal/mol; $\dfrac{dV}{dx} = 84.14$ kcal/mol-Å; $\dfrac{d^2V}{dx^2} = 197.0$ kcal/mol-Å^2

$\Delta r = -\dfrac{84.14}{197.0} = -0.427 => r' = 1.3 - 0.427 = 0.873$ Å

Second iteration

For $r = 0.873$ $\exp[-1.809(r - 1.1)] = \exp[-1.809(0.873 - 1.1)] = 1.508;$
$\exp[-2 \times 1.809(0.873 - 1.1)] = 2.274$

$V = 28.39$ kcal/mol; $\dfrac{dV}{dx} = -304.9$ kcal/mol-Å; $\dfrac{d^2V}{dx^2} = 2188.5$ kcal/mol-Å^2

$\Delta r = -\dfrac{-304.9}{2188.5} = 0.139 => r' = 0.873 + 0.139 = 1.012$ Å

Third iteration

For $r = 1.1012$ $\exp[-1.809(r - 1.1)] = \exp[-1.809(1.012 - 1.1)] = 1.173;$
$\exp[-2 \times 1.809(1.012 - 1.1)] = 1.376$

$V = 7.32$ kcal/mol; $\dfrac{dV}{dx} = -80.79$ kcal/mol-Å; $\dfrac{d^2V}{dx^2} = 1136.7$ kcal/mol-Å^2

$$\Delta r = -\frac{-80.79}{1136.7} = 0.071 => r' = 1.012 + 0.071 = 1.083 \text{ Å}$$

Fourth iteration

For $r = 1.083$ $\exp[-1.809(r - 1.1)] = \exp[-1.809(1.083 - 1.1)] = 1.031;$
$$\exp[-2 \times 1.809(1.083 - 1.1)] = 1.063$$

$V = 0.106$ kcal/mol; $\frac{dV}{dx} = -12.74$ kcal/mol-Å; $\frac{d^2V}{dx^2} = 788.3$ kcal/mol-Å²

$$\Delta r = -\frac{-12.74}{788.3} = 0.016 => r' = 1.083 + 0.016 = 1.099 \text{Å}$$

Fifth iteration

For $r = 1.099$ $\exp[-1.809(r - 1.1)] = \exp[-1.809(1.099 - 1.1)] = 1.0013;$
$$\exp[-2 \times 1.809(1.099 - 1.1)] = 1.0036$$

$V = 0.0004$ kcal/mol; $\frac{dV}{dx} = -0.716$ kcal/mol-Å; $\frac{d^2V}{dx^2} = 723.8$ kcal/mol-Å²

$$\Delta r = -\frac{-0.716}{723.8} = 0.00099 => r' = 1.099 + 0.00099 = 1.09999 \text{ Å}$$

Converged!

1.8. $R_{OH} = \sqrt{0 + 0.65^2 + 0.60^2} = 0.884 \text{ Å}$

For two equivalent OH bonds $V_r = 2 \times \frac{1}{2} 700 (0.884 - 0.93)^2 = 700 (0.046)^2$
$$= 1.481 \text{ kcal/mol}$$

For the bend $\theta = 2\cos^{-1}\left(\frac{0.60}{0.884}\right) = 94.51°$. Now $\theta - \theta_0$ (in rads)

$$= (94.51 - 104.5)\frac{\pi}{180°} = -0.1744$$

$V_\theta = \frac{1}{2} 100 \times (-0.1744)^2 = 1.52 \text{ kcal/mol}$

$V_{tot} = 1.481 + 1.52 = 3.00 \text{ kcal/mol}$

1.9.

Term	A	B	Δ
Stretch	0.15	0.16	0.01
Stretch–bend	0.05	0.07	0.02
Bend	0.29	0.63	0.41
Torsion	0.01	0.44	0.43
vdw	1.68	1.75	0.07
Total	2.18	3.05	0.87

Conformation **A** is lower in energy. The energetic difference is due to equal parts bend and torsion.

CHAPTER 2

2.5. Use Eq. (1.29) and the following data from Table 2.6:

Conformation 1	0.0
Conformation 2	0.11
Conformation 3	1.18
Conformation 4	5.30

$$P_1 = \frac{1}{1 + \exp(-0.11/0.590) + \exp(-1.18/0.590) + \exp(-5.30/0.590)}$$

$$P_2 = \frac{\exp(-0.11/0.590)}{1 + \exp(-0.11/0.590) + \exp(-1.18/0.590) + \exp(-5.30/0.590)}$$

$$P_3 = \frac{\exp(-1.18/0.590)}{1 + \exp(-0.11/0.590) + \exp(-1.18/0.590) + \exp(-5.30/0.590)}$$

$$P_4 = \frac{\exp(-5.30/0.590)}{1 + \exp(-0.11/0.590) + \exp(-1.18/0.590) + \exp(-5.30/0.590)}$$

$$P_1 = \frac{1}{d1 + 0.83 + 0.14 + 0.0001} = 0.508$$

$$P_2 = \frac{0.83}{d1 + 0.83 + 0.14 + 0.0001} = 0.421$$

$$P_3 = \frac{0.14}{d1 + 0.83 + 0.14 + 0.0001} = 0.071$$

$$P_4 = \frac{0.0001}{d1 + 0.83 + 0.14 + 0.0001} = 0.00005$$

2.9. The two enantiomers of lactic acid should have the strain–steric energies, the same bond distances, and the same bond angles?

2.10. The L with L and D with D combinations of lactic acid yield anhydrides that have the same energy but are not superimposable. The combinations L with D and D with L yield anhydrides that are superimposable, and hence the same.

2.11. The L with L and D with D combinations of lactic acid yield lactides that have the same energy but are not superimposable. The combinations L with D and D with L yield lactides that are superimposable, and hence the same.

2.12. As in Problem 2.11, the L with L and D with D combinations of lactic acid yield molecules that have the same energy but are not superimposable. Here the combinations L with D and D with L yield molecules with the same energy but are not superimposable. In this case the "superimposed" orientation of L–D lactic acid results in O being superimposed upon NH.

2.14. Your modeling software should be able to account for the steric differentiation in the first part of the problem, but not distinguish the differential electronic effect of a *p*-nitro group in the second part shown below.

X	Y	Z	$\Delta H\ddagger$	$\Delta\Delta H\ddagger$ H $=>$ NO$_2$
Cl	Me	H	11.8 ± 0.2	
Cl	Me	NO$_2$	13.4 ± 0.2	1.6
F	Me	H	5.5 ± 0.5	
F	Me	NO$_2$	4.7 ± 0.5	0.8

CHAPTER 3

3.1. F.3.1

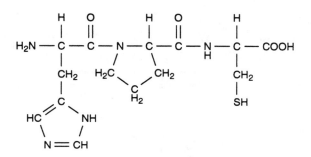

F.3.1

3.6. $V_{HB} = D_{IJ}\left\{-6\left[\dfrac{\rho_{IJ}}{\rho}\right]^{10} + 5\left[\dfrac{\rho_{IJ}}{\rho}\right]^{12}\right\}\cos^4\theta_{DHA}$

$V_{HB} = 4.25\left\{-6\left[\dfrac{2.75}{2.9}\right]^{10} + 5\left[\dfrac{2.75}{2.9}\right]^{12}\right\}\cos^4 150.0 =>$

$V_{HB} = 4.25\left\{-[0.9483]^{10} + 5[0.9483]^{12}\right\} 0.5625 =>$

$V_{HB} = 4.25\,(-6 \times 0.58811 + 5 \times 0.52887)\, 0.5625 = -2.11$ kcal/mol

3.7. $V_{vdw} = D_{IJ}\left\{-2\left[\dfrac{\rho_{IJ}}{\rho}\right]^{6} + \left[\dfrac{\rho_{IJ}}{\rho}\right]^{12}\right\}$

$V_{vdw} = 0.0152\left\{-2\left[\dfrac{3.195}{2.0}\right]^{6} + \left[\dfrac{3.195}{2.0}\right]^{12}\right\} =>$

$V_{vdw} = 0.0152\,(-2 \times 16.621 + 276.24) = 3.693$ kcal/mol

$V_{vdw} = 0.0152\left\{-2\left[\dfrac{3.195}{3.0}\right]^{6} + \left[\dfrac{3.195}{3.0}\right]^{12}\right\} =>$

$V_{vdw} = 0.0152\,(-2 \times 1.4591 + 2.1291 = -0.01199$ kcal/mol

3.10. The distance dependent dielectric parameter attempts to reproduce the shielding effect of water.

3.12. The trajectories should be correlated with temperature. You should see relative vibration of bonded atoms in the 150 K trajectory. You will not see the lysine side chain rotating at 150 K. Local unfolding or backbone conformational change occurs in about 10^5 ps. To do a conformational search on a peptide with dynamics you would need high temperatures and long-run times. The tuftsin dynamics described in the case study required a very high temperature (1000 K), and a long time (600 ps).

3.14. Tendamistat is called a β barrel because it looks a bit like a barrel made up of β-pleated sheets.

Yes, you should see interstrand NOEs.
Expected NOEs and assignments:

Atom	Residue	Atom	Residue	Measured Distance	NOE	NMR Distance[a]
HN	T 32	HC$_\alpha$	A 47	4.6	Weak	4.6
HN	T 32	HN	C 73	4.2	Weak	4.6
HC$_\alpha$	T 32	HN	Y 46	4.1	Weak	4.6
HC$_\alpha$	T 32	HC$_\alpha$	A 47	2.2	Very strong	2.3
HC$_\alpha$	T 32	HN	V 48	3.2	Medium	3.2

[a] The NMR distances are the NOE-derived constraints for tendamistat given in Kline, 1988.

3.15.

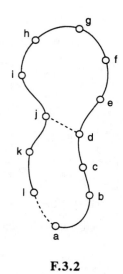

F.3.2

3.17. T, S (Both are hydrophillic, small-chain alkyl groups with an OH group, structurally similar.)

V, L, I (All three have small alkyl side chains and are hydrophobic.)

W, Y, F (All three have aromatic rings.)

N, D (Both are about the same size, either polar or charged, and both have a carbonyl group in the same position.)

Q, E (Both are about the same size, either polar or charged, and both have a carbonyl group in the same position.)

3.18. GTPAQ**TLNL**DF**DTGS**SD**LWVF**S**SETTA**
GT**TLNL**NF**DTGS**AD**LWVF**S**TELPA**SQQSGH
KST**S**ID GIA**DTGT**TLLYLPATVVS**A**Y

Residues where all three are identical are designated in boldface. Residues that are similar according to replacement rules are outlined.

CHAPTER 4

4.1. Inhibitor **4.43** is a hydrolysis transition state analog, and the C_2 symmetry of the inhibitor has been increased by reversing one-half of the molecule's N−C peptide polarity. The carbonyl oxygen of **4.44** mimics the internal HPR water.

4.2. Replacement of Ala10 by proline would cause the intramolecular hydrogen bond between residue 7 and 10 to be lost.

4.3. The $\log\left(\dfrac{1}{K_i}\right) = 0.59\pi + 6.49$.

4.4. The QSAR equation is

$$\log\left(\frac{1}{K_i}\right) = 0.95\text{MR}_{5'} + 0.89\text{MR}_{3'} + 0.80\text{MR}_4 - 0.21\text{MR}_4^2 + 1.58\pi_{3'}$$

$$- 1.77 \log(1.50 \times 10^{\pi 3'} + 1) + 6.65$$

For the hydrogen compound, $\text{MR}_{5'} = 0.10$, $\text{MR}_{3'} = 0.10$, $\text{MR}_4 = 0.10$, $\pi_{3'} = 0$. Do not forget that even though $\pi_{3'} = 0$, the log expression containing $\pi_{3'}$ is not equal to zero! Inserting these numbers into the equation yields $\log\left(\dfrac{1}{K_i}\right) = 6.21$. (The experimental number is 6.18.) For the CH_3 substituted compound, the CH_3 will go into the hydrophobic 3 position. Thus, $\text{MR}_{5'} = 0.10$, $\text{MR}_{3'} = 0.57$, $\text{MR}_4 = 0.10$, $\pi_{3'} = 0.52$. Inserting these numbers into the equation yields $\log\left(\dfrac{1}{K_i}\right) = 6.78$. (The experimental number is 6.70.) For the 3,4-$(OH)_2$ disubstituted compound, one OH will go into the hydrophilic 5 position, the other into the 4 position. $\text{MR}_{5'} = 0.29$, $\text{MR}_{3'} = 0.10$, $\text{MR}_4 = 0.29$, $\pi_{3'} = 0$. Inserting these numbers into the equation yields $\log\left(\dfrac{1}{K_i}\right) = 6.52$. (The experimental number is 6.46.)

4.5. Both allow fewer degrees of freedom by limiting the conformational flexibility.

CHAPTER 5

5.1.

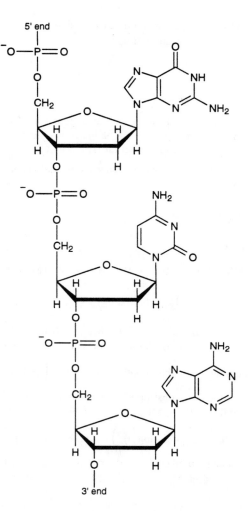

F.5.3

5.4. No. The conformation of the left-handed DNA is different: The helix is more slender, the numbr of bases per helical turn is higher, the sugar–phosphate skeleton is zigzag etc.

5.6. $V_{el} = 332.06 \dfrac{q_i q_j}{\varepsilon \rho_{IJ}} \Longrightarrow V_{el} = 332.06 \dfrac{0.27 \times 0.27}{\varepsilon \rho_{IJ}}$

$V_{el} = 332.06 \dfrac{0.27 \times 0.27}{80 \rho_{IJ}} \Longrightarrow$

$V_{el} = 332.06 \dfrac{0.27 \times 0.27}{80 \times 3.5} = 0.0865 \text{ kcal/mol}$

$$V_{el} = 332.06 \frac{0.27 \times 0.27}{80 \times 6.0} = 0.0504 \text{ kcal/mol}$$

$$V_{el} = 332.06 \frac{0.27 \times 0.27}{80 \times 2.0} = 0.0151 \text{ kcal/mol}$$

$$V_{el} = 332.06 \frac{0.27 \times 0.27}{1 \times 3.5} = 6.92 \text{ kcal/mol}$$

$$V_{el} = 332.06 \frac{0.27 \times 0.27}{3.5 \times 3.5} = 1.976 \text{ kcal/mol}$$

5.7. The distance dependent dielectric parameter attempts to reproduce the shielding effect of water. You should see atomic vibration only. It is not possible to see helix bending during a 2-ps calculation.

5.9. Yes. This molecule might bind to G-C base pair runs of DNA.

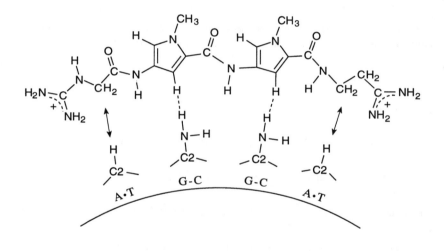

F.5.4

The van der Waals repulsions between pyrrole CH and guanine NH_2 in netropsin have been replaced by hydrogen bonding between the imidazole N and guanine NH_2 in the netropsin analog.

5.10. The new simulation should compare only the energetic components of the netropsin complexes; reactions of free DNA and netropsin will not be investigated. Dodecamers that differ as little as possible from each other should be bound to netropsin. Though you cannot compare perfect isomers as in Section 5.5, one example of two netropsin complexes that are alike at the ends and that differ only in the middle are netropsin bound to the central four base pairs of d(ATGC**ATAT**GCAT)$_2$ versus d(ATGC**GCGC**GCAT)$_2$.

5.11. The same factors that stabilize the stacking interactions between the bases in the double helix of free DNA also stabilize intercalators. These electronic factors are van der Waals interactions, dipole–dipole electrostatic interactions, and dipole-induced dipole polarization effects. van der Waals and electrostatic interactions are calculated in CHARMM and AMBER. Dipole-induced dipole interactions are not calculated in these force fields.

CHAPTER 6

6.1. For a 10 mer of polyethylene there are 19 C−C bonds 2 which are connected to terminal methyl groups so there are 17 bonds that can adopt gauche defect structures. For each of the 17 bonds there are g^+ and g^- possibilities, leading to 34 defect structures. This information along with Eq. (6.4) yields:

$$P_{trans} = \frac{1}{1 + 34 \exp(-0.5/0.397)} = \frac{1}{1 + 34 \times 0.284} = 0.094$$

CHAPTER 7

7.6. $\%ee = |\%R - \%S| => 95 = |\%R - \%S| => \%R = 97.5$ and $\%S = 2.5$
$\Delta\Delta G^{\ddagger} = -RT \ln(2.5/97.5) => \Delta\Delta G^{\ddagger} = 1.33$ kca/mol.

7.8. Quite likely the two diastereomers will differ by more than 1 or 2 kcal/mol, but an extensive conformational search would be necessary to obtain a good energy difference.

7.9. The three Rh−P distances are not the same due to the trans influence. Your modeling software may or may not treat this additional feature of inorganic complexes.

7.12. No!

CHAPTER 8

8.1. Use Eq. (8.9): $\mu = 4.80324\, r_{IJ}\, q$; $r_{IJ} = 1.15077$ Å and $\mu = 0.1587$

$$q = \frac{\mu}{4.80324\, r_{IJ}} = \frac{0.1587}{4.80324 \times 1.15077} = 0.0287$$

Compared to carbon monoxide at least the sign is consistent with electronegativities, the magnitude seems rather small.

8.2. (1) Experimental fitting; will not work directly, could interpolate.
(2) Theory: directly and interpolation.
(3) Rule based: interpolate the values for the rules.
(4) Assignment: directly.

8.3. An arithmetic average of C−Si and C−Sn gives C−Ge $r_{IJ} = \frac{1}{2}(1.88 + 2.147) = 2.01$ Å and $k_{IJ} = \frac{1}{2}(427.3 + 305.6) = 366$ kcal/mol-A².

A geometric average of C−Si and C−Sn gives C−Ge $r_{IJ} = \sqrt{1.88 \times 2.147} = 2.01$ Å and $k_{IJ} = \sqrt{427.3 \times 305.6} = 361$ kcal/mol-A².

The published parameters are C−Ge $r_{IJ} = 1.95$ Å, $k_{IJ} = 388.5$ kcal/mol-A².

8.4. Use $k = \mu\left(\dfrac{\omega}{108.593}\right)^2$, where k is the force constant in kilocalories per mole per square angstrom, μ is the reduced mass, and ω is the frequency in reciprocal centimeters.

Molecule	ω	μ	k (kcal/mol-Å²)
H_2	4401.213	0.5039	827.7
F_2	916.64	9.4992	676.8
Cl_2	559.72	17.4844	464.5
Br_2	325.321	39.45915	354.1
I_2	214.502	63.4522	125.3
HF	4138.32	0.9570	1389.9
HCl	2990.9463	0.9796	743.0
HBr	2648.975	0.9951	592.1
HI	2309.014	0.99986	452.1

Atomic weights: $^1H = 1.0078$, $^{19}F = 18.9984$, $^{35}Cl = 34.9688$, $^{79}Br = 78.9183$, $I = 126.9044$.

8.5. Use

$$k_{IJ} = 332.06\frac{Z^*_I Z^*_J}{r_{IJ}^3} => Z^*_I = \sqrt{\frac{k_{IJ}r_{IJ}^3}{332.60}}$$

Molecule	R (Å)	k	Z^*
H_2	0.74144	827.7	1.008
F_2	1.41193	676.8	2.395
Cl_2	1.9879	464.5	3.315
Br_2	2.28105	354.1	3.558
I_2	2.6663	125.3	2.674

$$k_{IJ} = 332.06\frac{Z^*_I Z^*_J}{r_{IJ}^3} \text{ and } \omega = 108.593\sqrt{\frac{k}{\mu}}$$

Molecule	R (Å)	k	ω
HF	0.916808	1040.3	3580.0
HCl	1.274552	535.9	2540.0
HBr	1.414435	420.9	2233.0
HI	1.60916	214.8	1592.0

Index